Feeling Good

The New Mood Therapy

David D. Burns, M.D.

Preface by Aaron T. Beck, M.D.

Collins
An Imprint of HarperCollinsPublishers

The ideas, procedures, and suggestions contained in this book are not intended as a substitute for consulting with your physician. All matters regarding your health require medical supervision.

A hardcover edition of this book was published in 1980 by William Morrow and Company, Inc.

HarperCollins books may be purchased for educational, business, or sales promotional use. For information, please e-mail the Special Markets Department at SPsales@harpercollins.com.

First WholeCare edition published 1999.
Reprinted in Quill 2000.

Library of Congress Cataloging-in-Publication Data is available.

ISBN 0-380-73176-2

20 21 21 LSC 60 59 58 57 56

This book is dedicated to Aaron T. Beck, M.D., in admiration of his knowledge and courage and in appreciation of his patience, dedication and empathy.

Acknowledgments

I am grateful to my wife, Melanie, for her editorial assistance and patience and encouragement on the many long evenings and weekends that were spent in the preparation of this book. I would also like to thank Mary Lovell for her enthusiasm and for her technical assistance in typing the manuscript.

The development of cognitive therapy has been a team effort involving many talented individuals. In the 1930s, Dr. Abraham Lowe, a physician, began a free-of-charge self-help movement for individuals with emotional difficulties, called "Recovery Incorporated," which is still in existence today. Dr. Lowe was one of the first health professionals to emphasize the important role of our thoughts and attitudes on our feelings and behavior. Although many people are not aware of his work, Dr. Lowe deserves a great deal of credit for pioneering many of the ideas that are still in vogue today.

In the 1950s, the noted New York psychologist, Dr. Albert Ellis, refined these concepts and created a new form of psychotherapy called Rational Emotive Therapy. Dr. Ellis published over fifty books that emphasize the role of negative self-talk (such as "shoulds" and "oughts") and irrational beliefs (such as "I must be perfect") in a wide variety of emotional problems. Like Dr. Lowe, his brilliant contributions are sometimes not sufficiently acknowledged by academic researchers and scholars. In fact, when I wrote the first edition of *Feeling Good*, I was not especially familiar with the work of Dr. Ellis and did not really appreciate the importance and magnitude of his contributions. I want to set the record straight here!

Finally, in the 1960s, my colleague at the University of Pennsylvania School of Medicine, Dr. Aaron Beck, adapted these ideas and treatment techniques to the problem of clinical depression. He described the depressed patient's negative view of the self, the world, and the future, and proposed a new form of "thinking therapy" for depression, which he called "cognitive therapy." The focus of cognitive therapy was helping the depressed patient change these negative thinking patterns. Dr. Beck's contributions, like those of Drs. Lowe and Ellis, have been substantial. His Beck Depression Inventory, published in 1964, allowed clinicians and researchers to measure depression for the first time. The idea that we could measure how severe a patient's depression was, and track changes in response to treatment, was revolutionary. Dr. Beck also emphasized the importance of systematic, quantitative research so we could get objective information on how well the different kinds of psychotherapy actually worked, and how effective they are in comparison to antidepressant drug therapy.

Since the time of those three early pioneers, many hundreds of gifted researchers and clinicians throughout the world have contributed to this new approach. In fact, there has probably been more published research on cognitive therapy than on any other form of psychotherapy ever developed, with the possible exception of behavior therapy. Clearly, I cannot mention all the individuals who have made important contributions to the development of cognitive therapy. In the early days of cognitive therapy, during the 1970s, I worked with several colleagues at the University of Pennsylvania School of Medicine who helped to create many of the treatment techniques still in use today.

They included Drs. John Rush, Maria Kovacs, Brian Shaw, Gary Emery, Steve Hollon, Rich Bedrosian, Ruth Greenberg, Ira Herman, Jeff Young, Art Freeman, Ron Coleman, Jackie Persons, and Robert Leahy.

Several individuals have given me permission to refer to their work in detail in this book, including Drs. Raymond Novaco, Arlene Weissman, and Mark K. Goldstein.

I would like to make special mention of Maria Guarnaschelli, the editor of this book, for her endless spark and vitality which have been a special inspiration to me.

During the time I was engaged in the training and research which led to this book, I was a Fellow of the Foundations' Fund for Research in Psychiatry. I would like to thank them for their support which made this experience possible.

And my thanks to Frederick K. Goodwin, M.D., a former chief at the National Institute of Mental Health, for his valuable consultation with regard to the role of biological factors and antidepressant drugs in treating mood disorders. Two Stanford colleagues, Drs. Greg Tarasoff and Joe Bellenoff, provided helpful feedback about the new drug chapters.

I would like to thank Arthur P. Schwartz for his encouragement and persistence. I would also like to thank Ann McKay Thoroman at Avon Books for editorial help on the new psychopharmacology chapters.

Finally, I would like to thank my daughter, Signe Burns, for extraordinarily helpful suggestions and meticulous editing of the new material in this 1999 edition.

Preface

I am pleased that David Burns is making available to the general public an approach to mood modification which has stimulated much interest and excitement among mental health professionals. Dr. Burns has condensed years of research conducted at the University of Pennsylvania on the causes and treatments of depression, and lucidly presents the essential self-help component of the specialized treatment that has derived from that research. The book is an important contribution to those who wish to give themselves a "top flight" education in understanding and mastering their moods.

A few words about the evolution of cognitive therapy may interest readers of *Feeling Good: The New Mood Therapy*. Soon after I began my professional career as an enthusiastic student and practitioner of traditional psychoanalytic psychiatry, I began to investigate the empirical support for the Freudian theory and therapy of depression. While such support proved elusive, the data I obtained in my quest suggested a new, testable theory about the causes of emotional disturbances. The research seemed to reveal that the depressed individual sees himself as a "loser," as an inadequate person doomed to frustration, deprivation, humiliation, and failure. Further experiments showed a marked difference between the depressed person's self-evaluation, expectations, the aspirations on the one hand and his actual achievements—often very striking—on the other. My conclusion was that depression must involve a disturbance in thinking: the depressed person thinks in idiosyncratic and negative ways about himself, his environment, and his future. The pessimistic mental set affects his mood, his motivation, and his relationships with others, and leads to the full spectrum of psychological and physical symptoms typical of depression.

We now have a large body of research data and clinical experience which suggests that people can learn to control painful mood swings and self-defeating behavior through the application of a few relatively simple principles and techniques. The promising results of this investigation have triggered interest in cognitive theory among psychiatrists, psychologists, and other mental health professionals. Many writers have viewed our findings as a major development in the scientific study of psychotherapy and personal change. The developing theory of the emotional disorders that underlies this research has become the subject of intensive investigations at academic centers around the world.

Dr. Burns clearly describes this advance in our understanding of depression. He presents, in simple language, innovative and effective methods for altering painful depressed moods and reducing debilitating anxiety. I expect that readers of this book will be able to apply to their own problems the principles and techniques evolved in our work with patients. While those individuals with more severe emotional disturbances will need the help of a mental health professional, individuals with more manageable problems can benefit by using the newly developed "common sense" coping skills which Dr. Burns delineates. Thus *Feeling Good* should prove to be an immensely useful step-by-step guide for people who wish to help themselves.

Finally, this book reflects the unique personal flair of its author, whose enthusiasm and creative energy have been his particular gifts to his patients and to his colleagues.

Aaron T. Beck, M.D.
Professor of Psychiatry,
University of Pennsylvania
School of Medicine

Contents

Introduction (Revised Edition, 1999)

I have been amazed by the interest in cognitive behavioral therapy that has developed since *Feeling Good* was first published in 1980. At that time, very few people had heard of cognitive therapy. Since that time, cognitive therapy has caught on in a big way among mental health professionals and the general public as well. In fact, cognitive therapy has become one of the most widely practiced and most intensely researched forms of psychotherapy in the world.

Why such interest in this particular brand of psychotherapy? There are at least three reasons. First, the basic ideas are very down-to-earth and intuitively appealing. Second, many research studies have confirmed that cognitive therapy can be very helpful for individuals suffering depression and anxiety and a number of other common problems as well. In fact, cognitive therapy appears to be at least as helpful as the best antidepressant medications (such as Prozac). And third, many successful self-help books, including my own *Feeling Good*, have created a strong popular demand for cognitive therapy in the United States and throughout the world as well.

Before I explain some of the exciting new developments, let me briefly explain what cognitive therapy is. A cognition is a thought or perception. In other words, your cognitions are the way you are thinking about things at any moment, including this very moment. These thoughts scroll across your mind automatically and often have a huge impact on how you feel.

For example, right now you are probably having some thoughts and feelings about this book. If you picked this book up because you have been feeling depressed and discouraged, you may be thinking about things in a negative, self-critical way: "I'm such a loser. What's wrong with me? I'll never get better. A stupid self-help book like this couldn't possibly help me. I don't have any problem with my *thoughts*. My problems are *real*." If you are feeling angry or annoyed you may be thinking: "This guy Burns is just a con artist and he's just trying to get rich. He probably doesn't even know what he's talking about." And if you are feeling optimistic and interested you may be thinking: "Hey, this is interesting. I may learn something really exciting and helpful." In each case, your thoughts create your feelings.

This example illustrates the powerful principle at the heart of cognitive therapy—your feelings result from the messages you give yourself. In fact, your thoughts often have much more to do with how you feel than what is actually happening in your life.

This isn't a new idea. Nearly two thousand years ago the Greek philosopher, Epictetus, stated that people are disturbed "not by things, but by the views we take of them." In the Book of Proverbs (23: 7) in the Old Testament you can find this passage: "For as he thinks within himself, so he is." And even Shakespeare expressed a similar idea when he said: "for there is nothing either good or bad, but thinking makes it so" (*Hamlet*, Act 2, Scene 2).

Although the idea has been around for ages, most depressed people do not really comprehend it. If you feel depressed, you may think it is because of bad things that have happened to you. You may think you are inferior and destined to be unhappy because you failed in your work or were rejected by someone you loved. You may think your feelings of inadequacy result from some personal defect—you may feel convinced you are not smart enough, successful enough, attractive enough, or talented enough to feel happy and fulfilled. You may think your negative feelings are the result of an unloving or traumatic childhood, or bad genes you inherited, or a chemical or hormonal imbalance of some type. Or you may blame others when you get upset: "It's these lousy stupid drivers that tick me off when I drive to work! If it weren't for these jerks, I'd be having a perfect day!" And nearly all depressed people are convinced that they are facing some special, awful truth about themselves and the world and that their terrible feelings are absolutely realistic and inevitable.

Certainly all these ideas contain an important germ of truth—bad things do happen, and life beats up on most of us at times. Many people do experience catastrophic losses and confront devastating personal problems. Our genes, hormones, and childhood experiences probably do have an impact on how we think and feel. And other people can be annoying, cruel, or thoughtless. But all these theories about the causes of our bad moods have the tendency to make us victims—because we think the causes result from something beyond our control. After all, there is little we can do to change the way people drive at rush hour, or the way we were treated when we were young, or our genes or body chemistry (save taking a pill). In contrast, you can learn to change the way you think about things, and you can also change your basic values and beliefs. And when you do, you will often experience profound and lasting changes in your mood, outlook, and productivity. That, in a nutshell, is what cognitive therapy is all about.

The theory is straightforward and may even seem overly simple—but don't write it off as pop psychology. I think you will discover that cognitive therapy can be surprisingly helpful—even if you feel pretty skeptical (as I did) when you first learn about it. I have personally conducted more than thirty thousand cognitive therapy sessions with hundreds of depressed and anxious individuals, and I am always surprised about how helpful and powerful this method can be.

The effectiveness of cognitive therapy has been confirmed by many outcome studies by researchers throughout the world during the past two decades. In a recent landmark article entitled "Psychotherapy vs. Medication for Depression: Challenging the Conventional Wisdom with Data," Drs. David O. Antonuccio and William G. Danton from the University of Nevada and Dr. Gurland Y. DeNelsky from the Cleveland Clinic reviewed many of the most carefully conducted studies on depression that have been published in scientific journals throughout the world.[1] The studies reviewed compared the antidepressant medications with psychotherapy in the treatment of depression and anxiety. Short-term studies as well as long-term follow-up studies were included in this review. The authors came to a number of startling conclusions that are at odds with the conventional wisdom:

- Although depression is conventionally viewed as a medical illness, research studies indicate that genetic influences appear to account for only about 16 percent of depression. For many individuals, life influences appear to be the most important causes.
- Drugs are the most common treatment for depression in the United States, and there is a widespread belief, popularized by the media, that drugs are the most effective treatment. However, this opinion is not consistent with the results of many carefully conducted outcome studies during the past twenty years. These studies show that the newer forms of psychotherapy, especially cognitive therapy, can be at least as effective as drugs, and for many patients appear to be more effective. This is good news for individuals who prefer to be treated without medications—due to personal preferences or health concerns. It is also good news for

the millions of individuals who have not responded adequately to antidepressants after years and years of treatment and who still struggle with depression and anxiety.

- Following recovery from depression, patients treated with psychotherapy are more likely to remain undepressed and are significantly less likely to relapse than patients treated with antidepressants alone. This is especially important because of the growing awareness that many people relapse following recovery from depression, especially if they are treated with antidepressant medications alone without any talking therapy.

Based on these findings, Dr. Antonuccio and his coauthors concluded that psychotherapy should not be considered a second-rate treatment but should usually be the initial treatment for depression. In addition, they emphasized that cognitive therapy appears to be one of the most effective psychotherapies for depression, if not the most effective.

Of course, medications can be helpful for some individuals—even life-saving. Medications can be combined with psychotherapy for maximum effect as well, especially when the depression is severe. It is extremely important to know that we have powerful new weapons to fight depression, and that drug-free treatments such as cognitive therapy can be highly effective.

Recent studies indicate that psychotherapy can be helpful not only for mild depressions, but also for severe depressions as well. These findings are at odds with the popular belief that "talking therapy" can only help people with mild problems, and that if you have a serious depression you need to be treated with drugs.

Although we are taught that depression may result from an imbalance in brain chemistry, recent studies indicate that cognitive behavioral therapy may actually change brain chemistry. In these studies, Drs. Lewis R. Baxter, Jr., Jeffrey M. Schwartz, Kenneth S. Bergman, and their colleagues at UCLA School of Medicine, used positron emission tomography (PET scanning) to evaluate changes in brain metabolism in two groups of patients before and after treatment.[2] One group received cognitive behavioral therapy and no drugs, and the other group received an antidepressant medication and no psychotherapy.

As one might expect, there were changes in brain chemistry in the patients in the drug therapy group who improved. These changes indicated that their brain metabolism had slowed down—in other words, the nerves in a certain region of the brain appeared to become more "relaxed." What came as quite a surprise was there were similar changes in the brains of the patients successfully treated with cognitive behavioral therapy. However, these patients received no medications. Further, there were *no significant differences* in the brain changes in the drug therapy and psychotherapy groups, or in the effectiveness of the two treatments. Because of these and other similar studies, investigators are starting for the first time to entertain the possibility that cognitive behavior therapy—the methods described in this book—may actually help people by changing the chemistry and architecture of the human brain!

Although no one treatment will ever be a panacea, research studies indicate that cognitive therapy can be helpful for a variety of disorders in addition to depression. For example, in several studies patients with panic attacks have responded so well to cognitive therapy without any medications that many experts now consider cognitive therapy alone to be the best treatment for this disorder. Cognitive therapy can also be helpful in many other forms of anxiety (such as chronic worrying, phobias, obsessive-compulsive disorder, and post-traumatic stress disorder), and is also being used with some success in the personality disorders, such as borderline personality disorder.

Cognitive therapy is gaining popularity in the treatment of many other disorders as well. At the 1998 Stanford Psychopharmacology Conference, I was intrigued by the presentation by a colleague from Stanford, Dr. Stuart Agras. Dr.

Agras is a renowned expert in eating disorders such as binge eating, anorexia nervosa, and bulimia. He presented the results of numerous recent studies on the treatment of eating disorders with antidepressant medications versus psychotherapy. These studies indicated that cognitive behavior therapy is the most effective treatment for eating disorders—better than any known drug or any other form of psychotherapy.*

We are also beginning to learn more about *how* cognitive therapy works. One important discovery is that self-help seems to be a key to recovery whether or not you receive treatment. In a series of five remarkable studies published in the prestigious *Journal of Consulting and Clinical Psychology* and in *The Gerontologist*, Dr. Forest Scogin and his colleagues at the University of Alabama studied the effects of simply reading a good self-help book like *Feeling Good*—without any other therapy. The name of this new type of treatment is "bibliotherapy" (reading therapy). They discovered that *Feeling Good* bibliotherapy may be as effective as a full course of psychotherapy or treatment with the best antidepressant drugs.[3–7] Given the tremendous pressures to cut health care costs, this is of considerable interest, since a paperback copy of the *Feeling Good* book costs less than two Prozac pills—and is presumably free of any troublesome side effects!

In a recent study, Dr. Scogin and his colleague, Dr. Christine Jamison, randomly assigned eighty individuals seeking treatment for a major depressive episode to one of two groups. The researchers gave the patients in the first group a copy of my *Feeling Good* and encouraged them to read it within four weeks. This group was called the Immediate Bibliotherapy Group. These patients also received a booklet containing blank copies of the self-help forms in the book in case they decided to do some of the suggested exercises in the book.

*No current treatment is a panacea, including cognitive therapy. Another new short-term therapy, called interpersonal therapy, has also shown some promise for patients with eating disorders. In the future, studies like those conducted by Dr. Agras and his colleagues will undoubtedly lead to more powerful and specific treatments for eating disorders.

Patients in the second group were told they would be placed on a four-week waiting list before beginning treatment. This group was called the Delayed Bibliotherapy Group because these patients were not given a copy of *Feeling Good* until the second four weeks of the study. The patients in the Delayed Bibliotherapy Group served as a control group to make sure that any improvement in the Immediate Bibliotherapy was not just due to the passage of time.

At the initial evaluation, the researchers administered two depression tests to all the patients. One was the Beck Depression Inventory (BDI), a time-honored self-assessment test that patients fill out on their own, and the second was the Hamilton Rating Scale for Depression (HRSD), which is administered by trained depression researchers. As you can see in Figure 1, there was no difference in the depression levels in the two groups at the initial evaluation. You can also see that the average scores for the patients in the Immediate Bibliotherapy Group and the Delayed Bibliotherapy group at the initial evaluation were both around 20 or above on the BDI and on the HRSD. These scores indicate that the depression levels in both groups were similar to the depression levels in most published studies of antidepressants or psychotherapy. In fact, the BDI score was nearly identical to the average BDI scores of approximately five hundred patients seeking treatment at my clinic in Philadelphia during the late 1980s.

Every week a research assistant called the patients in both groups and administered the BDI by telephone. The assistant also answered any questions patients had about the study and encouraged the patients in the Immediate Bibliotherapy Group to try to complete the book within four weeks. These calls were limited to ten minutes and no counseling was offered.

At the end of the four weeks, the two groups were compared. You can see in Figure 1 that the patients in the Immediate Bibliotherapy Group improved considerably. In fact, the average scores on both the BDI and HRSD were around 10 or below, scores in the range considered normal.

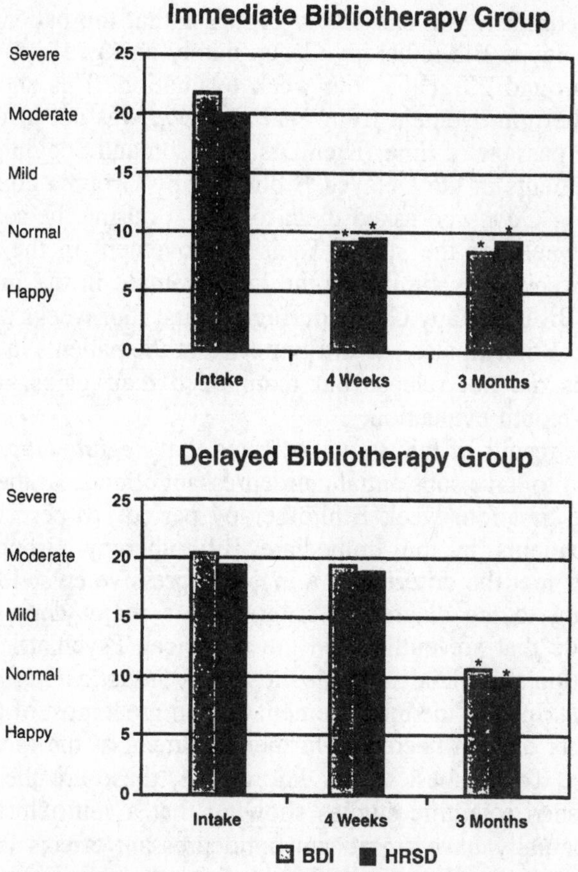

Figure 1. The patients in the Immediate Bibliotherapy Group (top chart) received *Feeling Good* at the intake evaluation. The patients in the Delayed Bibliotherapy Group (bottom chart) received *Feeling Good* at the four-week evaluation. BDI=Beck Depression Inventory. HRSD=Hamilton Rating Scale for Depression.

These changes in depression were very significant. You can also see that the patients maintained their gains at the three-month evaluation and did not relapse. In fact, there was a tendency for continued improvement following the completion of the bibliotherapy treatment; the scores on both depression tests were actually lower at the three-month evaluation.

In contrast, you can see in Figure 1 that the patients in the Delayed Bibliotherapy Group barely changed and were still around 20 at the four-week evaluation. This showed that the improvement from *Feeling Good* was not just due to the passage of time. Then Drs. Jamison and Scogin gave the patients in the Delayed Bibliotherapy Group a copy of *Feeling Good* and asked them to read it during the second four weeks of the study. Their improvement in the next four weeks was similar to the improvement in the Immediate Bibliotherapy Group during the first four weeks of the study. You can also see in Figure 1 that the patients in both groups did not relapse but maintained their gains at the three-month evaluation.

The results of this study indicated that *Feeling Good* appeared to have substantial antidepressant effects. At the end of the first four-week Bibliotherapy period, 70 percent of the patients in the Immediate Bibliotherapy Group no longer met the criteria for a major depressive episode, according to the diagnostic criteria for a major depressive episode that are outlined in the American Psychiatric Association's official *Diagnostic and Statistical Manual* (*DSM*). In fact, the improvement was so great most of these patients did not need any further treatment at the medical center. To the best of my knowledge, these are the first published scientific studies showing that a self-help book can actually have significant antidepressant effects in patients suffering from episodes of major depression.

In contrast, only 3 percent of the patients in the Delayed Bibliotherapy Group recovered during the first four weeks. In other words, the patients who did not read *Feeling Good* failed to improve. However, at the three-month evaluation, when both groups had read *Feeling Good*, 75 percent of the patients in the Immediate Bibliotherapy Group and 73 percent of the patients in the Delayed Bibliotherapy Group no longer qualified for a diagnosis of major depressive episode according to DSM criteria.

The researchers compared the magnitude of the improvement in these groups with the amount of improvement in published outcome studies using antidepressant medications or psychotherapy or both. In the large National Institute of Mental Health Collaborative Depression study, there was an average reduction of 11.6 points on the HRSD in patients who received cognitive therapy from highly trained therapists for twelve weeks. This was very similar to the 10.6-point change in the HRSD observed in the patients who read *Feeling Good* after just four weeks. However, the bibliotherapy treatment seemed to work significantly faster. My own clinical experience confirms this. In my private practice, very few patients have recovered during the first four weeks of treatment.

The percentage of patients who dropped out of the bibliotherapy therapy was also very small, around 10 percent. This is less than most published outcome studies using drugs or psychotherapy, which typically have dropout rates from 15 percent to over 50 percent. Finally, the patients developed significantly more positive attitudes and thinking patterns after reading *Feeling Good*. This was consistent with the premise of the book; namely, that you can defeat depression by changing the negative thinking patterns that cause it.

The researchers concluded that the bibliotherapy was effective for patients suffering from depression and might also have a significant role in public education and in depression prevention programs. They speculated that *Feeling Good* bibliotherapy might help prevent serious episodes of depression among individuals with a tendency toward negative thinking.

Finally, the researchers addressed another important concern: would the antidepressant effects of *Feeling Good* last? Skillful motivational speakers can get a crowd of people excited and optimistic for brief periods of time—but these brief mood-elevating effects often don't last. The same problem holds for the treatment of depression. Following successful treatment with drugs or psychotherapy, many patients feel tremendously improved—only to relapse back into depression after a period of time. These relapses can be devastating because patients feel so demoralized.

In 1997, the investigators reported the results of a three-year follow-up of the patients in the study I've just described.[7] The authors were Drs. Nancy Smith, Mark Floyd, and Forest Scogin from the University of Alabama and Dr. Christine Jamison from the Tuskegee Veterans Affairs Medical Center. The researchers contacted the patients three years after reading *Feeling Good* and administered the depression tests once again. They also asked the patients several questions about how they had been doing since the completion of the study. The researchers learned that the patients did not relapse but maintained their gains during this three-year period. In fact, the scores on the two depression tests at the three-year evaluation were actually slightly better than the scores at the completion of the bibliotherapy treatment. More than half of the patients said that their moods continued to improve following the completion of the initial study.

The diagnostic findings at the three-year evaluation confirmed this—72 percent of the patients still did not meet the criteria for a major depressive episode, and 70 percent did not seek or receive any further treatment with medications or psychotherapy during the follow-up period. Although they experienced the normal ups and downs we all feel from time to time, approximately half indicated that when they were upset, they opened up *Feeling Good* and reread the most helpful sections. The researchers speculated that these self-administered "booster sessions" may have been important in maintaining a positive outlook following recovery. Forty percent of the patients said that the best part of the book was that it helped them change their negative thinking patterns, such as learning to be less perfectionistic and to give up all-or-nothing thinking.

Of course, this study had limitations, like all studies. For one thing, not every patient was "cured" by reading *Feeling Good*. No treatment is a panacea. While it is encouraging that many patients seem to respond to reading *Feeling Good*, it is also clear that some patients with more severe or chronic depressions will need the help of a therapist and possibly an antidepressant medication as well. This is nothing to be ashamed of. Different individuals respond better to different approaches. It is good that we now have three types of effective treatment for depression: antidepressant medications, individual and group psychotherapy, and bibliotherapy.

Remember that you can use the cognitive bibliotherapy between therapy sessions to speed your recovery even if you are in treatment. In fact, when I first wrote *Feeling Good*, this is how I imagined the book would be used. I intended it to be a tool my patients could use between therapy sessions to speed up the treatment and never dreamed that it might someday be used alone as a treatment for depression.

It appears that more and more therapists are beginning to assign bibliotherapy to their patients as psychotherapy "homework" between therapy sessions. In 1994, the results of a nationwide survey about the use of bibliotherapy by mental health professionals were published in the *Authoritative Guide to Self-Help Books* (published by Guilford Press, New York). Drs. John W. Santrock and Ann M. Minnet from the University of Texas in Dallas and Barbara D. Campbell, a research associate at the university, conducted this study. These three researchers surveyed five hundred American mental health professionals from all fifty states and asked whether they "prescribed" books for patients to read between sessions to speed recovery. Seventy percent of the therapists polled indicated that they had recommended at least three self-help books to their patients during the previous year, and 86 percent reported that these books provided a positive benefit to their patients. The therapists were also asked which self-help books, from a list of one thousand, they most frequently recommended for their patients. *Feeling Good* was the number-one-rated book for depressed patients, and my *Feeling Good Handbook* (published as a Plume paperback in 1989) was rated number two.

I was not aware this survey was being conducted, and was thrilled to learn about the results of it. One of my goals when I wrote *Feeling Good* was to provide reading for my own patients to speed their learning and recovery between therapy sessions, but I never dreamed this idea would catch on in such a big way!

Should you expect to improve or recover after reading *Feeling Good*? That would be unreasonable. The research clearly indicates that while many people who read *Feeling Good* improved, others needed the additional help of a mental health professional. I have received many letters (probably more than ten thousand) from people who read *Feeling Good*. Many of them kindly described in glowing terms how *Feeling Good* had helped them, often after years and years of unsuccessful treatment with medications and even electroconvulsive therapy. Others indicated that they found the ideas in *Feeling Good* appealing but needed a referral to a good local therapist to make these ideas work for them. This is understandable—we are all different, and it would be unrealistic to think that any one book or form of therapy would be the answer for everyone.

Depression is one of the worst forms of suffering, because of the immense feelings of shame, worthlessness, hopelessness, and demoralization. Depression can seem worse than terminal cancer, because most cancer patients feel loved and they have hope and self-esteem. Many depressed patients have told me, in fact, that they yearned for death and prayed every night that they would get cancer, so they could die in dignity without having to commit suicide.

But no matter how terrible your depression and anxiety may feel, the prognosis for recovery is excellent. You may be convinced that your own case is so bad, so overwhelming and hopeless, that you are the one person who will never get well, no matter what. But sooner or later, the clouds have a way of blowing away and the sky suddenly clears and the sun begins to shine again. When this happens, the feelings of relief and joy can be overwhelming. And if you are now struggling with depression and low self-esteem, I believe this transformation can happen for you as well, no matter how discouraged or depressed you may feel.

Well, it's time to get on to Chapter 1 so we can start to work together. I want to wish you the very best as you read it, and hope you find these ideas and methods helpful!

David D. Burns, M.D.
Clinical Associate Professor of Psychiatry and Behavioral Sciences,
Stanford University School of Medicine

References

1. Antonuccio, D. O., Danton, W. G., & DeNelsky, G. Y. (1995). Psychotherapy versus medication for depression: Challenging the conventional wisdom with data. *Professional Psychology: Research and Practice*, 26(6), 574–585.

2. Baxter, L. R., Schwartz, J. M., & Bergman, K. S., et al. (1992). Caudate glucose metabolic rate changes with both drug and behavioral therapy for obsessive-compulsive disorders. *Archives of General Psychiatry*, 49, 681–689.

3. Scogin, F., Jamison, C., & Gochneaut, K. (1989). The comparative efficacy of cognitive and behavioral bibliotherapy for mildly and moderately depressed older adults. *Journal of Consulting and Clinical Psychology*, 57, 403–407.

4. Scogin, F., Hamblin, D., & Beutler, L. (1987). Bibliotherapy for depressed older adults: A self-help alternative. *The Gerontologist*, 27, 383–387.

5. Scogin, F., Jamison, C., & Davis, N. (1990). A two-year follow-up of the effects of bibliotherapy for depressed

older adults. *Journal of Consulting and Clinical Psychology*, 58, 665–667.

6. Jamison, C., & Scogin, F. (1995). Outcome of cognitive bibliotherapy with depressed adults. *Journal of Consulting and Clinical Psychology*, 63, 644–650.

7. Smith, N. M., Floyd, M. R., Jamison, C., & Scogin, F. (1997). Three-year follow-up of bibliotherapy for depression. *Journal of Consulting and Clinical Psychology*, 65(2), 324–327.

Part I
Theory and Research

Chapter 1

A Breakthrough in the Treatment of Mood Disorders

Depression has been called the world's number one public health problem. In fact, depression is so widespread it is considered the common cold of psychiatric disturbances. But there is a grim difference between depression and a cold. Depression can kill you. The suicide rate, studies indicate, has been on a shocking increase in recent years, even among children and adolescents. This escalating death rate has occurred in spite of the billions of antidepressant drugs and tranquilizers that have been dispensed during the past several decades.

This might sound fairly gloomy. Before you get even *more* depressed, let me tell you the good news. Depression is an illness and not a necessary part of healthy living. What's more important—you *can* overcome it by learning some simple methods for mood elevation. A group of psychiatrists and psychologists at the University of Pennsylvania School of Medicine has reported a significant breakthrough in the treatment and prevention of mood disorders. Dissatisfied with traditional methods for treating depression because they found them to be slow and ineffective, these doctors developed and systematically tested an entirely new and remarkably successful approach to depression and other emotional disorders. A series of recent studies confirms that these techniques reduce the symptoms of depression much more rapidly than conventional psychotherapy or drug therapy. The name of this revolutionary treatment is "cognitive therapy."

I have been centrally involved in the development of cognitive therapy, and this book is the first to describe these methods to the general public. The systematic application and scientific evaluation of this approach in treating clinical depression traces its origins to the innovative work of Drs. Albert Ellis and Aaron T. Beck, who began to refine their unique approach to mood transformation in the mid-1950's and early 1960's.* Their pioneering efforts began to emerge into prominence in the past decade because of the research that many mental-health professionals have undertaken to refine and evaluate cognitive therapy methods at academic institutions in the United States and abroad.

Cognitive therapy is a fast-acting technology of mood modification that you can learn to apply on your own. It can help you eliminate the symptoms and experience personal growth so you can minimize future upsets and cope with depression more effectively in the future.

The simple, effective mood-control techniques of cognitive therapy provide:

1. *Rapid Symptomatic Improvement*: In milder depressions, relief from your symptoms can often be observed in as short a time as twelve weeks.

---

*The idea that your thinking patterns can profoundly influence your moods has been described by a number of philosophers in the past 2500 years. More recently, the cognitive view of emotional disturbances has been explored in the writings of many psychiatrists and psychologists including Alfred Adler, Albert Ellis, Karen Horney, and Arnold Lazarus, to name just a few. A history of this movement has been described in Ellis, A., *Reason and Emotion in Psychotherapy*. New York: Lyle Stuart, 1962.

2. *Understanding*: A clear explanation of why you get moody and what you can do to change your moods. You will learn what causes your powerful feelings; how to distinguish "normal" from "abnormal" emotions; and how to diagnose and assess the severity of your upsets.
3. *Self-control*: You will learn how to apply safe and effective coping strategies that will make you feel better whenever you are upset. I will guide you as you develop a practical, realistic, step-by-step self-help plan. As you apply it, your moods can come under greater voluntary control.
4. *Prevention and Personal Growth*: Genuine and long-lasting prophylaxis (prevention) of future mood swings can effectively be based on a reassessment of some basic values and attitudes which lie at the core of your tendency toward painful depressions. I will show you how to challenge and reevaluate certain assumptions about the basis for human worth.

The problem-solving and coping techniques you learn will encompass every crisis in modern life, from minor irritations to major emotional collapse. These will include realistic problems, such as divorce, death, or failure, as well as those vague, chronic problems that seem to have no obvious external cause, such as low self-confidence, frustration, guilt, or apathy.

The question may now occur to you, "Is this just another self-help pop psychology?" Actually, cognitive therapy is one of the first forms of psychotherapy which has been shown to be effective through rigorous scientific research under the critical scrutiny of the academic community. This therapy is unique in having professional evaluation and validation at the highest academic levels. It is *not* just another self-help fad but a major development that has become an important part of the mainstream of modern psychiatric research and practice. Cognitive therapy's academic foundation has enhanced its impact and should give it staying power for years to come. But don't be turned off by the professional status that cognitive therapy has acquired. Unlike much traditional psychotherapy, it is not occult and anti-intuitive. It is practical and based on common sense, and you can make it work for you.

The first principle of cognitive therapy is that *all* your moods are created by your "cognitions," or thoughts. A cognition refers to the way you look at things—your perceptions, mental attitudes, and beliefs. It includes the way you interpret things—what you say about something or someone to yourself. You *feel* the way you do right now because of the *thoughts you are thinking at this moment*.

Let me illustrate this. How have you been feeling as you read this? You might have been thinking, "Cognitive therapy sounds too good to be true. It would never work for me." If your thoughts run along these lines, you are feeling skeptical or even discouraged. What causes you to feel that way? Your *thoughts*. You create those feelings by the dialogue you are having with yourself about this book!

Conversely, you may have felt a sudden uplift in mood because you thought, "Hey, this sounds like something which might finally help me." Your emotional reaction is generated *not* by the sentences you are reading but by the way you are *thinking*. The moment you have a certain thought and believe it, you will experience an immediate emotional response. Your thought actually *creates* the emotion.

The second principle is that when you are feeling depressed, your thoughts are dominated by a pervasive negativity. You perceive not only yourself but the entire world in dark, gloomy terms. What is even worse—you'll come to believe things *really are* as bad as you imagine them to be.

If you are substantially depressed, you will even begin to believe that things always have been and always will be negative. As you look into your past, you remember all the bad things that have happened to you. As you try to imagine the future, you see only emptiness or unending problems and anguish. This bleak vision creates a sense of hopelessness. This feeling is absolutely illogical, but it seems so real that you have convinced yourself that your inadequacy will go on forever.

The third principle is of substantial philosophical and therapeutic importance. Our research has documented that the negative thoughts which cause your emotional turmoil nearly *always* contain gross distortions. Although these thoughts appear valid, you will learn that they are irrational or just plain wrong, and that twisted thinking is a major *cause* of your suffering.

The implications are important. Your depression is probably not based on accurate perceptions of reality but is often the product of mental slippage.

Suppose you believe that what I've said has validity. What good will it do you? Now we come to the most important result of our clinical research. You can learn to deal with your moods more effectively if you master methods that will help you pinpoint and eliminate the mental distortions which cause you to feel upset. As you begin to think more objectively, you will begin to feel better.

How effective is cognitive therapy compared with other established and accepted methods for treating depression? Can the new therapy enable severely depressed individuals to get better without drugs? How rapidly does cognitive therapy work? Do the results last?

Several years ago a group of investigators at the Center for Cognitive Therapy at the University of Pennsylvania School of Medicine including Drs. John Rush, Aaron Beck, Maria Kovacs and Steve Hollon began a pilot study comparing cognitive therapy with one of the most widely used and effective antidepressant drugs on the market, Tofranil (imipramine hydrochloride). Over forty severely depressed patients were randomly assigned to two groups. One group was to receive individual cognitive therapy sessions and no drugs, while the other group would be treated with Tofranil and no therapy. This either-or research design was chosen because it provided the maximum opportunity to see how the treatments compared. Up to that time, no form of psychotherapy had been shown to be as effective for depression as treatment with an antidepressant drug. This is why antidepressants have experienced such a wave of interest from the media, and have come to be regarded by the professional community in the past two decades as the best treatment for most serious forms of depression.

Both groups of patients were treated for a twelve-week period. All patients were systematically evaluated with extensive psychological testing prior to therapy, as well as at several monthly intervals for one year after completion of treatment. The doctors who performed the psychological tests were not the therapists who administered the treatment. This ensured an objective assessment of the merits of each form of treatment.

The patients were suffering from moderate to severe depressive episodes. The majority had failed to improve in spite of previous treatment with two or more therapists at other clinics. Three quarters were suicidal at the time of their referral. The average patient had been troubled by chronic or intermittent depression for eight years. Many were absolutely convinced their problems were insoluble, and felt their lives were hopeless. Your own mood problems may not seem as overwhelming as theirs. A tough patient population was chosen so that the treatment could be tested under the most difficult, challenging conditions.

The outcome of the study was quite unexpected and encouraging. The cognitive therapy was at least as effective as, if not more effective than, the antidepressant drug therapy. As you can see (Table 1–1, page 15), fifteen of the nineteen patients treated with cognitive therapy had shown a substantial reduction of symptoms after twelve weeks of active treatment.* An additional two individuals had im-

*Table 1-1 was adapted from Rush, A. J., Beck, A. T., Kovacs, M., and Hollon, S. "Comparative Efficacy of Cognitive Therapy and Pharmacotherapy in the Treatment of Depressed Outpatients." *Cognitive Therapy and Research*, Vol. 1, No. I, March 1977, pp. 17–38.

Table 1–1. Status of 44 Severely Depressed Patients, 12 Weeks After Beginning Treatment

Number Who Entered Treatment	*Patients Treated with Cognitive Therapy Only* 19	*Patients Treated with Antidepressant Drug Therapy Only* 25
Number who had recovered completely*	15	5
Number who were considerably improved but still experienced borderline to mild depression	2	7
Number who were not substantially improved	1	5
Number who dropped out of treatment	1	8

*The superior improvement of the patients treated with cognitive therapy was statistically significant

proved, but were still experiencing borderline to mild depression. Only one patient had dropped out of treatment, and one had not yet begun to improve at the end of this period. In contrast, only five of the twenty-five patients assigned to antidepressant drug therapy had shown complete recovery by the end of the twelve-week period. Eight of these patients dropped out of therapy as a result of the adverse side effects of the medication, and twelve others showed no improvement or only partial improvement.

Of particular importance was the discovery that many patients treated with cognitive therapy improved more rapidly than those successfully treated with drugs. Within the first week or two, there was a pronounced reduction in suicidal thoughts among the cognitive therapy group. The effectiveness of cognitive therapy should be encouraging for individuals who prefer not to rely on drugs to raise their spirits, but prefer to develop an understanding of what is troubling them and do something to cope with it.

How about those patients who had not recovered by the end of twelve weeks? Like any form of treatment, this one is not a panacea. Clinical experience has shown that all individuals do not respond as rapidly, but most can nevertheless improve if they persist for a longer period of time. Sometimes this is hard work! One particularly encouraging development for individuals with refractory severe depressions is a recent study by Drs. Ivy Blackburn and her associates at the Medical Research Council at the University of Edinburgh in Scotland.* These investigators have shown that the combination of antidepressant drugs with cognitive therapy can be more effective than either modality above. In my experience the most crucial predictor of recovery is a persistent willingness to exert some effort to help yourself. Given this attitude, you will succeed.

Just how much improvement can you hope for? The average cognitively treated patient experienced a substantial elimination of symptoms by the end of treatment. Many reported they felt the happiest they had ever felt in their lives. They emphasized that the mood-training brought about a sense of self-esteem and confidence. No matter how miserable, depressed, and pessimistic you now feel, I am convinced that you can experience beneficial effects if you are willing to apply the methods described in this book with persistence and consistency.

How long do the effects last? The findings from follow-up studies during the year after completion of treatment are quite interesting. While many individuals from both groups had occasional mood swings at various times during the

*Blackburn, I. M., Bishop, S., Glen, A. I. M., Whalley, L. J. and Christie, J. E. "The Efficacy of Cognitive Therapy in Depression. A Treatment Trial Using Cognitive Therapy and Pharmacotherapy, Each Alone and in Combination." *British Journal of Psychiatry*, Vol. 139, January 1981, pp. 181–189.

year, both groups continued on the whole to maintain the gains they had demonstrated by the end of twelve weeks of active treatment.

Which group actually fared better during the follow-up period? The psychological tests, as well as the patients' own reports, confirmed that the cognitive therapy group continued to feel substantially better, and these differences were statistically significant. The relapse rate over the course of the year in the cognitive therapy group was less than half that observed in the drug patients. These were sizable differences that favored the patients treated with the new approach.

Does this mean that I can guarantee you will never again have the blues after using cognitive methods to eliminate your current depression? Obviously not. That would be like saying that once you have achieved good physical condition through daily jogging, you will never again be short of breath. Part of being human means getting upset from time to time, so I can guarantee you *will not* achieve a state of never-ending bliss! This means you will have to reapply the techniques that help you if you want to continue to master your moods. There's a difference between *feeling* better—which can occur spontaneously—and *getting* better—which results from systematically applying and reapplying the methods that will lift your mood whenever the need arises.

How has this work been received by the academic community? The impact of these findings on psychiatrists, psychologists, and other mental-health professionals has been substantial. It has now been twenty years since this chapter was first written. During that time, numerous well-controlled studies of the effectiveness of cognitive therapy have been published in scientific journals. These studies have compared the effectiveness of cognitive therapy with the effectiveness of antidepressant medications as well as other forms of psychotherapy in the treatment of depression, anxiety, and other disorders. The results of these studies have been quite encouraging. Researchers have confirmed our early impressions that cognitive therapy was at least as effective as medications, and often more effective, both in the short term and in the long term.

What does this all add up to? We are experiencing a crucial development in modern psychiatry and psychology—a promising new approach to understanding human emotions based on a cogent testable therapy. Large numbers of mental-health professionals are now showing a great interest in this approach, and the ground swell seems to be just beginning.

Since the first edition of *Feeling Good* in 1980, many thousands of depressed individuals have been successfully treated with cognitive therapy. Some had considered themselves hopelessly untreatable and came to us as a last-ditch effort before commmitting suicide. Many others were simply troubled by the nagging tensions of daily living and wanted a greater share of personal happiness. This book is a carefully thought-out practical application of our work, and it is designed for you. Good luck!

Chapter 2

How to Diagnose Your Moods: The First Step in the Cure

Perhaps you are wondering if you have in fact been suffering from depression. Let's go ahead and see where you stand. The Burns Depression Checklist (BDC) (see Table 2–1, page 20) is a reliable mood-measuring device that detects the presence of depression and accurately rates its severity.* This simple questionnaire will take only a few minutes to complete. After you have completed the BDC, I will show you how to make a simple interpretation of the results, based on your total score. Then you will know immediately whether or not you are suffering from a true depression and, if so, how severe it is. I will also lay out some important guidelines to help you determine whether you can safely and effectively treat your own blue mood using this book as your guide, or whether you have a more serious emotional disorder and might benefit from professional intervention in addition to your own efforts to help yourself.

As you fill out the questionnaire, read each item carefully and put a check (√) in the box that indicates how you have been feeling during the past few days. Make sure you check one answer for each of the twenty-five items.

If in doubt, make your best guess. Do not leave any questions unanswered. Regardless of the outcome, this can be your first step toward emotional improvement.

*Some readers may recall that I included the Beck Depression Inventory

Table 2–1. Burns Depression Checklist*

Instructions: Put a check (✔) to indicate how much you have experienced each symptom **during the past week,** including today. **Please answer all 25 items.**

	0—Not At All	1—Somewhat	2—Moderately	3—A Lot	4—Extremely
Thoughts and Feelings					
1. Feeling sad or down in the dumps					
2. Feeling unhappy or blue					
3. Crying spells or tearfulness					
4. Feeling discouraged					
5. Feeling hopeless					
6. Low self-esteem					
7. Feeling worthless or inadequate					
8. Guilt or shame					
9. Criticizing yourself or blaming yourself					
10. Difficulty making decisions					
Activities and Personal Relationships					
11. Loss of interest in family, friends or colleagues					
12. Loneliness					
13. Spending less time with family or friends					
14. Loss of motivation					
15. Loss of interest in work or other activities					
16. Avoiding work or other activities					
17. Loss of pleasure or satisfaction in life					
Physical Symptoms					
18. Feeling tired					
19. Difficulty sleeping or sleeping too much					
20. Decreased or increased appetite					
21. Loss of interest in sex					
22. Worrying about your health					

Burns Depression Checklist continued

	0—Not At All	1—Somewhat	2—Moderately	3—A Lot	4—Extremely
Suicidal Urges**					
23. Do you have any suicidal thoughts?					
24. Would you like to end your life?					
25. Do you have a plan for harming yourself?					
Please Total Your Score on Items 1 to 25 Here →					

**Anyone with suicidal urges should seek help from a mental health professional.

Interpreting the Burns Depression Checklist. Now that you have completed the test, add up the score for each of the twenty-five items and obtain the total. Since the highest score you can get on each of the twenty-five symptoms is 4, the highest score for the whole test would be 100. (This would indicate the most severe depression possible.) Since the lowest score for each item is 0, the lowest score for the test would be zero. (This would indicate no symptoms of depression at all.)

You can now evaluate your depression according to Table 2–2. As you can see, the higher the total score, the more severe your depression. In contrast, the lower the score, the better you are feeling.

Although the BDC is not difficult or time-consuming to fill out and score, don't be deceived by its simplicity. You have just learned to use a highly sophisticated tool for de-

(BDI) in the 1980 edition of *Feeling Good.* The BDI is a time-honored instrument that has been used in hundreds of research studies on depression. Dr. Aaron Beck, the creator of this test, deserves a great deal of credit for creating the BDI during the early 1960s. It was one of the first instruments for measuring depression in clinical and research settings, and I was grateful for his permission to reproduce it in the earlier edition of *Feeling Good.*

Table 2–2. Interpreting the Burns Depression Checklist

Total Score	*Level of Depression**
0–5	no depression
6–10	normal but unhappy
11–25	mild depression
26–50	moderate depression
51–75	severe depression
76–100	extreme depression

*Anyone with a persistent score above 10 may benefit from professional treatment. Anyone with suicidal feelings should seek an immediate consultation with a mental health professional.

tecting depression and measuring its severity. Research studies have demonstrated that the BDC is highly accurate and reliable. Studies in a variety of settings, such as psychiatric emergency rooms, have indicated that instruments of this type actually pick up the presence of depressive symptoms far more frequently than formal interviews by experienced clinicians.

---

*Mental health professionals may be interested to learn that the psychometric properties of the BDC are excellent. The reliability of the twenty-five-item BDC has been assessed in a group of ninety depressed outpatients seeking treatment at the Center for Cognitive Therapy in Oakland, California, and in a group of 145 outpatients seeking treatment at a Kaiser facility in Atlanta, Georgia. The reliability was extremely high and identical in both groups (Cronbach's coefficient alpha = 95%). The high correlation between the BDC and the BDI $r(68) = .88$, $p < .01$ in the Oakland group indicates that these two scales assess a similar if not identical construct. When both instruments were purged of errors of measurement using structural equation modeling techniques, the correlation between the scales was not significantly different from 1.0. The BDC was also normed against the widely used depression subscale of the Hopkins Symptom Checklist-90 in the Atlanta, Georgia, sample. The extremely high correlation between the two measures $r(131) = .90$, $p < .01$ further confirmed the validity of the BDC.

Extensive clinical experience with the BDC in a variety of treatment settings indicates it is well accepted by patients. Many have commented that the test is easy to complete and score and helpful for tracking changes in symptoms over time. A brief, five-item BDC with outstanding psychometric properties has also been developed. The brief BDC is ideal for testing patients on a session-by-session basis because patients can com-

You can use the BDC with confidence to monitor your progress as well. In my clinical work, I have insisted that every patient must fill out the test on his or her own between all sessions and report the score to me at the beginning of the next session. Changes in the score show me whether the patient is getting better, worse, or staying the same.

As you apply the various self-help techniques described in this book, take the BDC test at regular intervals to assess your progress objectively. I suggest a minimum of once a week. Compare it to weighing yourself regularly when you're on a diet. You will notice that various chapters in this book focus on different symptoms of depression. As you learn to overcome these symptoms, you will find that your total score will begin to fall. This will show that you are improving. When your score is under ten, you will be in the range considered normal. When it is under five, you will be feeling especially good. Ideally, I'd like to see your score under five the majority of the time. This is one aim of your treatment.

Is it safe for depressed individuals to try to help themselves using the principles and methods outlined in this book? The answer is—*definitely yes*! This is because the crucial decision *to try to help yourself* is the key that will allow you to feel better as soon as possible, regardless of how severe your mood disturbance might seem to be.

Under what conditions should you seek professional help? If your score is between 0 and 5, you are probably feeling good already. This is in the range of normal, and most people with scores this low feel pretty happily contented.

If your score was between 6 and 10, it is still in the range of normal, but you are probably feeling a bit on the "lumpy" side. There's room for improvement, a little mental "tune-up," if you will. The cognitive therapy techniques in this book can often be remarkably helpful in these instances. Problems

---

plete it in less than one minute. It has performed well with adults and adolescents in a variety of psychiatric and medical settings, including recently arrested juveniles in the California judicial system. Mental health professionals who are interesting in learning more about these and many other assessment instruments that can be used in clinical or research settings (including an electronic patient testing module) are cordially invited to visit my Web site at www.FeelingGood.com

in daily living bug all of us, and a change of perspective can often make a big difference in how you feel.

If your score was between 11 and 25, your depression, at least at this time, is mild and should not be a cause for alarm. You will definitely want to correct this problem, and you may be able to make substantial progress on your own. Systematic self-help efforts along the lines proposed in this book, combined with frank communication on a number of occasions with a trusted friend, may help a great deal. But if your score remains in this range for more than a few weeks, you should consider professional treatment. The help of a therapist or an antidepressant medication may considerably speed your recovery.

Some of the thorniest depressions I have treated were actually individuals whose scores were in the mild range. Often these individuals had been mildly depressed for years, sometimes for most of their entire life. A mild chronic depression that goes on and on is now called "dysthymic disorder." Although that is a big, fancy-sounding term, it has a simple meaning. All it means is, "this person is awfully gloomy and negative most of the time." You probably know someone who is like that, and you may have fallen into spells of pessimism yourself. Fortunately, the same methods in this book that have proven so helpful for severe depressions can also be very helpful for these mild, chronic depressions.

If you scored between 26 and 50 on the BDC, it means you are moderately depressed. But don't be fooled by the term, "moderate." A score in this range can indicate pretty intense suffering. Most of us can feel quite upset for brief periods, but we usually snap out of it. If your score remains in this range for more than two weeks, you should definitely seek professional treatment.

If your score was above 50, it indicates your depression is severe or even extreme. This degree of suffering can be almost unbearable, especially when the score is increased above 75. Your moods are apt to be intensely uncomfortable and possibly dangerous because the feelings of despair and hopelessness may even trigger suicidal impulses.

Fortunately, the prognosis for successful treatment is excellent. In fact, sometimes the most severe depressions respond the most rapidly. But it is not wise to try to treat a severe depression on your own. A professional consultation is a must. Seek out a trusted and competent counselor.

Even if you receive psychotherapy or antidepressant medications, I am convinced you can still benefit greatly by applying what I teach you. My research studies have indicated that the spirit of self-help greatly speeds up recovery, even when patients receive professional treatment.

In addition to evaluating your total score on the BDC, be sure you pay special attention to items 23, 24, and 25. These items ask about suicidal feelings, urges, and plans. If you had elevated scores on any of these items, I would strongly recommend that you obtain professional help right away.

Many depressed individuals have elevated scores on item 23, but zeros on items 24 and 25. This usually means they have suicidal thoughts, such as "I'd probably be better off dead," but no actual suicidal intentions or urges and no plans to commit suicide. This pattern is quite common. If your scores on item 24 or 25 are elevated, however, this is a cause for alarm. Seek treatment *immediately*!

I have provided some effective methods for assessing and reversing suicidal impulses in a later chapter, but you must consult a professional when suicide begins to appear to be a desirable or necessary option. Your conviction that you are hopeless is the reason to seek treatment, not suicide. The majority of seriously depressed individuals believe they are hopeless beyond any shadow of a doubt. This destructive delusion is merely a symptom of the illness, not a fact. Your feeling that you are hopeless is powerful evidence that you are actually not!

It is also important for you to look at item 22, which asks if you have been more worried about your health recently. Have you experienced any unexplained aches, pains, fever. weight loss, or other possible symptoms of medical illness? If so, it would be worthwhile to have a medical consultation, which would include a history, a complete physical examination, and laboratory tests. Your doctor will probably give you a clean bill of health. This will suggest that your uncomfortable physical symptoms are related to your emotional state. Depression can mimic a great number of medical disorders because your mood swings often create a wide variety of puzzling physical symptoms. These include, to name just a few, constipation, diarrhea, pain, insomnia or the tendency to sleep too much, fatigue, loss of sexual interest, light-headedness, trembling, and numbness. As your depression improves, these symptoms will in all likelihood vanish. However, keep in mind that many treatable illnesses may initially masquerade as depression, and a medical examination could reveal an early (and life-saving) diagnosis of a reversible organic disorder.

There are a number of symptoms that indicate—but do not prove—the existence of a serious mental disturbance, and these require a consultation with and possible treatment by a mental-health professional, *in addition to* the self-administered personal-growth program in this book. Some of the major symptoms include: the belief that people are plotting and conspiring against you in order to hurt you or take your life; a bizarre experience which the ordinary person cannot understand; the conviction that external forces are controlling your mind or body; the feeling that other people can hear your thoughts or read your mind; hearing voices from outside your head; seeing things that aren't there; and receiving personal messages broadcast from radio or television programs.

These symptoms are not a part of depressive illness, but represent major mental disorders. Psychiatric treatment is a must. Quite often, people with these symptoms are convinced that nothing is wrong with them, and may meet the suggestion to seek psychiatric therapy with suspicious resentment and resistance. In contrast, if you are harboring the deep fear that you are going insane and are experiencing episodes of panic in which you sense you are losing control or going over the deep end, it is a near certainty that you are not. These are typical symptoms of ordinary anxiety, a much less serious disorder.

Mania is a special type of mood disorder with which you should be familiar. Mania is the opposite of depression and requires prompt intervention by a psychiatrist who can prescribe lithium. Lithium stabilizes extreme mood swings and allows the patient to lead a normal life. However, until therapy is initiated, the disease can be emotionally destructive. The symptoms include an abnormally elated or irritable mood that persists for at least two days and is not caused by drugs or alcohol. The manic patient's behavior is characterized by impulsive actions which reflect poor judgment (such as irresponsible, excessive spending) along with a grandiose sense of self-confidence. Mania is accompanied by increased sexual or aggressive activity; hyperactive, continuous body movements; racing thoughts; nonstop, excited talking; and a decreased need to sleep. Manic individuals have the delusion that they are extraordinarily powerful and brilliant, and often insist they are on the verge of some philosophical or scientific breakthrough or lucrative money-making scheme. Many famous creative individuals suffer from this illness and manage to control it with lithium. Because the disease *feels* so good, individuals who are having their first attack often cannot be convinced to seek treatment. The first symptoms are so intoxicating that the victim resists accepting the idea that his or her sudden acquisition of self-confidence and inner ecstasy is actually just a manifestation of a destructive illness.

After a while, the euphoric state may escalate into uncontrollable delirium requiring involuntary hospitalization, or it may just as suddenly switch into an incapacitating depression with pronounced immobility and apathy. I want you to be familiar with the symptoms of mania because a significant percentage of individuals who experience a true major depressive episode will at some later time develop these symptoms. When this occurs, the personality of the afflicted individual undergoes a profound transformation over a period of days or weeks. While psychotherapy and a self-help program can be extremely helpful, concomitant treatment with lithium under medical supervision is a must for an optimal response. With such treatment the prognosis for manic illness is excellent.

Let's assume that you do *not* have a strong suicidal urge, hallucinations, or symptoms of mania. Instead of moping and feeling miserable, you can now proceed to get better, using the methods outlined in this book. You can start enjoying life and work, and use the energy spent in being depressed for vital and creative living.

Chapter 3

Understanding Your Moods: You Feel the Way You Think

As you read the previous chapter, you became aware of how extensive the effects of depression are—your mood slumps, your self-image crumbles, your body doesn't function properly, your willpower becomes paralyzed, and your actions defeat you. That's why you feel so *totally* down in the dumps. What's the key to it all?

Because depression has been viewed as an emotional disorder throughout the history of psychiatry, therapists from most schools of thought place a strong emphasis on "getting in touch" with your feelings. Our research reveals the unexpected: Depression is not an emotional disorder at all! The sudden change in the way you *feel* is of no more causal relevance than a runny nose is when you have a cold. Every bad feeling you have is the result of your distorted negative thinking. Illogical pessimistic attitudes play the central role in the development and continuation of all your symptoms.

Intense negative thinking *always* accompanies a depressive episode, or any painful emotion for that matter. Your moody thoughts are likely to be entirely different from those you have when you are not upset. A young woman, about to receive her Ph.D., expressed it this way:

> Every time I become depressed, I feel as if I have been hit with a sudden cosmic jolt, and I begin to *see* things differently. The change can come within less than an hour. My thoughts become negative and pessimistic. As I look into the past, I become convinced that everything that I've ever done is worthless. Any happy period seems like an illusion. My accomplishments appear as genuine as the false facade for the set of a Western movie. I become convinced that the real me is worthless and inadequate. I can't move forward with my work because I become frozen with doubt. But I can't stand still because the misery is unbearable.

You will learn, as she did, that the negative thoughts that flood your mind are the actual *cause* of your self-defeating emotions. These thoughts are what keep you lethargic and make you feel inadequate. Your negative thoughts, or cognitions, are the most frequently overlooked symptoms of your depression. These cognitions contain the key to relief and are therefore your most important symptoms.

Every time you feel depressed about something, try to identify a corresponding negative thought you had just prior to and during the depression. Because these thoughts have actually created your bad mood, by learning to restructure them, you can change your mood.

You are probably skeptical of all this because your negative thinking has become such a part of your life that it has become automatic. For this reason I call negative thoughts "automatic thoughts." They run through your mind automatically without the slightest effort on your part to put them there. They are as obvious and natural to you as the way you hold a fork.

The relationship between the way you *think* and the way you *feel* is diagramed in Figure 3–1. This illustrates the first major key to understanding your moods: Your emotions result entirely from the way you *look* at things. It is an obvious neurological fact that before you can experience any event, you must process it with your mind and give it

Figure 3–1. The relationship between the world and the way you feel. It is not the actual events but your perceptions that result in changes in mood. When you are *sad*, your thoughts will represent a realistic interpretation of negative events. When you are depressed or anxious, your thoughts will always be illogical, distorted, unrealistic, or just plain wrong.

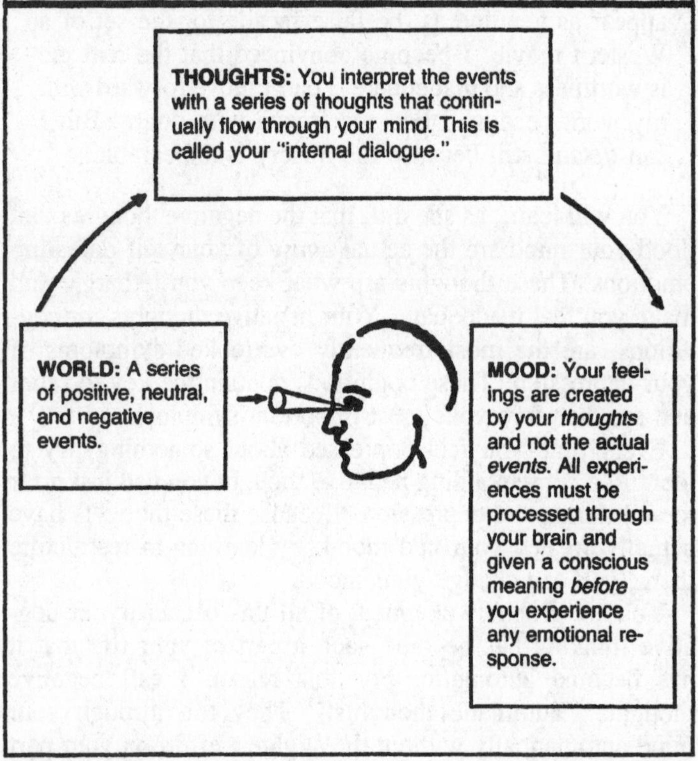

meaning. You must *understand* what is happening to you before you can *feel* it.

THOUGHTS: You interpret the events with a series of thoughts that continually flow through your mind. This is called your "internal dialogue."
WORLD: A series of positive, neutral, and negative events.
MOOD: Your feelings are created by your *thoughts* and not the actual *events*. All experiences must be processed through your brain and given a conscious meaning *before* you experience any emotional response.

If your understanding of what is happening is accurate, your emotions will be normal. If your perception is twisted and distorted in some way, your emotional response will be abnormal. Depression falls into this category. It is always the result of mental "static"—distortions. Your blue moods can be compared to the scratchy music coming from a radio that is not properly tuned to the station. The problem is *not* that the tubes or transistors are blown out or defective, or that the signal from the radio station is distorted as a result of bad weather. You just simply have to adjust the dials. When you learn to bring about this mental tuning, the music will come through clearly again and your depression will lift.

Some readers—maybe you—will experience a pang of despair when they read that paragraph. Yet there is *nothing upsetting* about it. If anything, the paragraph should bring hope. Then what caused your mood to plunge as you were reading? It was your thought, "For other people a little tuning may suffice. But I'm the radio that is broken beyond repair. My tubes are blown out. I don't care if ten thousand other depressed patients all get well—I'm convinced beyond any shadow of doubt that my problems are hopeless." I hear this statement fifty times a week! Nearly every depressed person seems convinced beyond all rhyme or reason that he or she is the special one who *really is* beyond hope. This delusion reflects the kind of mental processing that is at the very core of your illness!

I have always been fascinated by the ability certain people have to create illusions. As a child, I used to spend hours at the local library, reading books on magic. Saturdays I would hang out in magic stores for hours, watching the man behind the counter produce remarkable effects with cards and silks and chromium spheres that floated through the air, defying all the laws of common sense. One of my happiest childhood memories is when I was eight years old and saw "Blackstone—World's Greatest Magician" perform in Denver, Colorado. I was invited with several other children from the audience to come up on stage. Blackstone instructed us to place our hands on a two-feet by two-feet birdcage filled with live white doves until the top, bottom, and all four sides were enclosed entirely by our hands. He stood nearby and said, "Stare at the cage!" I did. My eyes were bulging and I refused to blink. He exclaimed, "Now

I'll clap my hands." He did. In that instant the cage of birds vanished. My hands were suspended in empty air. It was impossible! Yet it happened! I was stunned.

Now I know that his ability as an illusionist was no greater than that of the average depressed patient. This includes you. When you are depressed, you possess the remarkable ability to *believe*, and to get the people around you to believe, things which have no basis in reality. As a therapist, it is my job to *penetrate* your illusion, to teach you how to *look behind* the mirrors so you can see how you have been fooling yourself. You might even say that I'm planning to dis-illusion you! But I don't think you're going to mind at all.

Read over the following list of ten cognitive distortions that form the basis of all your depressions. Get a feel for them. I have prepared this list with great care; it represents the distilled essence of many years of research and clinical experience. Refer to it over and over when you read the how-to-do-it section of the book. When you are feeling upset, the list will be invaluable in making you aware of how you are fooling yourself.

Definitions of Cognitive Distortions

1. All-or-Nothing Thinking. This refers to your tendency to evaluate your personal qualities in extreme, black-or-white categories. For example, a prominent politician told me, "Because I lost the race for governor, I'm a zero." A straight-A student who received a B on an exam concluded, "Now I'm a total failure." All-or-nothing thinking forms the basis for perfectionism. It causes you to fear any mistake or imperfection because you will then see yourself as a complete loser, and you will feel inadequate and worthless.

This way of evaluating things is unrealistic because life is rarely completely either one way or the other. For example, no one is absolutely brilliant or totally stupid. Similarly, no one is either completely attractive or totally ugly. Look at the floor of the room you are sitting in now. Is it perfectly clean? Is every inch piled high with dust and dirt? Or is it partially clean? Absolutes do not exist in this universe. If you try to force your experiences into absolute categories, you will be constantly depressed because your perceptions will not conform to reality. You will set yourself up for discrediting yourself endlessly because whatever you do will *never* measure up to your exaggerated expectations. The technical name for this type of perceptual error is "dichotomous thinking." You see everything as black or white—shades of gray do not exist.

2. Overgeneralization. When I was eleven years old, I bought a deck of trick cards at the Arizona State Fair called the Svengali Deck. You may have seen this simple but impressive illusion yourself: I show the deck to you—every card is different. You choose a card at random. Let's assume you pick the Jack of Spades. Without telling me what card it is, you replace it in the deck. Now I exclaim, "Svengali!" As I turn the deck over, every card has turned into the Jack of Spades.

When you overgeneralize, this is performing the mental equivalent of Svengali. You arbitrarily conclude that one thing that happened to you once will occur over and over again, will multiply like the Jack of Spades. Since what happened is invariably unpleasant, you feel upset.

A depressed salesman noticed bird dung on his car window and thought, "That's just my luck. The birds are always crapping on *my* window!" This is a perfect example of overgeneralization. When I asked him about this experience, he admitted that in twenty years of traveling, he could not remember another time when he found bird dung on his car window.

The pain of rejection is generated almost entirely from overgeneralization. In its absence, a personal affront is temporarily disappointing but *cannot* be seriously disturbing. A shy young man mustered up his courage to ask a girl for a date. When she politely declined because of a previous engagement, he said to himself, "I'm never going to get a date. No girl would ever want a date with me. I'll be lonely and miserable all my life." In his distorted cognitions, he concluded that because she turned him down once, she would *always* do so, and that since all women have 100 percent identical tastes, he would be endlessly and repeatedly rejected by any eligible woman on the face of the earth. Svengali!

3. Mental Filter. You pick out a negative detail in any situation and dwell on it exclusively, thus perceiving that the whole situation is negative. For example, a depressed college student heard some other students making fun of her best friend. She became furious because she was thinking, "That's what the human race is basically like—cruel and insensitive!" She was overlooking the fact that in the previous months few people, if any, had been cruel or insensitive to her! On another occasion when she completed her first midterm exam, she felt certain she had missed approximately seventeen questions out of a hundred. She thought exclusively about those seventeen questions and concluded she would flunk out of college. When she got the paper back there was a note attached that read, "You got 83 out of 100 correct. This was by far the highest grade of any student this year. A+"

When you are depressed, you wear a pair of eyeglasses with special lenses that filter out anything positive. All that you allow to enter your conscious mind is negative. Because you are not aware of this "filtering process," you conclude that *everything* is negative. The technical name for this process is "selective abstraction." It is a bad habit that can cause you to suffer much needless anguish.

4. Disqualifying the Positive. An even more spectacular mental illusion is the persistent tendency of some depressed individuals to transform neutral or even positive experiences into negative ones. You don't just *ignore* positive experiences, you cleverly and swiftly turn them into their nightmarish opposite. I call this "reverse alchemy." The medieval alchemists dreamed of finding some method for transmuting the baser metals into gold. If you have been depressed, you may have developed the talent for doing the exact opposite—you can instantly transform golden joy into emotional lead. Not intentionally, however—you're probably not even aware of what you're doing to yourself.

An everyday example of this would be the way most of us have been conditioned to respond to compliments. When someone praises your appearance or your work, you might automatically tell yourself, "They're just being nice." With one swift blow you mentally disqualify their compliment. You do the same thing to them when you tell them, "Oh, it was nothing, really." If you constantly throw cold water on the good things that happen, no wonder life seems damp and chilly to you!

Disqualifying the positive is one of the most destructive forms of cognitive distortion. You're like a scientist intent on finding evidence to support some pet hypothesis. The hypothesis that dominates your depressive thinking is usually some version of "I'm second-rate." Whenever you have a negative experience, you dwell on it and conclude, "That proves what I've known all along." In contrast, when you have a positive experience, you tell yourself, "That was a fluke. It doesn't count." The price you pay for this tendency is intense misery and an inability to appreciate the good things that happen.

While this type of cognitive distortion is commonplace, it can also form the basis for some of the most extreme and intractable forms of depression. For example, a young woman hospitalized during a severe depressive episode told me, "No one could possibly care about me because I'm such an awful person. I'm a complete loner. Not one person on earth gives a damn about me." When she was discharged from the hospital, many patients and staff members expressed great fondness for her. Can you guess how she negated all this? "They don't count because they don't see me in the real world. A *real* person outside a hospital could never care about me.'' I then asked her how she reconciled this with the fact that she had numerous friends and family outside the hospital who *did* care about her. She replied, ''They don't count because they don't know the real me. You see Dr. Burns, inside I'm absolutely rotten. I'm the worst person in the world. It would be impossible for anyone to really like me for even one moment!'' By disqualifying positive experiences in this manner, she can maintain a negative belief which is clearly unrealistic and inconsistent with her everyday experiences.

While your negative thinking is probably not as extreme as hers, there may be many times every day when you do inadvertently ignore genuinely positive things that have happened to you. This removes much of life's richness and makes things appear needlessly bleak.

5. Jumping to Conclusions. You arbitrarily jump to a negative conclusion that is not justified by the facts of the situation. Two examples of this are ''mind reading'' and ''the fortune teller error.''

MIND READING: You make the assumption that other people are looking down on you, and you're so convinced about this that you don't even bother to check it out. Suppose you are giving an excellent lecture, and you notice that a man in the front row is nodding off. He was up most of the night on a wild fling, but you of course don't know this. You might have the thought, ''This audience thinks I'm a bore.'' Suppose a friend passes you on the street and fails to say hello because he is so absorbed in his thoughts he doesn't notice you. You might erroneously conclude, ''He is ignoring me so he must not like me anymore.'' Perhaps your spouse is unresponsive one evening because he or she was criticized at work and is too upset to want to talk about it. Your heart sinks because of the way you interpret the silence: ''He (or she) is mad at me. What did I do wrong?''

You may then respond to these imagined negative reactions by withdrawal or counterattack. This self-defeating behavior pattern may act as a self-fulfilling prophecy and set up a negative interaction in a relationship when none exists in the first place.

THE FORTUNE TELLER ERROR: It's as if you had a crystal ball that foretold only misery for you. You imagine that something bad is about to happen, and you take this prediction as a *fact* even though it is unrealistic. A high-school librarian repeatedly told herself during anxiety attacks, "I'm going to pass out or go crazy." These predictions were unrealistic because she had never once passed out (or gone crazy!) in her entire life. Nor did she have any serious symptoms to suggest impending insanity. During a therapy session an acutely depressed physician explained to me why he was giving up his practice: "I realize I'll be depressed forever. My misery will go on and on, and I'm absolutely convinced that this or any treatment will be doomed to failure." This negative prediction about his prognosis caused him to feel hopeless. His symptomatic improvement soon after initiating therapy indicated just how off-base his fortune telling had been.

Do you ever find yourself jumping to conclusions like these? Suppose you telephone a friend who fails to return your call after a reasonable time. You then feel depressed when you tell yourself that your friend probably got the message but wasn't interested enough to call you back. Your distorton?—mind reading. You then feel bitter, and decide not to call back and check this out because you say to yourself, "He'll think I'm being obnoxious if I call him back again. I'll only make a fool of myself." Because of these negative predictions (the fortune teller error), you avoid your friend and feel put down. Three weeks later you learn that your friend never got your message. All that stewing, it turns out, was just a lot of self-imposed hokum. Another painful product of your mental magic!

6. Magnification and Minimization. Another thinking trap you might fall into is called "magnification" and "minimization," but I like to think of it as the "binocular trick" because you are either blowing things up out of proportion or shrinking them. Magnification commonly occurs when you look at your own errors, fears, or imperfections and exaggerate their importance: "My God—I made a mistake. How terrible! How awful! The word will spread like wildfire! My reputation is ruined!" You're looking at your faults through the end of the binoculars that makes them appear gigantic and grotesque. This has also been called "catastrophizing" because you turn commonplace negative events into nightmarish monsters.

When you think about your strengths, you may do the opposite—look through the wrong end of the binoculars so that things look small and unimportant. If you magnify your imperfections and minimize your good points, you're guaranteed to feel inferior. But the problem isn't *you*—it's the crazy lenses you're wearing!

7. Emotional Reasoning. You take your emotions as evidence for the truth. Your logic: "I feel like a dud, therefore I *am* a dud." This kind of reasoning is misleading because your feelings reflect your thoughts and beliefs. If they are distorted—as is quite often the case—your emotions will have no validity. Examples of emotional reasoning include "I feel guilty. Therefore, I must have done something bad"; "I feel overwhelmed and hopeless. Therefore, my problems must be impossible to solve"; "I feel inadequate. Therefore, I must be a worthless person"; "I'm not in the mood to do anything. Therefore, I might as well just lie in bed"; or "I'm mad at you. This proves that you've been acting rotten and trying to take advantage of me."

Emotional reasoning plays a role in nearly all your depressions. Because things *feel* so negative to you, you assume they truly are. It doesn't occur to you to challenge the validity of the perceptions that create your feelings.

One usual side effect of emotional reasoning is procrastination. You avoid cleaning up your desk because you tell yourself, "I feel so lousy when I think about that messy desk, cleaning it will be impossible." Six months later you finally give yourself a little push and do it. It turns out to be quite gratifying and not so tough at all. You were fooling yourself all along because you are in the habit of letting your negative feelings guide the way you act.

8. Should Statements. You try to motivate yourself by saying, "I *should* do this" or "I *must* do that." These statements cause you to feel pressured and resentful. Paradoxically, you end up feeling apathetic and unmotivated. Albert Ellis calls this "*must*urbation." I call it the "shouldy" approach to life.

When you direct should statements toward others, you will usually feel frustrated. When an emergency caused me to be five minutes late for the first therapy session, the new patient thought, "He *shouldn't* be so self-centered and thoughtless. He *ought to be* prompt." This thought caused her to feel sour and resentful.

Should statements generate a lot of unnecessary emotional turmoil in your daily life. When the reality of your own behavior falls short of your standards, your shoulds and shouldn'ts create self-loathing, shame, and guilt. When the all-too-human performance of other people falls short of your expectations, as will inevitably happen from time to time, you'll feel bitter and self-righteous. You'll either have to change your expectations to approximate reality or always feel let down by human behavior. If you recognize this bad *should* habit in yourself, I have outlined many effective "should and shouldn't" removal methods in later chapters on guilt and anger.

9. Labeling and Mislabeling. Personal labeling means creating a completely negative self-image based on your errors. It is an extreme form of overgeneralization. The philosophy behind it is "The measure of a man is the mistakes he makes." There is a good chance you are involved in a personal labeling whenever you describe your mistakes with sentences beginning with "*I'm a . . .*" For example, when you miss your putt on the eighteenth hole, you might say, "*I'm a* born loser" instead of "I goofed up on my putt." Similarly, when the stock you invested in goes down instead of up, you might think, "*I'm* a failure" instead of "I made a mistake."

Labeling yourself is not only self-defeating, it is irrational. Your *self* cannot be equated with any *one* thing you do. Your life is a complex and ever-changing flow of thoughts, emotions, and actions. To put it another way, you are more like a river than a statue. Stop trying to define yourself with negative labels—they are overly simplistic and wrong. Would you think of yourself exclusively as an "eater" just because you eat, or a "breather" just because you breathe? This is nonsense, but such nonsense becomes painful when you label yourself out of a sense of your own inadequacies.

When you label other people, you will invariably generate hostility. A common example is the boss who sees his occasionally irritable secretary as "an uncooperative bitch." Because of this label, he resents her and jumps at every chance to criticize her. She, in turn, labels him an "insensitive chauvinist" and complains about him at every opportunity. So, around and around they go at each other's throats, focusing on every weakness or imperfection as proof of the other's worthlessness.

Mislabeling involves describing an event with words that are inaccurate and emotionally heavily loaded. For example, a woman on a diet ate a dish of ice cream and thought, "How disgusting and repulsive of me. I'm a *pig*." These thoughts made her so upset she ate the whole quart of ice cream!

10. Personalization. This distortion is the mother of guilt! You assume responsibility for a negative even when there is no basis for doing so. You arbitrarily conclude that what happened was your fault or reflects your inadequacy, even when you were not responsible for it. For example, when a patient didn't do a self-help assignment I had suggested, I felt guilty because of my thought, "I must be a lousy therapist. It's my fault that she isn't working harder to help herself. It's my responsibility to make sure she gets well." When a mother saw her child's report card, there was a note from the teacher indicating the child was not working well. She immediately decided, "I must be a bad mother. This shows how I've failed."

Personalization causes you to feel crippling guilt. You suffer from a paralyzing and burdensome sense of responsibility that forces you to carry the whole world on your shoulders. You have confused *influence* with *control* over others. In your role as a teacher, counselor, parent, physician, salesman, executive, you will certainly influence the people you interact with, but no one could reasonably expect you to control them. What the other person does is ultimately his or her responsibility, not yours. Methods to help you overcome your tendency to personalize and trim your sense of responsibility down to manageable, realistic proportions will be discussed later on in this book.

The ten forms of cognitive distortions cause many, if not all, of your depressed states. They are summarized in Table 3–1 on page 42. Study this table and master these concepts; try to become as familiar with them as with your phone number. Refer to Table 3–1 over and over again as you learn about the various methods for mood modification. When you become familiar with these ten forms of distortion, you will benefit from this knowledge all your life.

I have prepared a simple self-assessment quiz to help you test and strengthen your understanding of the ten distortions. As you read each of the following brief vignettes, imagine you are the person who is being described. Circle one or more answers which indicate the distortions contained in the negative thoughts. I will explain the answer to the first question. The answer key to subsequent questions is given at the end of this chapter. But don't look ahead! I'm *certain* you will be able to identify at least *one* distortion in the first question—and that will be a start!

Table 3–1. Definitions of Cognitive Distortions

1. ALL-OR-NOTHING THINKING: You see things in black-and-white categories. If your performance falls short of perfect, you see yourself as a total failure.
2. OVERGENERALIZATION: You see a single negative event as a never-ending pattern of defeat.
3. MENTAL FILTER: You pick out a single negative detail and dwell on it exclusively so that your vision of all reality becomes darkened, like the drop of ink that colors the entire beaker of water.
4. DISQUALIFYING THE POSITIVE: You reject positive experiences by insisting they "don't count" for some reason or other. In this way you can maintain a negative belief that is contradicted by your everyday experiences.
5. JUMPING TO CONCLUSIONS: You make a negative interpretation even though there are no definite facts that convincingly support your conclusion.

   a. *Mind reading.* You arbitrarily conclude that someone is reacting negatively to you, and you don't bother to check this out.
   b. *The Fortune Teller Error.* You anticipate that things will turn out badly, and you feel convinced that your prediction is an already-established fact.

6. MAGNIFICATION (CATASTROPHIZING) OR MINIMIZATION: You exaggerate the importance of things (such as your goof-up or someone else's achievement), or you inappropriately shrink things until they appear tiny (your own desirable qualities or the other fellow's imperfections). This is also called the "binocular trick."
7. EMOTIONAL REASONING: You assume that your negative emotions necessarily reflect the way things really are: "I feel it, therefore it must be true."
8. SHOULD STATEMENTS: You try to motivate yourself with shoulds and shouldn'ts, as if you had to be whipped and punished before you could be expected to do anything. "Musts" and "oughts" are also offenders. The emotional consequence is guilt. When you direct should statements toward others, you feel anger, frustration, and resentment.

Table 3–1. cont.

9. LABELING AND MISLABELING: This is an extreme form of overgeneralization. Instead of describing your error, you attach a negative label to yourself: "I'm a *loser*." When someone else's behavior rubs you the wrong way, you attach a negative label to him: "He's a goddam louse." Mislabeling involves describing an event with language that is highly colored and emotionally loaded.
10. PERSONALIZATION: You see yourself as the cause of some negative external event which in fact you were not primarily responsible for.

1. You are a housewife, and your heart sinks when your husband has just complained disgruntledly that the roast beef was overdone. The following thought crosses your mind: "I'm a total failure. I can't stand it! I *never* do *anything* right. I work like a slave and this is all the thanks I get! The jerk!" These thoughts cause you to feel sad and angry. Your distortions include one or more of the following:

a. all-or-nothing thinking;
b. overgeneralization;
c. magnification;
d. labeling;
e. all the above.

Now I will discuss the correct answers to this question so you can get some immediate feedback. Any answer(s) you might have circled was (were) correct. So if you circled *anything*, you were right! Here's why. When you tell yourself, "I'm a *total* failure," you engage in *all-or-nothing* thinking. Cut it out! The meat was a little dry, but that doesn't make your entire life a total failure. When you think, "I *never* do *anything* right," you are *overgeneralizing*. Never? Come on now! Not *anything*? When you tell yourself, "I can't stand it," you are *magnifying* the pain you are feeling. You're blowing it way out of proportion because you *are* standing it, and if you *are*, you *can*. Your husband's grumbling is not exactly what you like to hear, but it's not a reflection of your worth. Finally, when you proclaim, "I work like a slave and this is all the thanks I get! The jerk!" you are *labeling* both of you. He's not a *jerk*, he's just being irritable and insensitive. Jerky behavior exists, but jerks do not. Similarly, it's silly to label yourself a *slave*. You're just letting his moodiness sour your evening.

Okay, now let's continue with the quiz.

2. You have just read the sentence in which I informed you that you would have to take this self-assessment quiz. Your heart suddenly sinks and you think, "Oh no, not another test! I always do lousy on tests. I'll have to skip this section of the book. It makes me nervous, so it wouldn't help anyway." Your distortions include:

a. jumping to conclusions (fortune teller error);
b. overgeneralization;
c. all-or-nothing thinking;
d. personalization;
e. emotional reasoning.

3. You are a psychiatrist at the University of Pennsylvania. You are attempting to revise your manuscript on depression after meeting with your editor in New York. Although your editor seemed extremely enthusiastic, you notice you are feeling nervous and inadequate due to your thoughts, "They made a terrible mistake when they chose my book! I won't be able to do a good job. I'll never be able to make the book fresh, lively, and punchy. My writing is too drab, and my ideas aren't good enough." Your cognitive distortions include:

a. all-or-nothing thinking;
b. jumping to conclusions (negative prediction);
c. mental filter;
d. disqualifying the positive;
e. magnification.

4. You are lonely and you decide to attend a social affair for singles. Soon after you get there, you have the urge to leave because you feel anxious and defensive. The following thoughts run through your mind: "They probably aren't very interesting people. Why torture myself? They're just a bunch of losers. I can tell because I feel so bored. This party will be a drag." Your errors involve:
   a. labeling;
   b. magnification;
   c. jumping to conclusions (fortune teller error and mind reading);
   d. emotional reasoning;
   e. personalization.

5. You receive a layoff notice from your employer. You feel mad and frustrated. You think, "This proves the world is no damn good. I never get a break." Your distortions include:
   a. all-or-nothing thinking;
   b. disqualifying the positive;
   c. mental filter;
   d. personalization;
   e. should statement.

6. You are about to give a lecture and you notice that your heart is pounding. You feel tense and nervous because you think, "My God, I'll probably forget what I'm supposed to say. My speech isn't any good anyway. My mind will blank out. I'll make a fool of myself." Your thinking errors involve:
   a. all-or-nothing thinking;
   b. disqualifying the positive;
   c. jumping to conclusions (fortune teller error);
   d. minimization;
   e. labeling.

7. Your date calls you at the last minute to cancel out because of illness. You feel angry and disappointed because you think, "I'm getting jilted. What did I do to foul things up?" Your thinking errors include:

a. all-or-nothing thinking;
b. should statements;
c. jumping to conclusions (mind reading);
d. personalization;
e. overgeneralization.

8. You have put off writing a report for work. Every night when you try to get down to it, the whole project seems so difficult that you watch TV instead. You begin to feel overwhelmed and guilty. You are thinking the following: "I'm so lazy I'll never get this done. I just can't do the darn thing. It would take forever. It won't turn out right anyway." Your thinking errors include:

a. jumping to conclusions (fortune teller error);
b. overgeneralization;
c. labeling;
d. magnification;
e. emotional reasoning.

9. You've read this entire book and after applying the methods for several weeks, you begin to feel better. Your BDC score went down from twenty-six (moderately depressed) to eleven (borderline depression). Then you suddenly begin to feel worse, and in three days your score has gone back up to twenty-eight. You feel disillusioned, hopeless, bitter, and desperate due to thinking, "I'm not getting anywhere. These methods won't help me after all. I should be well by now. That 'improvement' was a fluke. I was fooling myself when I thought I was feeling better. I'll never get well." Your cognitive distortions include:

a. disqualifying the positive;
b. should statement;
c. emotional reasoning;

d. all-or-nothing thinking;
e. jumping to conclusions (negative prediction).

10. You've been trying to diet. This weekend you've been nervous, and, since you didn't have anything to do, you've been nibbling, nibbling. After your fourth piece of candy, you tell yourself, "I just can't control myself. My dieting and jogging all week have gone down the drain. I must look like a balloon. I shouldn't have eaten that. I can't stand this. I'm going to pig out all weekend!" You begin to feel so guilty you push another handful of candy into your mouth in an abortive effort to feel better. Your distortions include:

a. all-or-nothing thinking;
b. mislabeling;
c. negative prediction;
d. should statement;
e. disqualifying the positive.

ANSWER KEY

1.	A B C D E	6.	A C D E
2.	A B C E	7.	C D
3.	A B D E	8.	A B C D E
4.	A B C D	9.	A B C D E
5.	A C	10.	A B C D E

Feelings Aren't Facts

At this point you may be asking yourself, "Okay. I understand that my depression results from my negative thoughts because my outlook on life changes enormously when my moods go up or down. But if my negative thoughts are so distorted, how do I continually get fooled? I can think as clearly and realistically as the next person, so if what I am telling myself is irrational, why does it seem so right?"

Even though your depressing thoughts may be distorted, they nevertheless create a powerful illusion of truth. Let me expose the basis for the deception in blunt terms—your feelings are not facts! In fact, your feelings, per se, don't even count—except as a mirror of the way you are thinking. If your perceptions make no sense, the feelings they create will be as absurd as the images reflected in the trick mirrors at an amusement park. But these abnormal emotions *feel* just as valid and realistic as the genuine feelings created by undistorted thoughts, so you automatically attribute truth to them. This is why depression is such a powerful form of mental black magic.

Once you invite depression through an "automatic" series of cognitive distortions, your feelings and actions will reinforce each other in a self-perpetuating vicious cycle. Because you *believe* whatever your depressed brain tells you, you find yourself feeling negative about almost everything. This reaction occurs in milliseconds, too quickly for you even to be aware of it. The negative emotion *feels* realistic and in turn lends an aura of credibility to the distorted thought which created it. The cycle goes on and on, and you are eventually trapped. The mental prison is an illusion, a hoax you have inadvertently created, but it *seems* real because it *feels* real.

What is the key to releasing yourself from your emotional prison? Simply this: Your thoughts create your emotions; therefore, your emotions cannot prove that your thoughts are accurate. Unpleasant feelings merely indicate that you are thinking something negative and believing it. Your emotions *follow* your thoughts just as surely as baby ducks follow their mother. But the fact that the baby ducks follow faithfully along doesn't prove that the mother knows where she is going!

Let's examine your equation, "I feel, therefore I am." This attitude that emotions reflect a kind of self-evident, ultimate truth is not unique to depressed people. Most psychotherapists today share the conviction that becoming more *aware* of your feelings and expressing them more openly represent emotional maturity. The implication is that your feelings represent a higher reality, a personal integrity, a truth beyond question.

My position is quite different. Your feelings, per se, are not necessarily special at all. In fact, to the extent that your negative emotions are based on mental distortions—as is all too often the case—they can hardly be viewed as desirable.

Do I mean you should get rid of *all* emotions? Do I want you to turn into a robot? No. I want to teach you to avoid painful feelings based on mental distortions, because they are neither valid nor desirable. I believe that once you have learned how to perceive life more realistically you will experience an enhanced emotional life with a greater appreciation for genuine sadness—which lacks distortion—as well as joy.

As you go on to the next sections of this book, you can learn to correct the distortions that fool you when you are upset. At the same time, you will have the opportunity to reevaluate some of the basic values and assumptions that create your vulnerability to destructive mood swings. I have outlined the necessary steps in detail. The modifications in illogical thinking patterns will have a profound effect on your moods and increase your capacity for productive living. Now, let's go ahead and see how we can turn your problems around.

Part II
Practical Applications

Chapter 4

Start by Building Self-Esteem

When you are depressed, you invariably believe that you are worthless. The worse the depression, the more you feel this way. You are not alone. A survey by Dr. Aaron Beck revealed that over 80 percent of depressed patients expressed self-dislike.* Furthermore, Dr. Beck found that depressed patients see themselves as deficient in the very qualities they value most highly: intelligence, achievement, popularity, attractiveness, health, and strength. He said a depressed self-image can be characterized by the four D's: You feel Defeated, Defective, Deserted, and Deprived.

Almost all negative emotional reactions inflict their damage *only* as a result of low self-esteem. A poor self-image is the magnifying glass that can transform a trivial mistake or an imperfection into an overwhelming symbol of personal defeat. For example, Eric, a first-year law student, feels a sense of panic in class. "When the professor calls on me, I'll probably goof up." Although Eric's fear of "goofing up" was foremost on his mind, my dialogue with him re-

*Beck, Aaron T. *Depression: Clinical, Experimental, & Theoretical Aspects*. New York: Hoeber, 1967. (Republished as *Depression: Causes and Treatment*. Philadelphia: University of Pennsylvania Press, 1972, pp. 17–23.)

vealed that a sense of personal inadequacy was the real cause of the problem:

DAVID: Suppose you did goof up in class. Why would that be particularly upsetting to you? Why is that so tragic?

ERIC: Then I would make a fool of myself.

DAVID: Suppose you did make a fool of yourself. Why would that be upsetting?

ERIC: Because then everyone would look down on me.

DAVID: Suppose people did look down on you? What then?

ERIC: Then I would feel miserable.

DAVID: Why? Why is it that you would have to feel miserable if people were looking down on you?

ERIC: Well, that would mean I wouldn't be a worthwhile person. Furthermore, it might ruin my career. I'd get bad grades, and maybe I could never be an attorney.

DAVID: Suppose you didn't become an attorney. Let's assume for the purposes of discussion that you did flunk out. Why would that be particularly upsetting to you?

ERIC: That would mean that I had failed at something I've wanted all my life.

DAVID: And what would that mean to you?

ERIC: Life would be empty. It would mean I was a failure. It would mean I was worthless.

In this brief dialogue, Eric showed that he believed it would be terrible to be disapproved of or to make a mistake or to fail. He seemed convinced that if one person looked down on him then everyone would. It was as if the word REJECT would suddenly be stamped on his forehead for everyone to see. He seemed to have no sense of self-esteem that was not contingent upon approval and/or success. He measured himself by the way others looked at him and by what he had achieved. If his cravings for approval and accomplishment were not satisfied, Eric sensed he would be nothing because there would be no true support from within.

If you feel that Eric's perfectionistic drive for achievement and approval is self-defeating and unrealistic, you are right. But to Eric, this drive was *realistic* and *reasonable*. If you are now depressed or have ever been depressed, you may find it much harder to recognize the illogical thinking patterns which cause you to look down on yourself. In fact, you are probably convinced that you really are inferior or worthless. And any suggestion to the contrary is likely to sound foolish and dishonest.

Unfortunately, when you are depressed you may not be alone in your conviction about your personal inadequacy. In many cases you will be so *persuasive* and *persistent* in your maladaptive belief that you are defective and no good, you may lead your friends, family, and even your therapist into accepting this idea of yourself. For many years psychiatrists have tended to "buy into" the negative self-evaluation system of depressed individuals without probing the validity of what the patients are saying about themselves. This is illustrated in the writings of such a keen observer as Sigmund Freud in his treatise "Mourning and Melancholia," which forms the basis for the orthodox psychoanalytic approach to treating depression. In this classic study Freud said that when the patient says he is worthless, unable to achieve, and morally despicable, he *must be right*. Consequently, it was fruitless for the therapist to disagree with the patient. Freud believed the therapist should agree that the patient is, in fact, uninteresting, unlovable, petty, self-centered, and dishonest. These qualities describe a human being's true self, according to Freud, and the disease process simply makes the truth more obvious:

> The patient represents his ego to us as worthless, incapable of any achievement and morally despicable; he reproaches himself, vilifies himself and expects to be cast out and punished. . . . It would be equally fruitless from a scientific and therapeutic point of view to contradict a patient who brings these accusations against his ego. He must *surely be right in some way* [emphasis mine] and be describing something that is as it seems to him to be. Indeed we must at once confirm some of his statements without reservation. *He really is as lacking in interest and incapable of love and achievement as he says* [emphasis mine]. . . . He also seems to us justified in certain other self-accusations; *it is merely that he has a keener eye for the truth than other people who are not melancholic* [emphasis mine]. When in his heightened self-criticism he describes himself as petty, egoistic, dishonest, lacking in independence, one whose sole aim has been to hide the weaknesses of his own nature, it may be so far as we know, that *he has come pretty near to understanding himself* [emphasis mine]; we only wonder why a man has to be ill before he can be accessible to truth of this kind.
>
> —SIGMUND FREUD, "*Mourning and Melancholia*"*

The way a therapist handles your feelings of inadequacy is crucial to the cure, as your sense of worthlessness is a key to depression. The question also has considerable philosophical relevance—is human nature *inherently* defective? Are depressed patients actually facing the ultimate truth about themselves? And what, in the final analysis, is the source of genuine self-esteem? This, in my opinion, is the most important question you will ever confront.

First, you *cannot earn* worth through what you do.

---

*Freud, S. *Collected Papers*, 1917. (Translated by Joan Riviere, Vol. IV, Chapter 8, "Mourning and Melancholia," pp. 155–156. London: Hogarth Press Ltd., 1952.)

Achievements can bring you satisfaction but not happiness. Self-worth based on accomplishments is a "pseudo-esteem," not the genuine thing! My many successful but depressed patients would all agree. Nor can you base a valid sense of self-worth on your looks, talent, fame, or fortune. Marilyn Monroe, Mark Rothko, Freddie Prinz, and a multitude of famous suicide victims attest to this grim truth. Nor can love, approval, friendship, or a capacity for close, caring human relationships add one iota to your inherent worth. The great majority of depressed individuals are in fact very much loved, but it doesn't help one bit because *self*-love and *self*-esteem are missing. At the bottom line, only your own sense of self-worth determines how you feel.

"So," you may now be asking with some exasperation, "how *do* I get a sense of self-worth? The fact is, I *feel* damn inadequate, and I'm convinced I'm really not as good as other people. I don't believe there's anything I can do to change those rotten feelings because that's the way I basically am."

One of the cardinal features of cognitive therapy is that it stubbornly refuses to buy into your sense of worthlessness. In my practice I lead my patients through a systematic re-evaluation of their negative self-image. I raise the same question over and over again: "Are you really *right* when you insist that somewhere inside you are essentially a loser?"

The first step is to take a close look at what you say about yourself when you insist you are no good. The evidence you present in defense of your worthlessness will usually, if not always, make no sense.

This opinion is based on a recent study by Drs. Aaron Beck and David Braff which indicated that there is actually a formal thinking disturbance in depressed patients. Depressed individuals were compared with schizophrenic patients and with undepressed persons in their ability to interpret the meaning of a number of proverbs, such as "A stitch in time saves nine." Both the schizophrenic and depressed patients made many logical errors and had difficulty in extracting the meaning of the proverbs. They were overly concrete and couldn't make accurate generalizations. Although the severity of the defect was obviously less profound and bizarre in the depressed patients than in the schizophrenic group, the depressed individuals were clearly abnormal as compared with the normal subjects.

In practical terms the study indicated that during periods of depression you lose some of your capacity for clear thinking; you have trouble putting things into proper perspective. Negative events grow in importance until they dominate your entire reality—and you can't really tell that what is happening is distorted. It all seems very *real* to you. The illusion of hell you create is *very convincing*.

The more depressed and miserable you feel, the more twisted your thinking becomes. And, conversely, in the absence of mental distortion, you *cannot* experience low self-worth or depression!

What types of mental errors do you make most generally when you look down on yourself? A good place to begin is with the list of distortions you began to master in Chapter 3. The most usual mental distortion to look out for when you are feeling worthless is all-or-nothing thinking. If you see life only in such extreme categories, you will believe your performance will be either great or terrible—nothing else will exist. As a salesman told me, "Achieving 95 percent or better of my goal for monthly sales is acceptable. Ninety-four percent or below is the equivalent of total failure."

Not only is this all-or-nothing system of self-evaluation highly unrealistic and self-defeating, it creates overwhelming anxiety and frequent disappointment. A depressed psychiatrist who was referred to me noticed a lack of sexual drive and a difficulty in maintaining erections during a two-week period when he was feeling blue. His perfectionistic tendencies had dominated not only his illustrious professional career but also his sexual life. Consequently, he had intercourse regularly with his wife every other day precisely on schedule for the twenty years of their married life. In spite of his decreased sex drive—which is a common symptom of depression—he told himself, "I *must* continue to perform intercourse on schedule." This thought created such anxiety that he became increasingly unable to achieve a satisfactory erection. Because his perfect intercourse track record was broken, he now began clubbing himself with the "nothing" side of his all-or-nothing system and concluded, "I'm not a full marriage partner anymore. I'm a failure as a husband. I'm not even a man. I'm a worthless nothing." Although he was a competent (and some might even say brilliant) psychiatrist, he confided to me tearfully, "Dr. Burns, you and I both know it is an undeniable fact that I will never be able to have intercourse again." In spite of his years of medical training, he could actually convince himself of such a thought.

Overcoming the Sense of Worthlessness

By now you might be saying, "Okay, I can see that there is a certain illogic which lurks behind the sense of worthlessness. At least for *some* people. But they are basically winners; they're not like me. You seem to be treating famous physicians and successful businessmen. Anyone could have told you that their lack of self-esteem was illogical. But I really *am* a mediocre nothing. Others *are*, in fact, better looking and more popular and successful than I am. So what can I do about it? Nothing, that's what! My feeling of worthlessness is very valid. It's based on reality, so there is little consolation in being told to *think* logically. I don't think there's any way to make these awful feelings go away unless I try to fool myself, and you and I both know that won't work." Let me first show you a couple of popular approaches, used by many therapists, which I feel do *not* represent satisfactory solutions to your problem of worthlessness. Then I'll show you some approaches that will make sense and help you.

In keeping with the belief that there is some deep truth in your conviction you are basically worthless, some psychotherapists may allow you to ventilate these feelings of inadequacy during a therapy session. There is undoubtedly some benefit to getting such feelings off your chest. The cathartic release may sometimes, but not always, result in a temporary mood elevation. However, if the therapist does not provide objective feedback about the validity of your self-evaluation, you may conclude that he agrees with you. And you may be right! You may, in fact, have fooled him as well as yourself! As a result you probably will feel even more inadequate.

Prolonged silences during therapy sessions may cause you to become more upset and preoccupied with your critical internal voice—much like a sensory-deprivation experiment. This kind of nondirective therapy, in which the therapist adopts a passive role, frequently produces greater anxiety and depression for the patient. And even when you do feel better as a result of achieving emotional release with an empathetic and caring therapist, the sense of improvement is likely to be short-lived if you haven't significantly transformed the way you evaluate yourself and your life. Unless you substantially reverse your self-defeating thinking and behavior patterns, you are likely to slip back again into depression.

Just as emotional ventilation for its own sake is usually not enough to overcome the sense of worthlessness, insight and psychological interpretation generally don't help either. For example, Jennifer was a writer who came for treatment for panic she experienced before publication of her novel. In the first session she told me, "I have been to several therapists. They have told me that my problem is *perfectionism* and impossible expectations and demands on myself. I also have learned that I probably picked up this trait from my mother, who is compulsive and perfectionistic. She can find nineteen things wrong with an incredibly clean room. I always tried to please her, but rarely felt I succeeded no matter how well I did. Therapists have told me, 'Stop seeing everyone as your mother! Stop being so perfectionistic.' But how do I *do* this? I'd like to, I want to, but no one ever was able to tell me how to go about it."

Jennifer's complaint is one I hear nearly every day in my practice. Pinpointing the nature or origin of your problem may give you insight, but usually fails to change the way you act. That is not surprising. You have been practicing for years and years the bad mental habits that helped create your low self-esteem. It will take systematic and ongoing effort to turn the problem around. Does a stutterer stop stuttering because of his insight into the fact that he doesn't vocalize properly? Does a tennis player's game improve just because the coach tells him he hits the ball into the net too often?

Since ventilation of emotions and insight—the two staples of the standard psychotherapeutic diet—won't help, what will? As a cognitive therapist, I have three aims in dealing with your sense of worthlessness: a rapid and decisive transformation in the way you *think*, *feel*, and *behave*. These results will be brought about in a systematic training program that employs simple concrete methods you can apply on a daily basis. If you are willing to commit some regular time and effort to this program, you can expect success proportionate to the effort you put in.

Are you willing? If so, we've come to the beginning. You're about to take the first crucial step toward an improved mood and self-image.

I have developed many specific and easily applied techniques that can help you develop your sense of worth. As you read the following sections, keep in mind that simply reading them is not guaranteed to bolster your self-esteem—at least not for long. You will have to work at it and practice the various exercises. In fact, I recommend that you set some time aside each day to work at improving your self-image because *only* in this way can you experience the fastest and most enduring personal growth.

Specific Methods for Boosting Self-Esteem

1. Talk Back to That Internal Critic! A sense of worthlessness is created by your internal self-critical dialogue. It is self-degrading statements, such as "I'm no damn good," "I'm a shit," "I'm inferior to other people," and so on, that create and feed your feelings of despair and poor self-esteem. In order to overcome this bad mental habit, three steps are necessary:

a. Train yourself to recognize and write down the self-critical thoughts as they go through your mind;
b. Learn why these thoughts are distorted; and
c. Practice talking back to them so as to develop a more realistic self-evaluation system.

One effective method for accomplishing this is the "triple-column technique." Simply draw two lines down the center of a piece of paper to divide it into thirds (see Figure 4–1, page 63). Label the left-hand column "Automatic Thoughts (Self-Criticism)," the middle column "Cognitive Distortion," and the right-hand column "Rational Response (Self-Defense)." In the left-hand column write down all those hurtful self-criticisms you make when you are feeling worthless and down on yourself.

Suppose, for example, you suddenly realize you're late for an important meeting. Your heart sinks and you're gripped with panic. Now ask yourself, "What thoughts are going through my mind right now? What am I saying to myself? Why is this upsetting me?" Then write these thoughts down in the left-hand column.

You might have been thinking, "I never do anything right," and "I'm always late." Write these thoughts down in the left-hand column and number them (see Figure 4–1). You might also have thought, "Everyone will look down at me. This shows what a jerk I am." Just as fast as these thoughts cross your mind, jot them down. Why? Because they are the very *cause* of your emotional upset. They rip away at you like knives tearing into your flesh. I'm sure you know what I mean because you've *felt* it.

Figure 4–1. The "triple-column technique" can be used to restructure the way you think about yourself when you have goofed up in some way. The aim is to substitute more objective rational thoughts for the illogical, harsh self-criticisms that automatically flood your mind when a negative event occurs.

Automatic Thought	*Cognitive Distortion*	*Rational Response*
(SELF-CRITICISM)		(SELF-DEFENSE)
1. I never do anything right.	1. Overgeneralization	1. Nonsense! I do a lot of things right.
2. I'm always late.	2. Overgeneralization	2. I'm not *always* late. That's ridiculous. Think of all the times I've been on time. If I'm late more often than I'd like, I'll work on this problem and develop a method for being more punctual.
3. Everyone will look down on me.	3. Mind reading Overgeneralization All-or-nothing thinking Fortune teller error	3. Someone may be disappointed that I'm late but it's not the end of the world. Maybe the meeting won't even start on time.
4. This shows what a jerk I am.	4. Labeling	4. Come on, now, I'm not "a jerk."
5. I'll make a fool of myself.	5. Labeling Fortune teller error	5. Ditto. I'm not "a fool" either. I may appear foolish if I come in late, but this doesn't make me a fool. Everyone is late sometimes.

What's the second step? You already began to prepare for this when you read Chapter 3. Using the list of ten cognitive distortions (page 42), see if you can identify the thinking errors in each of your negative automatic thoughts. For instance, "I never do anything right" is an example of overgeneralization. Write this down in the middle column. Continue to pinpoint the distortions in your other automatic thoughts, as shown in Figure 4–1.

You are now ready for the crucial step in mood transformation—substituting a more rational, less upsetting thought in the right-hand column. You do not try to cheer yourself up by rationalizing or saying things you do not believe are objectively valid. Instead, try to recognize *the truth*. If what you write down in the Rational Response column is not convincing and realistic, it won't help you one bit. Make sure you believe in your rebuttal to self-criticism. This rational response can take into account what was illogical and erroneous about your self-critical automatic thought.

For example, in answer to "I never do anything right," you could write, "Forget that! I do some things right and some wrong, just like everyone else. I fouled up on my appointment, but let's not blow this up out of proportion."

Suppose you cannot think of a rational response to a particular negative thought. Then just forget about it for a few days and come back to it later. You will usually be able to see the other side of the coin. As you work at the triple-column technique for fifteen minutes every day over a period of a month or two, you will find it gets easier and easier. Don't be afraid to ask other people how they would answer an upsetting thought if you can't figure out the appropriate rational response on your own.

One note of caution: Do *not* use words describing your emotional reactions in the Automatic Thought column. Just write the thoughts that created the emotion. For example, suppose you notice your car has a flat tire. Don't write "I feel crappy" because you can't disprove that with a rational response. The fact is, you *do* feel crappy. Instead, write down the thoughts that automatically flashed through your mind the moment you saw the tire; for example, "I'm so stupid—I should have gotten a new tire this last month," or "Oh, hell! This is just my rotten luck!" Then you can substitute rational responses such as "It might have been better to get a new tire, but I'm not stupid and no one can predict the future with certainty." This process won't put air in the tire, but at least you won't have to change it with a deflated ego.

While it's best not to describe your emotions in the Automatic Thought column, it can be quite helpful to do some "emotional accounting" before and after you use the triple-column technique to determine how much your feelings actually improve. You can do this very easily if you record how upset you are between 0 and 100 percent before you pinpoint and answer your automatic thoughts. In the previous example, you might note that you were 80 percent frustrated and angry at the moment you saw the flat tire. Then, once you complete the written exercise, you can record how much relief you experienced, say, to 40 percent or so. If there's a decrease, you'll know that the method has worked for you.

A slightly more elaborate form developed by Dr. Aaron Beck called the Daily Record of Dysfunctional Thoughts allows you to record not only your upsetting thoughts but also your feelings and the negative event that triggered them (see Figure 4–2, page 66).

For example, suppose you are selling insurance and a potential customer insults you without provocation and hangs up on you. Describe the actual event in the Situation column, but *not* in the Automatic Thought(s) column. Then write down your feelings and the negative distorted thoughts that created them in the appropriate column. Finally, talk back to these thoughts and do your emotional accounting. Some individuals prefer to use the Daily Record of Dysfunctional Thoughts because it allows them to analyze neg-

Figure 4–2. Daily Record of Dysfunctional Thoughts*

Situation Briefly describe the actual event leading to the unpleasant emotion.	*Emotion(s)* 1. Specify sad/ anxious/ angry, etc. 2. Rate degree of emotion, 1–100%.	*Automatic Thought(s)* Write the automatic thought(s) that accompany the emotion(s).	*Cognitive Distortion(s)* Identify the distortion(s) present in each automatic thought.	*Rational Response(s)* Write rational response(s) to the automatic thought(s).	*Outcome* Specify and rate subsequent emotions, 0–100%.
Potential customer hangs up on me when I call to describe our new insurance program. He said, "Get out of my goddam hair!"	Angry, 99% Sad, 50%	1. I'll never sell a policy. 2. I'd like to strangle the bastard. 3. I must have said the wrong thing.	1. Overgeneralization 2. Magnification; labeling 3. Jumping to conclusions; personalization	1. I've sold a lot of policies. 2. He acted like a pain in the butt. We all do at times. Why let this get to me? 3. I actually didn't do anything different from the way I usually approach a new customer. So why sweat it?	Angry, 50% Sad, 10%

Explanation: When you experience an unpleasant emotion, note the situation that seemed to stimulate it. Then, note the automatic thought associated with the emotion. In rating degree of emotion, 1 = a trace; 100 = the most intense possible.

*Copyright 1979, Aaron T. Beck.

ative events, thoughts, and feelings in a systematic way. Be sure to use the technique that feels most comfortable to you.

Writing down your negative thoughts and rational responses may strike you as simplistic, ineffective, or even gimmicky. You might even share the feelings of some patients who initially refused to do this, saying, "What's the point? It won't work—it couldn't work because I really am hopeless and worthless."

This attitude can only serve as a self-fulfilling prophecy. If you are unwilling to pick up the tool and use it, you won't be able to do the job. Start by writing down your automatic thoughts and rational responses for fifteen minutes every day for two weeks and see the effect this has on your mood, as measured by the Burns Depression Checklist. You may be surprised to note the beginning of a period of personal growth and a healthy change in your self-image.

This was the experience of Gail, a young secretary whose sense of self-esteem was so low that she felt in constant danger of being criticized by friends. She was so sensitive to her roommate's request to help clean up their apartment after a party that she felt rejected and worthless. She was initially so pessimistic about her chances for feeling better that I could barely persuade her to give the triple-column technique a try. When she reluctantly decided to try it, she was surprised to see how her self-esteem and mood began to undergo a rapid transformation. She reported that *writing down* the many negative thoughts that flowed through her mind during the day helped her gain objectivity. She stopped taking these thoughts so seriously. As a result of Gail's daily written exercises, she began to feel better, and her interpersonal relationships improved by a quantum leap. An excerpt from her written homework is included in Figure 4–3.

Gail's experience is not unusual. The simple exercise of answering your negative thoughts with rational responses on a daily basis is at the heart of the cognitive method. It is one of the most important approaches to changing your

Figure 4–3. Excerpts from Gail's daily written homework using the "triple-column technique." In the left column she recorded the negative thoughts that automatically flowed through her mind when her roommate asked her to clean up the apartment. In the middle column she identified her distortions, and in the right-hand column she wrote down more realistic interpretations. This daily written exercise greatly accelerated her personal growth and resulted in substantial emotional relief.

Automatic Thoughts (SELF-CRITICISM)	*Cognitive Distortion*	*Rational Response* (SELF-DEFENSE)
1. Everyone knows how disorganized and selfish I am.	Jumping to conclusions (mind reading); overgeneralization	1. I'm disorganized at times and I'm organized at times. Everybody doesn't think the same way about me.
2. I'm completely self-centered and thoughtless. I'm just no good.	All-or-nothing thinking	2. I'm thoughtless at times, and at times I can be quite thoughtful. I probably do act overly self-centered at times. I can work on this. I may be imperfect but I'm not "no good!"
3. My roommate probably hates me. I have no real friends.	Jumping to conclusions (mind reading); all-or-nothing thinking	3. My friendships are just as real as anyone's. At times I take criticism as rejection of *me*, Gail, the person. But others are usually not rejecting *me*. They're just expressing dislike for what I *did* (or said)—and they still accept me afterward.

thinking. It is crucial to *write down* your automatic thoughts and rational responses; do not try to do the exercise in your head. Writing them down forces you to develop much more objectivity than you could ever achieve by letting responses swirl through your mind. It also helps you locate the mental errors that depress you. The triple-column technique is not limited to problems of personal inadequacy, but can be applied to a great range of emotional difficulties in which distorted thinking plays a central role. You can take the major sting out of problems you would ordinarily assume are entirely "realistic," such as bankruptcy, divorce, or severe mental illness. Finally, in the section on prophylaxis and personal growth, you will learn how to apply a slight variation of the automatic-thought method to penetrate to the part of your psyche where the causes of mood swings lurk. You will be able to expose and transform those "pressure points" in your mind that cause you to be vulnerable to depression in the first place.

2. Mental Biofeedback. A second method which can be very useful involves monitoring your negative thoughts with a wrist counter. You can buy one at a sporting-goods store or a golf shop; it looks like a wristwatch, is inexpensive, and every time you push the button, the number changes on the dial. Click the button each time a negative thought about yourself crosses your mind; be on the constant alert for such thoughts. At the end of the day, note your daily total score and write it down in a log book.

At first you will notice that the number increases; this will continue for several days as you get better and better at identifying your critical thoughts. Soon you will begin to notice that the daily total reaches a plateau for a week to ten days, and then it will begin to go *down*. This indicates that your harmful thoughts are diminishing and that you are getting better. This approach usually requires three weeks.

It is not known with certainty why such a simple technique works so well, but systematic self-monitoring frequently helps develop increased self-control. As you learn to stop haranguing yourself, you will begin to feel much better.

In case you decide to use a wrist counter, I want to emphasize it is not intended to be a substitute for setting aside ten to fifteen minutes each day to write down your distorted negative thoughts and answering them as outlined in the previous pages. The written method cannot be bypassed because it exposes to the light of day the illogical nature of the thoughts that trouble you. Once you are doing this regularly, you can then use your wrist counter to nip your painful cognitions in the bud at other times.

3. Cope, Don't Mope!—The Woman Who Thought She Was a "Bad Mother." As you read the previous sections, the following objection may have occurred to you: "All this deals with is my *thoughts*. But what if my problems are realistic? What good will it do me to think differently? I have some real inadequacies that need to be dealt with."

Nancy is a thirty-four-year-old mother of two who felt this way. Six years ago she divorced her first husband and has just recently remarried. She is completing her college degree on a part-time basis. Nancy is usually animated and enthusiastic and quite committed to her family. However, she has experienced episodic depressions for many years. During those low periods she becomes extremely critical of herself and others, and expresses self-doubt and insecurity. She was referred to me during such a period of depression.

I was struck by the vehemence of her self-reproach. She had received a note from her son's teacher stating that he was having some difficulty in school. Her immediate reaction was to mope and blame herself. The following is an excerpt from our therapy session:

NANCY: I should have worked with Bobby on his homework because now he is disorganized and not ready for school. I spoke to Bobby's teacher, who said Bobby lacks self-confidence and doesn't follow directions adequately. Consequently, his schoolwork has been deteriorating.

> I had a number of self-critical thoughts after the call and I felt suddenly dejected. I began to tell myself that a good mother spends time with her kids on some activity every night. I'm responsible for his poor behavior—lying, not doing well in school. I just can't figure out how to handle him. I'm really a bad mother. I began to think he was stupid and about to flunk and how it was all my fault.

My first strategy was to teach her how to attack the statement "I am a bad mother," because I felt this self-criticism was hurtful and unrealistic, creating a paralyzing internal anguish which would not help her in her efforts to guide Bobby through his crisis.

DAVID: Okay. What's wrong with this statement, "I am a bad mother"?

NANCY: Well . . .

DAVID: Is there any such thing as a "bad mother"?

NANCY: Of course.

DAVID: What is your definition of a "bad mother"?

NANCY: A bad mother is one who does a bad job of raising her kids. She isn't as effective as other mothers, so her kids turn out bad. It seems obvious.

DAVID: So you would say a "bad mother" is one who is low on mothering skills? That's your definition?

NANCY: Some mothers lack mothering skills.

DAVID: But all mothers lack mothering skills to some extent.

NANCY: They do?

DAVID: There's no mother in this world who is perfect in all mothering skills. So they all lack mothering skills in some part. According to your definition, it would seem that all mothers are bad mothers.

NANCY: I feel that *I'm* a bad mother, but not everybody is.

DAVID: Well, define it again. What is a "bad mother"?

NANCY: A bad mother is someone who does not understand her children or is constantly making damaging errors. Errors that are detrimental.

DAVID: According to this new definition, you're not a "bad mother," and there are no "bad mothers" because no one constantly makes damaging errors.

NANCY: No one . . . ?

DAVID: You said that a bad mother *constantly* makes damaging errors. There is no such person who constantly makes damaging errors twenty-four hours a day. Every mother is capable of doing *some* things right.

NANCY: Well, there can be abusive parents who are always punishing, hitting—you read about them in the papers. Their children end up battered. That could certainly be a bad mother.

DAVID: There are parents who resort to abusive behavior, that's true. And these individuals could improve their behavior, which might make them feel better about themselves and their children. But it's not realistic to say that such parents are *constantly* doing abusing or damaging things, and it's not going to help matters by attaching the label "bad" to them. Such individuals do have a problem with aggression and need training in self-control, but it would only make matters worse if you tried to convince them that their problem was badness. They usually already believe they are rotten human beings, and that is part of their problem. Labeling them as "bad mothers" would be inaccurate, and it would also be irresponsible, like trying to put out a fire by throwing gasoline on it.

At this point I was trying to show Nancy that she was just defeating herself by labeling herself as a "bad mother." I hoped to show her that no matter how she defined "bad mother," the definition would be unrealistic. Once she gave up the destructive tendency to mope and label herself as worthless, we could then go on to coping strategies for helping her son with his problems at school.

NANCY: But I still have the feeling I am a "bad mother."

DAVID: Well, once again, what is your definition?

NANCY: Someone who doesn't give her child enough attention, positive attention. I'm so busy in school. And when I do pay attention, I'm afraid it may be all negative attention. Who knows? That's what I'm saying.

DAVID: A "bad mother" is one who doesn't give her child enough attention, you say? Enough for what?

NANCY: For her child to do well in life.

DAVID: Do well in *everything*, or in some things?

NANCY: In some things. No one can do well at everything.

DAVID: Does Bobby do well at some things? Does he have any redeeming virtues?

NANCY: Oh yes. There are many things he enjoys and does well at.

DAVID: Then you can't be a "bad mother" according to your definition because your son does well at many things.

NANCY: Then why do I feel like a bad mother?

DAVID: It seems that you're labeling yourself as a "bad mother" because you'd like to spend more time with your son, and because you sometimes feel inadequate, and because there is a clear-cut need to improve your communication with Bobby. But it won't help you solve these problems if you conclude automatically you are a "bad mother." Does that make sense to you?

NANCY: If I paid more attention to him and gave him more help, he could do better at school and he could be a whole lot happier. I feel it's my fault when he doesn't do well.

DAVID: So you are willing to take the blame for his mistakes?

NANCY: Yes, it's my fault. So I'm a bad mother.

DAVID: And you also take the credit for his achievements? And for his happiness?

NANCY: No—*he* should get the credit for that, not me.

DAVID: Does that make sense? That you're responsible for his faults but not his strengths?

NANCY: No.

DAVID: Do you understand the point I'm trying to make?

NANCY: Yep.

DAVID: "Bad mother" is an abstraction; there is no such thing as a "bad mother" in this universe.

NANCY: Right. But mothers can do bad things.

DAVID: They're just people, and people do a whole variety of things—good, bad, and neutral. "Bad mother" is just a fantasy; there's no such thing. The chair is a thing. A "bad mother" is an abstraction. You understand that?

NANCY: I got it, but some mothers are more experienced and more effective than others.

DAVID: Yes, there are all degrees of effectiveness at parenting skills. And most everyone has plenty of room for improvement. The meaningful question is not "Am I a good or bad mother?" but rather "What are my relative skills and weaknesses, and what can I do to improve?"

NANCY: I understand. That approach makes more sense and it feels much better. When I label myself "bad mother," I just feel inadequate and depressed, and I don't do anything productive. Now I see what you've been driving at. Once I give up criticizing myself, I'll feel better, and maybe I can be more helpful to Bobby.

DAVID: Right! So when you look at it that way, you're talking about coping strategies. For example, what are your parenting skills? How can you begin to improve on those skills? Now that's the type of thing I would suggest with regard to Bobby. Seeing yourself as a "bad mother" eats up emotional energy and distracts you from the task of improving your mothering skills. It's irresponsible.

NANCY: Right. If I can stop punishing myself with that statement, I'll be much better off and I can start working toward helping Bobby. The moment I stop calling myself a bad mother, I'll start feeling better.

DAVID: Yes, now what can you say to yourself when you have the urge to say "I'm a bad mother"?

NANCY: I can say I don't have to hate my whole self if there is a particular thing I find I dislike about Bobby, or if he has a problem at school. I can try to *define* that problem, and *attack* that problem, and work toward solving it.

DAVID: Right. Now, that's a positive approach. I like it. You refute the negative statement and then add a positive statement. I like that.

We then worked on answering several "automatic thoughts" she had written down after the call from Bobby's teacher (see Figure 4–4, below). As Nancy learned to refute her self-critical thoughts, she experienced much-needed emotional relief. She was then able to develop some specific coping strategies designed to help Bobby with his difficulties.

Figure 4–4. Nancy's written homework concerning Bobby's difficulties at school. This is similar to the "triple-column technique," except that she did not find it necessary to write down the cognitive distortions contained in her automatic thoughts.

Automatic Thought (SELF-CRITICISM)	*Rational Response* (SELF-DEFENSE)
1. I didn't pay attention to Bobby.	1. I really spend *too much* time with him; I'm overprotective.
2. I should have worked with him on his homework, and now he is disorganized and not ready for school.	2. Homework is his responsibility, not mine. I can explain to him how to get organized. What are my responsibilities? a. Check homework; b. Insist it be done at a certain time; c. Ask if he's having any difficulties; d. Set up a reward system.
3. A good mother spends time with her kids on some activity every night.	3. Not true. I spend time when I can and want to, but it isn't feasible always. Besides, his schedule is his.
4. I'm responsible for his poor behavior and not doing well in school.	4. I can only guide Bobby. It's up to him to do the rest.

Figure 4–4. cont.

Automatic Thought (SELF-CRITICISM)	*Rational Response* (SELF-DEFENSE)
5. He wouldn't have gotten into trouble at school if I had helped him. If I had supervised his homework earlier, this problem wouldn't have occurred.	5. That is not so. Problems will occur even if I'm around to oversee things.
6. I'm a bad mother. I'm the cause of his problems.	6. I'm not a bad mother; I try. I can't control what goes on in all areas of his life. Maybe I can talk to him and his teacher and find out how to help him. Why punish myself whenever someone I love has a problem?
7. All other mothers work with their kids, but I don't know how to get along with Bobby.	7. Overgeneralization! Not true. Stop moping and start coping.

The first step of her coping plan was to talk to Bobby about the difficulties he had been having so as to find out what the real problem was. Was he having difficulties as his teacher had suggested? What was his understanding of the problem? Was it true that he was feeling tense and low in confidence? Had his homework been particularly hard for him recently? Once Nancy had obtained this information and defined the real problem, she realized she would then be in a position to work toward an appropriate solution. For example, if Bobby said he found some of his courses particularly difficult, she could develop a reward system at home to encourage him to do extra homework. She also decided to read several books on parenting skills. Her relationship with Bobby improved, and his grades and behavior at school underwent a rapid turnabout.

Nancy's mistake had been to view herself in a global way, making the moralistic judgment that she was a bad mother. This type of criticism incapacitated her because it created the impression that she had a personal problem so big and bad that no one could do anything about it. The emotional upset this labeling caused prevented her from *defining* the real problem, *breaking it down* into its specific parts, and *applying appropriate solutions*. If she had continued to mope, there was the distinct possibility that Bobby would have continued to do poorly, and she would have become increasingly ineffectual.

How can you apply what Nancy learned to your own situation? When you are down on yourself, you might find it helpful to ask what you actually mean when you try to define your true identity with a negative label such as "a fool," "a sham," "a stupid dope," etc. Once you begin to pick these destructive labels apart, you will find they are arbitrary and meaningless. They actually cloud the issue, creating confusion and despair. Once rid of them, you can define and cope with any real problems that exist.

Summary. When you are experiencing a blue mood, the chances are that you are telling yourself you are inherently inadequate or just plain "no good." You will become convinced that you have a bad core or are essentially worthless. To the extent that you believe such thoughts, you will experience a severe emotional reaction of despair and self-hatred. You may even feel that you'd be better off dead because you are so unbearably uncomfortable and self-denigrating. You may become inactive and paralyzed, afraid and unwilling to participate in the normal flow of life.

Because of the negative emotional and behavioral consequences of your harsh thinking, the first step is to stop telling yourself you are worthless. However, you probably won't be able to do this until you become absolutely convinced that these statements are *incorrect* and *unrealistic*.

How can this be accomplished? You must first consider that a human life is an ongoing process that involves a constantly changing physical body as well as an enormous number of rapidly changing thoughts, feelings, and behaviors. Your life therefore is an evolving experience, a continual flow. You are not a thing; that's why any label is constricting, highly inaccurate, and global. Abstract labels such as "worthless" or "inferior" *communicate nothing* and *mean nothing*.

But you may still be convinced you are second-rate. What is your evidence? You may reason, "I feel inadequate. Therefore, I must *be* inadequate. Otherwise, why would I be filled with such unbearable emotions?" Your error is in emotional reasoning. Your feelings do not determine your worth, simply your relative state of comfort or discomfort. Rotten, miserable internal states do not prove that you are a rotten, worthless person, merely that you think you are; because you are in a temporarily depressed mood, you are thinking illogically and unreasonably about yourself.

Would you say that states of mood elevation and happiness prove you are great or especially worthy? Or do they simply mean that you are feeling good?

Just as your feelings do not determine your worth, neither do your thoughts or behaviors. Some may be positive, creative, and enhancing; the great majority are neutral. Others may be irrational, self-defeating, and maladaptive. These can be modified if you are willing to exert the effort, but they certainly do not and cannot mean that you are no good. There is no such thing in this universe as a worthless human being.

"Then how can I develop a sense of self-esteem?" you may ask. The answer is—you don't have to! You don't have to do anything especially worthy to create or deserve self-esteem; all you have to do is turn off that critical, haranguing, inner voice. Why? *Because that critical inner voice is wrong!* Your internal self-abuse springs from illogical, distorted thinking. Your sense of worthlessness is not based on truth, it is just the abscess which lies at the core of depressive illness.

So remember three crucial steps when you are upset:

1. Zero in on those automatic negative thoughts and write them down. Don't let them buzz around in your head; snare them on paper!
2. Read over the list of ten cognitive distortions. Learn precisely how you are twisting things and blowing them out of proportion.
3. Substitute a more objective thought that puts the lie to the one which made you look down on yourself. As you do this, you'll begin to feel better. You'll be boosting your self-esteem, and your sense of worthlessness (and, of course, your depression) will disappear.

Chapter 5

Do-Nothingism: How to Beat It

In the last chapter you learned that you can change your mood by changing how you *think*. There is a second major approach to mood elevation that is enormously effective. People are not only thinkers, they are doers, so it is not surprising that you can substantially change the way you feel by changing the way you act. There's only one hitch— when you're depressed, you don't feel like doing much.

One of the most destructive aspects of depression is the way it paralyzes your willpower. In its mildest form you may simply procrastinate about doing a few odious chores. As your lack of motivation intensifies, virtually any activity appears so difficult that you become overwhelmed by the urge to do nothing. Because you accomplish very little, you feel worse and worse. Not only do you cut yourself off from your normal sources of stimulation and pleasure, but your lack of productivity aggravates your self-hatred, resulting in further isolation and incapacitation.

If you don't recognize the emotional prison in which you are trapped, this situation can go on for weeks, months, or even years. Your inactivity will be all the more frustrating if you once took pride in the energy you had for life. Your do-nothingism can also affect your family and friends, who, like yourself, cannot understand your behavior. They may say that you must want to be depressed or else you'd "get off your behind." Such a comment only worsens your anguish and paralysis.

Do-nothingism represents one of the great paradoxes of human nature. Some people naturally throw themselves into life with great zest, while others always hang back, defeating themselves at every turn as if they were involved in a plot against themselves. Do you ever wonder why?

If a person were condemned to spend months in isolation, cut off from all normal activities and interpersonal relationships, a substantial depression would result. Even young monkeys slip into a retarded, withdrawn state if they are separated from their peers and confined to a small cage. Why do you voluntarily impose a similar punishment on yourself? Do you want to suffer? Using cognitive techniques, you can discover the precise reasons for your difficulties in motivating yourself.

In my practice I find that the great majority of the depressed patients referred to me improve substantially if they try to help themselves. Sometimes it hardly seems to matter what you do as long as you do something with the attitude of self-help. I know of two presumably "hopeless" cases who were helped enormously simply by putting a mark on a piece of paper. One patient was an artist who had been convinced for years that he couldn't even draw a straight line. Consequently he didn't even try to draw. When his therapist suggested he test his conviction by actually attempting to draw a line, it came out so straight he began drawing again and soon was symptom-free! And yet many depressed individuals will go through a phase in which they *stubbornly refuse* to do anything to help themselves. The moment this crucial motivational problem has been solved, the depression typically begins to diminish. You can therefore understand why much of our research has been directed to locating the causes of this paralysis of the will. Using this knowledge, we have developed some specific methods to help you deal with procrastination.

Let me describe two perplexing patients I treated recently.

You might think their do-nothingism is extreme and wrongly conclude they must be "crazies" with whom you would have little in common. In fact, I believe their problems are caused by attitudes similar to yours, so don't write them off.

Patient A, a twenty-eight-year-old woman, has done an experiment to see how her mood would respond to a variety of activities. It turns out that she feels substantially better when she does nearly *anything*. The list of things that will reliably give her a mood lift includes cleaning the house, playing tennis, going to work, practicing her guitar, shopping for dinner, etc. Only one thing makes her feel reliably worse; this single activity nearly always makes her intensely miserable. Can you guess what it is? DO-NOTHINGISM: lying around in bed all day long, staring at the ceiling and courting negative thoughts. And guess what she does weekends. Right! She crawls right into bed on Saturday morning and begins her descent into inner hell. Do you think she really wants to suffer?

Patient B, a physician, gives me a clear, definite message early in her therapy. She says she understands that the speed of improvement is dependent on her willingness to work between sessions, and insists she wants to get well more than anything else in the world, having been wracked by depression for over sixteen years. She emphasizes she'll be happy to come to therapy sessions, but I must not ask her to lift one finger to help herself. She says that if I push her to spend five minutes on self-help assignments, she'll kill herself. As she describes in detail the lethal, gruesome method of self-destruction she had carefully planned in her hospital's operating room, it becomes obvious that she is deadly serious. Why is she so determined not to help herself?

I know your procrastination is probably less severe and only deals with minor things, like paying bills, a trip to the dentist, etc. Or maybe you've had trouble finishing a relatively straightforward report that is crucial to your career. But the perplexing question is the same—why do we frequently behave in ways that are not in our self-interest?

Procrastinating and self-defeating behavior can seem funny, frustrating, puzzling, infuriating, or pathetic, depending on your perspective. I find it a very human trait, so widespread that we all bump into it nearly every day. Writers, philosophers, and students of human nature throughout history have tried to formulate some explanation for self-defeating behavior, including such popular theories as:

1. You're basically lazy; it's just your "nature."
2. You *want* to hurt yourself and suffer. You either like feeling depressed, or you have a self-destructive drive, a "death wish."
3. You're passive-aggressive, and you want to frustrate the people around you by doing nothing.
4. You must be getting some "payoff" from your procrastination and do-nothingism. For example, you enjoy getting all that attention when you are depressed.

Each of these famous explanations represents a different psychological theory, and each is inaccurate! The first is a "trait" model; your inactivity is seen as a fixed personality trait and stems from your "lazy streak." The problem with this theory is that it just labels the problem without explaining it. Labeling yourself as "lazy" is useless and self-defeating because it creates the false impression that your lack of motivation is an irreversible, innate part of your makeup. This kind of thinking does not represent a valid scientific theory, but is an example of a cognitive distortion (labeling).

The second model implies you want to hurt yourself and suffer because there is something enjoyable or desirable about procrastination. This theory is so ludicrous I hesitate to include it, except that it is widespread and vigorously supported by a substantial percentage of psychotherapists. If you have the hunch that you or someone else likes being depressed and doing nothing, then remind yourself that depression is the most agonizing form of human suffering. Tell me—what is so great about it? I haven't yet met a patient who really enjoys the misery.

If you aren't convinced but think you really do enjoy pain and suffering, then give yourself the paper-clip test. Straighten out one end of a paper clip and push it under your fingernail. As you push harder and harder, you may notice how the pain becomes more and more excruciating. Now ask yourself—is this really enjoyable? Do I *really* like to suffer?

The third hypothesis—you're "passive-aggressive"—represents the thinking of many therapists, who believe that depressive behavior can be explained on the basis of "internalized anger." Your procrastination could be seen as an expression of that pent-up hostility because your inaction often annoys the people around you. One problem with this theory is that most depressed or procrastinating individuals simply do not feel particularly angry. Resentment can sometimes contribute to your lack of motivation, but is usually not central to the problem. Although your family may feel frustrated about your depression, you probably do not intend them to react this way. In fact, it is more often the case that you *fear* displeasing them. The implication that you are *intentionally* doing nothing in order to frustrate them is insulting and untrue; such a suggestion will only make you feel worse.

The last theory—you must be getting some "payoff" from procrastination—reflects more recent, behaviorally oriented psychology. Your moods and actions are seen as the result of rewards and punishments from your environment. If you are feeling depressed and doing nothing about it, it follows that your behavior is being rewarded in some way.

There is a grain of truth in this; depressed people do sometimes receive substantial support and reassurance from others who try to help them. However, the depressed person rarely enjoys all the attention he receives because of his profound tendency to disqualify it. If you are depressed and someone tells you they like you, you will probably think, "He doesn't know how rotten I am. I don't deserve this praise." Depression and lethargy have no real rewards. Theory number four bites the dust with the others.

How can you find the real cause of motivational paralysis? The study of mood disorders gives us the unique opportunity to observe extraordinary transformations in levels of personal motivation within short periods of time. The same individual who ordinarily bursts with creative energy and optimism may be reduced during an episode of depression to pathetic, bedridden immobility. By tracing dramatic mood swings, we can gather valuable clues that unlock many of the mysteries of human motivation. Simply ask yourself, "When I think about that undone task, what thoughts immediately come to mind?" Then write those thoughts down on a piece of paper. What you write will reflect a number of maladaptive attitudes, misconceptions, and faulty assumptions. You will learn that the feelings that impede your motivation, such as apathy, anxiety, or the sense of being overwhelmed, are the result of distortions in your thinking.

Figure 5–1 shows a typical Lethargy Cycle. The thoughts on this patient's mind are negative; he says to himself, "There's no point in doing anything because I am a born loser and so I'm bound to fail." Such a thought sounds very convincing when you are depressed, immobilizing you and making you feel inadequate, overwhelmed, self-hating, and helpless. You then take these negative emotions as proof that your pessimistic attitudes are valid, and you begin to change your approach to life. Because you are convinced you will botch up anything, you don't even try; you stay in bed instead. You lie back passively and stare at the ceiling, hoping to drift into sleep, painfully aware you are letting your career go down the drain while your business dwindles into bankruptcy. You may refuse to answer the phone for fear of hearing bad news; life becomes a treadmill of boredom, apprehension, and misery. This vicious cycle can go on indefinitely unless you know how to beat it.

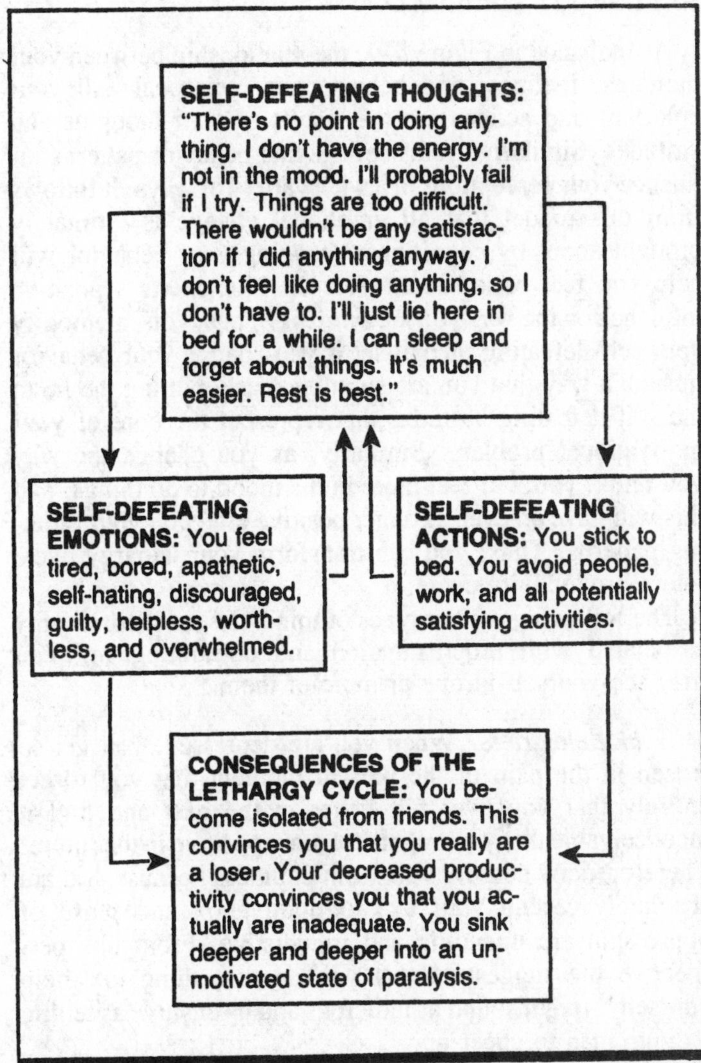

SELF-DEFEATING THOUGHTS: "There's no point in doing anything. I don't have the energy. I'm not in the mood. I'll probably fail if I try. Things are too difficult. There wouldn't be any satisfaction if I did anything anyway. I don't feel like doing anything, so I don't have to. I'll just lie here in bed for a while. I can sleep and forget about things. It's much easier. Rest is best."
SELF-DEFEATING EMOTIONS: You feel tired, bored, apathetic, self-hating, discouraged, guilty, helpless, worthless, and overwhelmed.
SELF-DEFEATING ACTIONS: You stick to bed. You avoid people, work, and all potentially satisfying activities.
CONSEQUENCES OF THE LETHARGY CYCLE: You become isolated from friends. This convinces you that you really are a loser. Your decreased productivity convinces you that you actually are inadequate. You sink deeper and deeper into an unmotivated state of paralysis.

Figure 5–1. The Lethargy Cycle. Your self-defeating negative thoughts make you feel miserable. Your painful emotions in turn convince you that your distorted, pessimistic thoughts are actually valid. Similarly, self-defeating thoughts and actions reinforce each other in a circular manner. The unpleasant consequences of do-nothingism make your problems even worse.

As indicated in Figure 5–1, the relationship between your thoughts, feelings, and behaviors is reciprocal—all your emotions and actions are the results of your thoughts and attitudes. Similarly, your feelings and behavior patterns influence your perceptions in a wide variety of ways. It follows from this model that all emotional change is ultimately brought about by cognitions; changing your behavior will help you feel better about yourself if it exerts a positive influence on the way you are *thinking*. Thus, you can modify your self-defeating mental set if you change your behavior in such a way that you are simultaneously putting the lie to the self-defeating attitudes that represent the core of your motivational problem. Similarly, as you change the way you think, you will feel more in the mood to do things, and this will have an even stronger positive effect on your thinking patterns. Thus, you can transform your lethargy cycle into a productivity cycle.

The following are the types of mind-sets most commonly associated with procrastination and do-nothingism. You may see yourself in one or more of them.

1. Hopelessness. When you are depressed, you get so frozen in the pain of the present moment that you forget entirely that you ever felt better in the past and find it inconceivable that you might feel more positive in the future. Therefore, any activity will seem pointless because you are absolutely certain your lack of motivation and sense of oppression are unending and irreversible. From this perspective the suggestion that you do something to "help yourself" might sound as ludicrous and insensitive as telling a dying man to cheer up.

2. Helplessness. You can't possibly do anything that will make yourself feel better because you are convinced that your moods are caused by factors beyond your control, such as fate, hormone cycles, dietary factors, luck, and other people's evaluations of you.

3. Overwhelming Yourself. There are several ways you may overwhelm yourself into doing nothing. You may magnify a task to the degree that it seems impossible to tackle. You may assume you must do everything at once instead of breaking each job down into small, discrete, manageable units which you can complete one step at a time. You might also inadvertently distract yourself from the task at hand by obsessing about endless other things you haven't gotten around to doing yet. To illustrate how irrational this is, imagine that every time you sat down to eat, you thought about all the food you would have to eat during your lifetime. Just imagine for a moment that all piled up in front of you are tons of meat, vegetables, ice cream, and thousands of gallons of fluids! And you have to eat every bit of this food before you die! Now, suppose that before every meal you said to yourself, "This meal is just a drop in the bucket. How can I ever get all that food eaten? There's just no point in eating one pitiful hamburger tonight." You'd feel so nauseated and overwhelmed your appetite would vanish and your stomach would turn into a knot. When you think about all the things you are putting off, you do this very same thing without being aware of it.

4. Jumping to Conclusions. You sense that it's not within your power to take effective action that will result in satisfaction because you are in the habit of saying, "I can't," or "I would but . . ." Thus when I suggested that a depressed woman bake an apple pie, she responded, "I can't cook anymore." What she really meant to say was, "I have the feeling I wouldn't enjoy cooking and it seems like it would be awfully difficult." When she tested these assumptions by attempting to bake a pie, she found it surprisingly satisfying and not at all difficult.

5. Self-labeling. The more you procrastinate, the more you condemn yourself as inferior. This saps your self-confidence further. The problem is compounded when you label yourself "a procrastinator" or "a lazy person." This causes you to see your lack of effective action as the "real you" so that you automatically expect little or nothing from yourself.

6. Undervaluing the Rewards. When you are depressed you may fail to initiate any meaningful activity not only because you conceive of any task as terribly difficult, but also because you feel the reward simply wouldn't be worth the effort.

"Anhedonia" is the technical name for a diminished ability to experience satisfaction and pleasure. A common thinking error—your tendency to "disqualify the positive"—may be at the root of this problem. Do you recall what this thinking error consists of?

A businessman complained to me that nothing he did all day was satisfying. He explained that in the morning he had attempted to return a call from a client, but found the line was busy. As he hung up, he told himself, "That was a waste of time." Later in the morning he successfully completed an important business negotiation. This time he told himself, "Anyone in our firm could have handled it just as well or better. It was an easy problem, and so my role wasn't really important." His lack of satisfaction results from the fact that he always finds a way to discredit his efforts. His bad habit of saying "It doesn't count" successfully torpedoes any sense of fulfillment.

7. Perfectionism. You defeat yourself with inappropriate goals and standards. You will settle for nothing short of a magnificent performance in anything you do, so you frequently end up having to settle for just that—*nothing*.

8. Fear of Failure. Another mind-set which paralyzes you is the fear of failure. Because you imagine that putting in the effort and not succeeding would be an overwhelming personal defeat, you refuse to try at all. Several thinking errors are involved in the fear of failure. One of the most common is overgeneralization. You reason, "If I fail at this, it means I will fail at anything." This, of course, is impossible. Nobody can fail at everything. We all have our share of victories and defeats. While it is true that victory tastes sweet and defeat is often bitter, failing at any task need not be a fatal poison, and the bad taste will not linger forever.

A second mind-set that contributes to the fear of defeat is when you evaluate your performance exclusively on the outcome regardless of your individual effort. This is illogical and reflects a "product orientation" rather than a "process orientation." Let me explain this with a personal example. As a psychotherapist I can control only what I say and how I interact with each patient. I cannot control how any particular patient will respond to my efforts during a given therapy session. What I say and how I interact is the process; how each individual reacts is the product. In any given day, several patients will report that they have benefited greatly from that day's session, while a couple of others will tell me that their session was not particularly helpful. If I evaluated my work exclusively on the outcome or product, I would experience a sense of exhilaration whenever a patient did well, and feel defeated and defective whenever a patient reacted negatively. This would make my emotional life a roller coaster, and my self-esteem would go up and down in an exhausting and unpredictable manner all day long. But if I admit to myself that all I can control is the input I provide in the therapeutic process, I can pride myself on good consistent work regardless of the outcome of any particular session. It was a great personal victory when I learned to evaluate my work based on the process rather than on the product. If a patient gives me a negative report, I try to learn from it. If I did make an error, I attempt to correct it, but I don't need to jump out the window.

9. *Fear of Success.* Because of your lack of confidence, success may seem even more risky than failure because you are certain it is based on chance. Therefore, you are con-

vinced you couldn't keep it up, and you feel your accomplishments will falsely raise the expectations of others. Then when the awful truth that you are basically "a loser" ultimately comes out, the disappointment, rejection, and pain will be all the more bitter. Since you feel sure you will eventually fall off the cliff, it seems safer not to go mountain climbing at all.

You may also fear success because you anticipate that people will make even greater demands on you. Because you are convinced you *must* and *can't* meet their expectations, success would put you into a dangerous and impossible situation. Therefore, you try to maintain control by avoiding any commitment or involvement.

10. Fear of Disapproval or Criticism. You imagine that if you try something new, any mistake or flub will be met with strong disapproval and criticism because the people you care about won't accept you if you are human and imperfect. The risk of rejection seems so dangerous that to protect yourself you adopt as low a profile as possible. If you don't make any effort, you can't goof up!

11. Coercion and Resentment. A deadly enemy of motivation is a sense of coercion. You feel under intense pressure to perform—generated from within and without. This happens when you try to motivate yourself with moralistic "shoulds" and "oughts." You tell yourself, "I *should* do this" and "I *have* to do that." Then you feel obliged, burdened, tense, resentful, and guilty. You feel like a delinquent child under the discipline of a tyrannical probation officer. Every task becomes colored with such unpleasantness that you can't stand to face it. Then as you procrastinate, you condemn yourself as a lazy, no-good bum. This further drains your energies.

12. Low Frustration Tolerance. You assume that you should be able to solve your problems and reach your goals rapidly and easily, so you go into a frenzied state of panic and rage when life presents you with obstacles. Rather than persist patiently over a period of time, you may retaliate against the "unfairness" of it all when things get tough, so you give up completely. I also call this the "entitlement syndrome" because you feel and act as if you were entitled to success, love, approval, perfect health, happiness, etc.

Your frustration results from your habit of comparing reality with an ideal in your head. When the two don't match, you condemn reality. It doesn't occur to you that it might be infinitely easier simply to change your expectations than to bend and twist reality.

This frustration is frequently generated by should statements. While jogging, you might complain, "For all the miles I've gone, I should be in better shape by now." Indeed? Why should you? You may have the illusion that such punishing, demanding statements will help you by driving you on to try harder and to put out more effort. It rarely works this way. The frustration just adds to your sense of futility and increases your urge to give up and do nothing.

13. Guilt and Self-blame. If you are frozen in the conviction you are bad or have let others down, you will naturally feel unmotivated to pursue your daily life. I recently treated a lonely elderly woman who spent her days in bed in spite of the fact that she felt better when she shopped, cooked, and socialized with her friends. Why? This sweet woman was holding herself responsible for her daughter's divorce five years earlier. She explained, "When I visited them, I should have sat down and talked things over with my son-in-law. I should have asked him how things were going. Maybe I could have helped. I wanted to and yet I didn't take the opportunity. Now I feel I failed them." After we reviewed the illogic in her thinking, she felt better immediately and became active again. Because she was human and not God, she could not have been expected to predict the future or to know precisely how to intervene.

By now you may be thinking, "So what? I know that my do-nothingism is in a way illogical and self-defeating. I can see myself in several of the mental sets you've described. But I feel like I'm trying to wade through a pool of molasses. I just can't get myself going. You may say all this oppression just results from my attitudes, but it feels like a ton of bricks. So what can I do about it?"

Do you know why virtually *any* meaningful activity has a decent chance of brightening your mood? If you do nothing, you will become preoccupied with the flood of negative, destructive thoughts. If you do something, you will be temporarily distracted from that internal dialogue of self-denigration. What is even more important, the sense of mastery you will experience will disprove many of the distorted thoughts that slowed you down in the first place.

As you review the following self-activation techniques, choose a couple that appeal most to you and work at them for a week or two. Remember you don't have to master them all! One man's salvation can be another's curse. Use the methods that seem the most tailored to your particular brand of procrastination.

The Daily Activity Schedule. The Daily Activity Schedule (see Figure 5–2, page 95) is simple but effective, and can help you get organized in your fight against lethargy and apathy. The schedule consists of two parts. In the Prospective column, write out an hour-by-hour plan for what you would like to accomplish each day. Even though you may actually carry out only a portion of your plan, the simple act of creating a method of action every day can be immensely helpful. Your plans need not be elaborate. Just put one or two words in each time slot to indicate what you'd like to do, such as "dress," "eat lunch," "prepare résumé," etc. It should not require more than five minutes to do this.

At the end of the day, fill out the Retrospective column. Record in each time slot what you actually did during the day. This may be the same as or different from what you

Figure 5–2. Daily Activity Schedule.

	PROSPECTIVE: Plan your activities on an hour-by-hour basis at the start of the day.	*RETROSPECTIVE:* At the end of the day, record what you actually did and rate each activity with an M for mastery or a P for pleasure.*
Date_______		
TIME		
8–9		
9–10		
10–11		
11–12		
12–1		
1–2		
2–3		
3–4		
4–5		
5–6		
6–7		
7–8		
8–9		
9–12		

*Mastery and pleasure activities must be rated from 0 to 5: the higher the number, the greater the sense of satisfaction.

actually planned; nevertheless, even if it was just staring at the wall, write it down. In addition, label each activity with the letter M for mastery or the letter P for pleasure. Mastery activities are those which represent some accomplishment, such as brushing your teeth, cooking dinner, driving to work, etc. Pleasure might include reading a book, eating, going to a movie, etc. After you have written M or P for each activity, estimate the actual amount of pleasure, or the degree of difficulty in the task by using a zero to five rating. For example, you could give yourself a score of M-1 for particularly easy tasks like getting dressed, while M-4 or M-5 would indicate you did something more difficult and challenging, such as not eating too much or applying for a job. You can rate the pleasure activities in a similar manner. If any activity was pleasurable in the past when you were not depressed, but today it was nearly or totally devoid of pleasure, put a P-½ or a P-0. Some activities, such as cooking dinner, can be labeled M and P.

Why is this simple activity schedule likely to be helpful? First, it will undercut your tendency to obsess endlessly about the value of various activities and to debate counterproductively about whether or not to do something. Accomplishing even a part of your scheduled activities will in all probability give you some satisfaction and will combat your depression.

As you plan your day, develop a balanced program that provides for enjoyable leisure activities as well as work. If you are feeling blue, you may want to put a special emphasis on fun, even if you doubt you can enjoy things as much as usual. You may be depleted from having asked too much of yourself, causing an imbalance in your "give-and-get" system. If so, take a few days of "vacation" and schedule only those things you *want* to do.

If you adhere to the schedule, you will find your motivation increasing. As you start doing things, you will begin to disprove your belief that you are incapable of functioning effectively. As one procrastinator reported, "By scheduling my day and comparing the results, I have become aware of how I spend my time. This has helped me take charge of my life once again. I realize that I can be in control if I want to."

Keep this Daily Activity Schedule for at least a week. As you review the activities in which you participated during the previous week, you will see that some have given you a greater sense of mastery and pleasure, as indicated by higher scores. As you continue planning each upcoming day, use this information to schedule more of those activities, and avoid others which are associated with lower satisfaction levels.

The Daily Activity Schedule can be especially helpful for a common syndrome I call the "weekend/holiday blues." This is a pattern of depression most often reflected in people who are single and have their greatest emotional difficulties when alone. If you fit this description, you probably assume these periods are bound to be unbearable, so you do very little to care for yourself creatively. You stare at the walls and mope, or lie in bed all day Saturday and Sunday; or, for good times, you watch a boring TV show and eat a meager dinner of a peanut-butter sandwich and a cup of instant coffee. No wonder your weekends are tough! Not only are you depressed and alone but you treat yourself in a way that can only inflict pain. Would you treat someone else in such a sadistic manner?

These weekend blues can be overcome by using the Daily Activity Schedule. On Friday night, schedule some plans for Saturday on an hourly basis. You may resist this, saying, "What's the point? I'm all alone." The fact that you are all alone is the very reason for using the schedule. Why assume you're bound to be miserable? This prediction can function only as a self-fulfilling prophecy! Put it to the test by adopting a productive approach. Your plans need not be elaborate in order to be helpful. You can schedule going to the hairdresser, shopping, visiting an art museum, reading a book, or walking through the park. You will discover that laying out and adhering to a simple plan for the day can go a long way toward lifting your mood. And who knows—if you are willing to care for yourself, you may suddenly notice that others will act more interested in you as well!

At the end of the day before you go to bed, write down what you actually did each hour and rate each activity for Mastery and Pleasure. Then make out a new schedule for the following day. This simple procedure may be the first step toward a sense of self-respect and genuine self-reliance.

The Antiprocrastination Sheet. In Figure 5–3 is a form I have found effective in breaking the habit of procrastination. You may be avoiding a particular activity because you predict it will be too difficult and unrewarding. Using the Antiprocrastination Sheet, you can train yourself to test these negative predictions. Each day write down in the appropriate column one or more tasks you have been putting off. If the task requires substantial time and effort, it is best to break it down into a series of small steps so that each one can be completed in fifteen minutes or less. Now write down in the next column how difficult you predict each step of the task will be, using a 0-to-100 percent scale. If you imagine the task will be easy, you can write down a low estimate such as 10 to 20 percent; for harder tasks, use 80 to 90 percent. In the next column, write down your prediction of how satisfying and rewarding it will be to complete each phase of the task, again using the percentage system. Once you've recorded these predictions, go ahead and complete the first step of the task. After you've completed each step, take note of how difficult it actually turned out to be, as well as the amount of pleasure you gained from doing it. Record this information in the last two columns, again using the percentage system.

Figure 5–3 shows how a college professor used this form to overcome several months of putting off writing a letter applying for a teaching position opening up at another university. As you can see, he anticipated that writing the letter would be difficult and unrewarding. After he recorded his pessimistic predictions, he became curious to outline the letter and prepare a rough draft to see if it would be as

Figure 5–3. A professor procrastinated for several months in writing a letter because he imagined it would be difficult and unrewarding. He decided to break the task down into small steps and to predict on a 0-to-100 percent scale how difficult and rewarding each step would be (see the appropriate columns). After completing each step, he wrote down how difficult and rewarding it actually was. He was amazed to see how off-base his negative expectations really were.

The Antiprocrastination Sheet

(Write down the predicted difficulty and satisfaction *before* you attempt the task. Write down the actual difficulty and satisfaction *after* you have completed each step.)

Date	*Activity (Break each task down into small steps)*	*Predicted Difficulty (0–100%)*	*Predicted Satisfaction (0–100%)*	*Actual Difficulty (0–100%)*	*Actual Satisfaction (0–100%)*
6/10/99	1. Outline letter.	90	10	10	60
	2. Write rough draft.	90	10	10	75
	3. Type up final draft.	75	10	5	80
	4. Address the envelope and mail the letter.	50	5	0	95

tedious and unrewarding as he thought. He found to his great surprise that it turned out to be easy and satisfying, and he felt sufficiently motivated that he went on to complete the letter. He recorded this data in the last two columns. The information gained from this experiment so greatly astonished him that he used the Antiprocrastination Sheet in many other areas in his life. Consequently, his productivity and self-confidence underwent a dramatic increase, and his depression disappeared.

Daily Record of Dysfunctional Thoughts. This record, introduced in Chapter 4, can be used to great advantage when you are overwhelmed by the urge to do nothing. Simply write down the thoughts that run through your mind when you think about a particular task. This will immediately show you what your problem is. Then write down appropriate rational responses that show these thoughts are unrealistic. This will help you mobilize enough energy to take that first difficult step. Once you've done that, you will gain momentum and be on your way.

An example of this approach is indicated in Figure 5–4. Annette is an attractive, young single woman who owns and operates a successful boutique (she is Patient A, described on page 83). She does well during the week because of all the bustle at her store. On weekends she tends to hide away in bed unless she has social activities lined up. The moment she gets into bed, she becomes despondent, yet claims it is beyond her control to get out of bed. As Annette recorded her automatic thoughts one Sunday evening (Figure 5–4), it became obvious what her problems were: She was waiting around until she felt the desire, interest, and energy to do something; she was assuming that there was no point in doing anything since she was alone; and she was persecuting and insulting herself because of her inactivity.

When she talked back to her thoughts, she reported that the clouds lifted just a bit so that she was able to get up, take a shower, and get dressed. She then felt even better

Figure 5–4. Daily Record of Dysfunctional Thoughts.

Date	*Situation*	*Emotions*	*Automatic Thoughts*	*Rational Responses*	*Outcome*
7/15/99	I stayed in bed all day Sunday—slept off and on—no desire or energy to get up or do anything productive.	Depressed Exhausted Guilty Self-hatred Lonely	I have no desire to do anything.	That's because I'm doing nothing. Remember motivation follows action!	Felt some relief and decided to get up and take a shower at least.
			I don't have the energy to get out of bed.	I *can* get out of bed; I'm not crippled.	
			I'm a failure as a person.	I do succeed at things when I want to. Doing nothing makes me depressed and bored, but it doesn't mean I'm "a failure as a person" because there *is no such thing*!	
			I have no real interests.	I do have interests, but not when I'm doing nothing. If I get started at something, I'll probably get more interested.	

Figure 5–4. cont.

Date	*Situation*	*Emotions*	*Automatic Thoughts*	*Rational Responses*	*Outcome*
			I'm self-centered because I don't care about anything that's going on around me.	I do care about other things when I'm feeling really good. It's natural to be less interested when you're depressed.	
			Most people are out enjoying themselves.	So what does that have to do with me? I'm free to do anything I want to.	
			I don't enjoy anything.	I enjoy things when I feel good. If I do something I'll probably enjoy it once I get started, even though it doesn't seem that way when I'm lying in bed.	

I'll never have a normal energy level.	I have no proof of that; I'm working on it now and seeing some results. When I feel good, I'm full of energy. When I get involved in things, I get more energetic.
I don't want to talk to anyone or see anyone.	So don't! No one's forcing me to talk. So, decide to do something on my own. At least I can get out of bed and start doing things.

and arranged to meet a friend for dinner and a movie. As she predicted in the Rational Responses column, the more she did, the better she felt.

If you decide to use this method, be sure you actually write down upsetting thoughts. If you try to figure them out in your head, you will in all probability get nowhere because the thoughts that stymie you are slippery and complex. When you try to talk back to them, they'll come at you even harder from all angles with such speed that you won't even know what hit you. But when you write them down, they become exposed to the light of reason. This way you can reflect on them, pinpoint the distortions, and come up with some helpful answers.

The Pleasure-Predicting Sheet. One of Annette's self-defeating attitudes is her assumption that there is no point in doing anything productive if she is alone. Because of this belief, she does nothing and feels miserable, which just confirms her attitude that it's terrible to be alone.

Solution: Test your belief that there is no point in doing anything by using the Pleasure-Predicting Sheet shown in Figure 5–5, page 105. Over a period of weeks, schedule a number of activities that contain a potential for personal growth or satisfaction. Do some of them by yourself and some with others. Record who you did each activity with in the appropriate column, and predict how satisfying each will be—between 0 and 100 percent. Then go and do them. In the Actual Satisfaction column, write down how enjoyable each activity really turned out to be. You may be surprised to learn that things you do on your own are more gratifying than you thought.

Make sure that the things you do by yourself are of equal quality as those you do with others so that your comparisons will be valid. If you choose to eat a TV dinner alone, for example, don't compare it with the fancy French restaurant dinner you share with a friend!

Figure 5–5 shows the activities of a young man who learned that his girlfriend (who lived 200 miles away) had

Figure 5–5. The Pleasure-Predicting Sheet.

Date	*Activity for Satisfaction. (Sense of Achievement or Pleasure)*	*Who Did You Do This With? (If Alone, Specify Self)*	*Predicted Satisfaction (0–100%). (Write This* Before *the Activity)*	*Actual Satisfaction (0–100%). (Record This* After *the Activity)*
8/2/99	Reading (1 hour)	self	50%	60%
8/3/99	Dinner + bar w/Ben	Ben	80%	90%
8/4/99	Susan's party	self	80%	85%
8/5/99	N.Y.C. and Aunt Helen	parents and grandma	40%	30%
8/5/99	Nancy's house	Nancy and Joelle	75%	65%
8/6/99	Dinner at Nancy's	12 people	60%	80%
8/6/99	Luci's party	Luci + 5 people	70%	70%
8/7/99	Jogging	self	60%	90%
8/8/99	Theater	Luci	80%	70%
8/9/99	Harry's	Harry, Jack, Ben and Jim	60%	85%
8/10/99	Jogging	self	70%	80%
8/10/99	Phillies game	Dad	50%	70%
8/11/99	Dinner	Susan and Ben	70%	70%
8/12/99	Art museum	self	60%	70%
8/12/99	Peabody's	Fred	80%	85%
8/13/99	Jogging	self	70%	80%

a new boyfriend and didn't want to see him. Instead of moping in self-pity, he became involved with life. You will notice in the last column that the satisfaction levels he experienced by himself ranged from 60 to 90 percent, while those with other people ranged from 30 to 90 percent. This knowledge strengthened his self-reliance because he realized that he wasn't condemned to misery because he lost his girl, and that he didn't need to depend on others to enjoy himself.

You can use the Pleasure-Predicting Sheet to test a number of assumptions you might make that lead to procrastination. These include:

1. I can't enjoy anything when I'm alone.
2. There's no point in doing anything because I failed at something important to me (e.g., I didn't get the job or promotion I had my heart set on).
3. Since I'm not rich, successful, or famous, I can't really enjoy things to the hilt.
4. I can't enjoy things unless I'm the center of attention.
5. Things won't be particularly satisfying unless I can do them perfectly (or successfully).
6. I wouldn't feel very fulfilled if I did just a part of my work. I've got to get it *all* done today.

All of these attitudes will produce a round of self-fulfilling prophecies if you don't put them to the test. If, however, you check them out using the Pleasure-Predicting Sheet, you may be amazed to learn that life can offer you enormous fulfillment. Help yourself!

A question that commonly comes up about the Pleasure-Predicting Sheet is: "Suppose I do schedule a number of activities, and I find out they are just as unpleasant as I had anticipated?" This might happen. If so, try noting your negative thoughts and write them down, answering them with the Daily Record of Dysfunctional Thoughts. For example, suppose you go to a restaurant on your own and feel tense. You might be thinking, "These people probably think I'm a loser because I'm here all alone."

How would you answer this? You might remind yourself that other people's thoughts do not affect your mood one iota. I have demonstrated this to patients by telling them I will think two thoughts about them for fifteen seconds each. One thought will be extremely positive, and the other will be intensely negative and insulting. They are to tell me how each of my thoughts affects them. I close my eyes and think, "Jack here is a fine person and I like him." Then I think, "Jack is the worst person in Pennsylvania." Since Jack doesn't know which thought is which, they have no effect on him!

Does that brief experiment strike you as trivial? It's not—because only *your* thoughts can ever affect you. For example, if you are in a restaurant feeling miserable because you are alone, you really have no idea what people are thinking. It's your thoughts and only yours that are making you feel terrible; *you're the only person in the world who can effectively persecute yourself*. Why do you label yourself a "loser" because you're in a restaurant alone? Would you be so cruel to someone else? Stop insulting yourself like that! Talk back to that automatic thought with a rational response: "Going to a restaurant alone doesn't make me a loser. I have just as much right to be here as anyone else. If someone doesn't like it, so what? As long as I respect myself, I don't need to be concerned with others' opinions."

How to Get off Your "But"—the But Rebuttal. Your "but" may represent the greatest obstacle to effective action. The moment you think of doing something productive, you give yourself excuses in the form of buts. For example, "I *could* go out and jog today, BUT . . ."

1. I'm really too tired to;
2. I'm just too lazy;
3. I'm not particularly in the mood, etc.

Figure 5–6. The But-Rebuttal Method. The zigzag arrows trace your thinking pattern as you debate the issue in your mind.

But Column	*But Rebuttal*
I really *should* mow the lawn, but I'm just not in the mood.	I'll feel more like it once I get started. When I'm done I'll feel terrific.
But now it's so long it would take forever.	It won't take that much extra time with the power mower. I can always do a part of it now.
But I'm too tired.	So just do some of it and rest.
I'd rather rest now or watch TV.	I can, but I won't feel very good about it knowing this chore is hanging over my head.
But I'm just too lazy to do it today.	That can't be true—I've done it on numerous occasions in the past.

Here's another example. "I *could* cut down on my smoking, BUT . . ."

1. I don't have that kind of self-discipline;
2. I don't really feel like going cold turkey, and cutting down gradually would be slow torture;
3. I've been too nervous lately.

If you really want to motivate yourself, you'll have to learn how to get off your but. One way to do this is with the "But-Rebuttal Method" shown in Figure 5–6. Suppose it's Saturday and you've scheduled mowing the lawn. You've procrastinated for three weeks, and it looks like a jungle. You tell yourself, "I really should, BUT I'm just not in the mood." Record this in the But column. Now fight back by writing a But Rubuttal: "I'll feel more like it once I get started. When I'm done, I'll feel terrific." Your next impulse will probably be to dream up a new objection: "BUT it's so long it will take forever." Now fight back with a new rebuttal, as shown in Figure 5–6, and continue this process until you've run out of excuses.

Learn to Endorse Yourself. Do you frequently convince yourself that what you do doesn't count? If you have this bad habit, you will naturally feel that you never do anything worthwhile. It won't make any difference if you are a Nobel laureate or a gardener—life will seem empty because your sour attitude will take the joy out of all your endeavors and defeat you before you even begin. No wonder you feel unmotivated!

To reverse this destructive tendency, a good first step would be to pinpoint the self-downing thoughts that cause you to feel this way in the first place. Talk back to these thoughts and replace them with ones that are more objective and self-endorsing. Some examples of this are shown in Figure 5–7. Once you get the knack of it, practice consciously endorsing yourself all day long for the things you do even if they seem trivial. You may not feel a pleasant emotional lift in the beginning, but keep practicing even if it seems mechanical. After a few days you will begin to experience some mood lift, and you will feel more pride about what you're doing.

You may object, "Why should I have to pat myself on the back for everything I do? My family, friends, and business associates should be more appreciative of me." There are several problems here. In the first place, even if people are overlooking your efforts, you are guilty of the same crime if you also neglect yourself, and pouting won't improve the situation.

Even when someone does stroke you, you can't absorb the praise unless you decide to believe and therefore validate what is being said. How many genuine compliments fall on

Figure 5–7.

Self-Downing Statement	*Self-Endorsing Statement*
Anybody could wash these dishes.	If it's a routine, boring job, I deserve extra credit for doing it.
There was no point in washing these dishes. They'll just get dirty again.	That's just the point. They'll be clean when we need them.
I could have done a better job straightening up.	Nothing in the universe is perfect, but I did make the room look a hell of a lot better.
It was just luck the way my speech turned out.	It wasn't a matter of luck. I prepared well and delivered my talk effectively. I did a darn good job.
I waxed the car, but it still doesn't look as good as my neighbor's new car.	The car looks a heck of a lot better than it did. I'll enjoy driving it around.

your deaf ears because you mentally discredit them? When you do this, other people feel frustrated because you don't respond positively to what they are saying. Naturally, they give up trying to combat your self-downing habit. Ultimately, only what you think about what you do will affect your mood.

It can be helpful simply to make a written or mental list of the things you do each day. Then give yourself a mental credit for each of them, however small. This will help you focus on what you *have* done instead of what you haven't gotten around to doing. It may sound simplistic, but it works!

TIC-TOC Technique. If you are procrastinating about getting down to a specific task, take note of the way you are thinking about it. These TICs, or Task-Interfering Cognitions, will lose much of their power over you if you simply write them down and substitute more adaptive TOCs, or Task-Oriented Cognitions, using the double-column technique. A number of examples are shown in Figure 5–8. When you record your TIC-TOCs, be sure to pinpoint the distortion in the TIC that defeats you. You may find, for example, that your worst enemy is all-or-nothing thinking or disqualifying the positive, or you may be in the bad habit of making arbitrary negative predictions. Once you become aware of the type of distortion that most commonly thwarts you, you will be able to correct it. Your procrastination and time-wasting will give way to action and creativity.

You can also apply this principle to mental images and daydreams as well as to thoughts. When you avoid a task, you probably automatically fantasize about it in a negative, defeatist fashion. This creates unnecessary tension and apprehension, which impairs your performance and increases the likelihood that your dreaded fear will actually come true.

For example, if you have to give a speech to a group of associates, you may fret and worry for weeks ahead of time because in your mind's eye you *see* yourself forgetting what you have to say or reacting defensively to a pushy question from the audience. By the time you give the speech, you have effectively programmed yourself to behave just this way, and you're such a nervous wreck it turns out just as badly as you had imagined!

If you dare to give it a try, here's a solution: For ten minutes every night before you go to sleep, practice fantasizing that you deliver the speech in a positive way. Imagine that you appear confident, that you present your material in an energetic manner, and that you handle all questions from the audience warmly and capably. You may be surprised that this simple exercise can go a long way to improving how you feel about what you do. Obviously there is no guarantee things will always come out exactly as you imagine, but there's *no* doubt that your expectations and mood will profoundly *influence* what actually does happen.

Figure 5–8. The TIC-TOC Technique. In the left-hand column, record the thoughts that inhibit your motivation for a specific task. In the right-hand column, pinpoint the distortions and substitute more objective, productive attitudes.

TICs (*Task-interfering Cognitions*)	*TOCs* (*Task-oriented Cognitions*)
Housewife: I'll never be able to get the garage cleaned out. The junk's been piling up for years.	Overgeneralization; all-or-nothing thinking. Just do a little bit and get started. There's no reason I have to do it all today.
Bank Clerk: My work isn't very important or exciting.	Disqualifying the positive. It may seem routine to me, but it's quite important to the people who use the bank. When I'm not depressed, it can be very enjoyable. Many people do routine work but this doesn't make them unimportant human beings. Maybe I could do something more exciting in my free time.
Student: Writing this term paper is pointless. The subject is boring.	All-or-nothing thinking. Just do a routine job. It doesn't have to be a masterpiece. I might learn something, and it will make me feel better to get it done.
Secretary: I'll probably flub typing this and make a bunch of typos. Then my boss will yell at me.	Fortune teller error. I don't have to type perfectly. I can correct the errors. If he's overly critical, I can disarm him, or tell him I'd do better if he were more supportive and less demanding.
Politician: If I lose this race for governor, I'll be a laughing stock.	Fortune teller error; labeling. It's not shameful to lose a political contest. A lot of people respect me for trying and taking an honest stand on some

Figure 5–8. cont.

TICs (Task-interfering Cognitions)	*TOCs (Task-oriented Cognitions)*
	important issues. Unfortunately, the best man often doesn't win, but I can believe in myself whether or not I come out on top.
Insurance Salesman:	Mind reading.
What's the point in calling this guy back? He didn't sound interested.	I have no way of knowing. Give it a try. At least he asked me to call back. Some people will be interested and I have to sift the chaff from the wheat. I can feel productive even when someone turns me down. I'll sell one policy on the average for every five people who turn me down, so it's to my advantage to get as many turndowns as possible! The more turndowns, the more sales!
Shy Single Man:	Fortune teller error; overgeneralization.
If I call up an attractive girl, she'll just dump on me, so what's the point? I'll just wait around until some girl makes it real obvious that she likes me. Then I won't have to take a risk.	They can't *all* turn me down, and it's not shameful to try. I can learn from any rejection. I've got to start practicing to improve my style, so take the big plunge! It took courage to jump off the high dive the first time, but I did it and survived. I can do this too!
Author:	All-or-nothing thinking.
This chapter has to be great. But I don't feel very creative.	Just prepare an adequate draft. I can improve it later.
Athlete:	Disqualifying the positive; all-or-nothing thinking.
I can't discipline myself. I have no self-control. I'll never get in shape.	I must have self-control because I've done well. Just work out for a while and call it quits if I get exhausted.

Little Steps for Little Feet. A simple and obvious self-activation method involves learning to break any proposed task down into its tiny component parts. This will combat your tendency to overwhelm yourself by dwelling on all the things you have to do.

Suppose your job involves attending lots of meetings, but you find it difficult to concentrate due to anxiety, depression, or daydreaming. You can't concentrate effectively because you think, "I don't understand this as I should. Gosh, this is boring. I'd really prefer to be making love or fishing right now."

Here's how you can beat the boredom, defeat the distraction, and increase your ability to concentrate: Break the task down into its smallest component parts! For example, decide to listen for only three minutes, and then take a one-minute break to daydream intensively. At the end of this mental vacation, listen for another three minutes, and do not entertain any distracting thoughts for this brief period. Then give yourself another one-minute break to daydream.

This technique will enable you to maintain a more effective level of overall concentration. Giving yourself permission to dwell on distracting thoughts for short periods will diminish their power over you. After a while, they will seem ludicrous.

An extremely useful way to divide a task into manageable units is through time limitation. Decide how much time you will devote to a particular task, and then stop at the end of the allotted time and go on to something more enjoyable, whether or not you're finished. As simple as this sounds, it can work wonders. For example, the wife of a political VIP spent years harboring resentment toward her husband for his successful, glamorous life. She felt her life consisted of an oppressive load of child-rearing and housecleaning. Because she was compulsive she never felt she had enough time to complete her dreary chores. Life was a treadmill. She was straddled by depression, and had been unsuccessfully treated by a long string of famous therapists for over a decade as she looked in vain for the elusive key to personal happiness.

After consulting twice with one of my colleagues (Dr. Aaron T. Beck), she experienced a rapid mood swing out of her depression (his therapeutic wizardry never ceases to astonish me). How did he perform this seeming miracle? Easy. He suggested to her that her depression was due in part to the fact that she wasn't pursuing goals that were meaningful to her because she didn't believe in herself. Instead of acknowledging and confronting her fear of taking risks, she blamed her lack of direction on her husband and complained about all the undone housework.

The first step was to decide how much time she felt she wanted to spend on the housework each day; she was to spend no more than this amount even if the house wasn't perfect, and she was to budget the rest of the day to pursue activities that interested her. She decided that one hour of housework would be fair, and enrolled in a graduate program so she could develop her own career. This gave her a feeling of liberation. Like magic, the depression vanished along with the anger she harbored toward her husband.

I don't want to give you the idea that depression is usually so easy to eliminate. Even in the above case, this patient will probably have to fight off a number of depressive recurrences. She may at times fall back temporarily into the same trap of trying to do too much, blaming others, and feeling overwhelmed. Then she will have to apply the same solution again. The important thing is—she has found a method that works for her.

The same approach might work for you. Do you tend to bite off bigger pieces than you can comfortably chew? *Dare* to put modest time limits on what you do! *Have the courage* to walk away from an unfinished task! You may be amazed that you will experience a substantial increase in your productivity and mood, and your procrastination may become a thing of the past.

Motivation Without Coercion. A possible source of your procrastination is an inappropriate system for self-motivation. You may inadvertently undermine what you attempt by flagellating yourself with so many "oughts," "shoulds," and "musts" that you end up drained of any desire to get moving. You are defeating yourself by the *way* you *kill* yourself to get moving! Dr. Albert Ellis describes this mental trap as "*must*erbation."

Reformulate the way you tell yourself to do things by eliminating those coercive words from your vocabulary. An alternative to pushing yourself to get up in the morning would be to say, "It will make me feel better to get out of bed, even though it will be hard at first. Although I'm not *obliged* to, I might end up being glad I did. If, on the other hand, I'm really benefiting from the rest and relaxation, I may as well go ahead and enjoy it!" If you translate shoulds into wants, you will be treating yourself with a sense of respect. This will produce a feeling of freedom of choice and personal dignity. You will find that a reward system works better and lasts longer than a whip. Ask yourself, "What do I *want* to do? What course of action would be to my best advantage?" I think you will find that this way of looking at things will enhance your motivation.

If you still have the desire to lie in bed, mope, and feel doubtful that getting up is really what you want to do, make a list of the advantages and disadvantages of staying in bed for another day. For example, an accountant who was far behind in his work around tax time found it hard to get up each day. His customers began to complain about the undone work, and in order to avoid these embarrassing confrontations, he lay in bed for weeks trying to escape, not even answering the phone. Many customers fired him, and his business began to fail.

His mistake was in telling himself, "I know I *should* go to work but I don't want to. And I don't have to either! So I won't!" Essentially, the word "should" created the illusion that the only reason for him to get out of bed was to please a bunch of angry, demanding customers. This was

Figure 5–9.

Advantages of Lying in Bed	*Disadvantages of Lying in Bed*
1. It's easy.	1. While it seems easy, it gets awfully boring and painful after a while. It's actually not so easy to do nothing and to lie here moping and criticizing myself hour after hour.
2. I won't have to do anything or face my problems.	2. I won't be obliged to do anything if I get out of bed either, but it might feel better. If I avoid my problems they won't go away, they'll just get worse, and I won't have the satisfaction of trying to solve them. The short-term discomfort of facing up to things is probably less depressing than the endless anguish of staying in bed.
3. I can sleep and escape.	3. I can't sleep forever, and I really don't need any more sleep since I have been sleeping nearly sixteen hours a day. I will probably feel less fatigued if I get up and get my arms and legs moving rather than lie around in bed like a cripple waiting for my arms and legs to rot!

so unpleasant that he *resisted*. The absurdity of what he was doing to himself became apparent when he made a list of the advantages and disadvantages of staying in bed (Figure 5–9, above). After preparing this list, he realized it was to his advantage to get out of bed. As he subsequently became more involved with his work, his mood rapidly improved in spite of the fact that he had lost many accounts during the period of inactivity.

Disarming Technique. Your sense of paralysis will be intensified if your family and friends are in the habit of pushing and cajoling you. Their nagging should statements reinforce the insulting thoughts already echoing through your head. Why is their pushy approach doomed to failure? It's a basic law of physics that for every action there's an equal and opposite reaction. Any time you feel shoved, whether by someone's hand actually on your chest or by someone trying to boss you around, you will naturally tighten up and resist so as to maintain your equilibrium and balance. You will attempt to exert your self-control and preserve your dignity by refusing to do the thing that you are being pushed to do. The paradox is that you often end up hurting yourself.

It can be very confusing when someone obnoxiously insists you do something that actually would be to your advantage. This puts you in a "can't win" situation because if you refuse to do what the person tells you, you end up defeating yourself just in order to spite him or her. In contrast, if you do what the person tells you to do, you feel had. Because you gave in to those pushy demands, you get the feeling the individual controlled you, and this robs you of self-respect. No one likes to be coerced.

For example, Mary is a woman in her late teens who was referred to us by her parents after many years of depression. Mary was a real "hibernator," and had the capacity to sit alone in her room watching TV soap operas for months at a time. This was due in part to her irrational belief that she looked "peculiar," and that people would stare at her if she went out in public, and also by her feeling of being coerced by her domineering mother. Mary admitted that doing things might help her feel better, but this would mean giving in to her mother, who kept telling her to get off her duff and do something. The harder Mom pushed, the more stubbornly Mary resisted.

It is an unfortunate fact of human nature that it can be extremely difficult to do something when you sense you are being forced into it. Fortunately, it's very easy to learn how to handle people who nag and harangue you and try to run your life. Suppose you are Mary, and after thinking things over, you decide you would be better off if you got involved in doing a number of things. You've just made this decision when your mother comes into your bedroom and announces, "Don't you lie around any longer! Your life is going down the drain. Get moving! Get involved in things the way the other girls your age do!" At that moment, in spite of the fact that you already have decided to do just that, you develop a tremendous aversion to it!

The disarming technique is an assertive method that will solve this problem for you (other applications of this verbal maneuver will be described in the next chapter). The essence of the disarming technique is to agree with your mother, but to do so in a way that you remind her you are agreeing with her based on your own decision, and not because she was telling you what to do. So, you might answer this way: "Yes, Mom, I just thought the situation over myself and decided it *would* be to my advantage to get moving on things. Because of *my own* decision, I'm going to do it." Now you can start doing things and not feel had. Or if you wish to put more of a barb in your comments, you can always say, "Yes, Mom, I *have* in fact decided to get out of bed in spite of the fact that you've been telling me to!"

Visualize Success. A powerful self-motivation method involves making a list of the advantages of a productive action you've been avoiding because it requires more self-discipline than you have been able to muster. Such a list will train you to look at the positive consequences of doing it. It's only human to go after what you want. Furthermore, clubbing yourself into effective action doesn't usually work nearly as well as a fat, fresh carrot.

Suppose, for example, you want to quit smoking. You may be reminding yourself about cancer and all the other dangers of smoking. These fear tactics make you so nervous that you immediately reach for another cigarette; they don't work. Here's a three-step method that *does work*.

The first step is to make a list of all the positive consequences that will result when you become a nonsmoker. List as many as you can think of, including:

1. Improved health.
2. I'll respect myself.
3. I'll have greater self-discipline. With my new self-confidence, I may be able to do a whole lot of other things I've been putting off.
4. I will be able to run and dance actively, and still feel good about my body. I'll have lots of stamina and extra energy.
5. My lungs and heart will become strong. My blood pressure will go down.
6. My breath will be fresh.
7. I'll have extra spending money.
8. I'll live longer.
9. The air around me will be clean.
10. I'll be able to tell people that I've become a nonsmoker.

Once you have prepared the list, you're ready for the second step. Every night before you go to sleep, fantasize you are in your favorite spot—walking through the woods in the mountains, on a crisp autumn day, or maybe lying on a quiet beach near a crystal-blue ocean, with the sun warming your skin. Whatever fantasy you choose, visualize every enjoyable detail as vividly as possible, and let your body relax and let go. Allow every muscle to unwind. Let the tension flow out of your arms and legs and leave your body. Notice how your muscles begin to feel limp and loose. Notice how peaceful you feel. Now you are ready for the third step.

Fantasize that you are still in that scene, and you have

become a nonsmoker. Go through your list of benefits and repeat each one to yourself in the following way: "Now I have improved health and I like it. I can run along the beach, and I want this. The air around me is clean and fresh, and I feel good about myself. I respect myself. Now I have greater self-discipline, and I can take on other challenges if I want to. I have extra spending money," etc.

This method of habit management through the power of positive suggestion works amazingly well. It enabled me and many of my patients to quit smoking after a single treatment session. You can do it easily, and you'll find it's well worth your efforts. It can be used for self-improvement in losing weight, lawn mowing, getting up on time in the morning, adhering to a jogging routine, or for any other habit you'd like to modify.

Count What Counts. A three-year-old boy named Stevie stood by the edge of the children's pool, afraid to jump in. His mother sat in the water in front of him, urging him to take the leap. He held back; she cajoled. The power struggle went on for thirty minutes. Finally, he jumped. The water felt fine. It wasn't so difficult, and there was actually nothing to fear. But his mother's efforts backfired. The unfortunate message imprinted on Stevie's mind was "I have to be *pushed* before I can do anything risky. I don't have the gumption to jump in on my own like the other kids." His mother and father got the same idea; they began to think, "Left to his own devices, Stevie would never dare go into the water at all. If he isn't constantly pushed, he'll do nothing by himself. Raising him is going to be a long, hard struggle."

Sure enough, as Stevie grew up, the drama was repeated over and over. He had to be *persuaded* and *pushed* to go to school, to join the baseball team, to go to parties, and so on. He rarely initiated any action on his own. By the time he was referred to me at age twenty-one, he was chronically depressed, living with his parents, and not doing much with his life. He was still waiting around for people to tell him what to do and how to do it. But by now his parents were fed up trying to motivate him.

After each therapy session, he would leave the office charged with my enthusiasm to follow through on whatever self-help assignment we had discussed. For example, one week he decided to smile or say hello to three people he didn't know as a small first step in breaking his isolation. But the next week he would come into my office with a drooping head and a sheepish look that let me know he had "forgotten" to say hello to anyone. Another week, his assignment was to read a three-page article I had written for a singles magazine on how an unmarried man learned to overcome his loneliness. Steve came back the next week and said he had lost the manuscript before having a chance to read it. Each week as he left, he would feel a grea' surge of eagerness to help himself, but by the time he was in the elevator, he would "know" in his heart of hearts that the week's assignment, however simple, would just be too *hard* to do!

What was Stevie's problem? The explanation goes back to that day at the swimming pool. He still carries in his mind the powerfully imprinted idea that "I really can't do anything on my own. I'm the kind of guy who's got to be pushed." Because it never occurred to him to challenge this belief, it continued to function as a self-fulfilling prophecy, and he had over fifteen years of procrastination to back up his belief that he "really was" like that.

What was the solution? First Stevie had to become aware of the two mental errors that were the key to his problem: mental filter and labeling. His mind was dominated by thoughts about the various things he put off doing, and he *ignored* the hundreds of things he did each week that did *not* involve his being pushed by someone else.

"All of that is well and good," Stevie said after we discussed this. "You seem to have explained my problem, and I think that's correct. But how can I *change* the situation?"

The solution turned out to be simpler than he anticipated. I suggested he obtain a wrist counter (as discussed in the last chapter), so that each day he could count the things he did on his own without prodding or encouragement from anyone. At the end of the day he was to write down the total number of clicks he scored and keep a daily log.

Over a several-week period, he began to notice that his daily score increased. Every time he clicked the counter, he reminded himself that *he* was in control of his life, and in this way he trained himself *to notice what he did do*. Stevie began to feel increased self-confidence, and to view himself as a more capable human being.

Does it sound simple? It is! Will it work for you? You probably don't think so. But why not put it to the test? If you have a negative reaction and are convinced the wrist counter won't work for you, why not evaluate your pessimistic prediction with an experiment? Learn to count what counts; you may be surprised at the results!

Test Your "Can'ts." An important key to successful self-activation involves learning to adopt a scientific attitude toward the self-defeating predictions you make about your performance and abilities. If you put these pessimistic thoughts to the test, you can discover what the truth is.

One common self-defeating thought pattern when you are depressed or procrastinating is to "can't" yourself every time you think of something productive to do. Perhaps this stems from your fear of being blamed for your do-nothingism. You try to save face by creating the illusion that you are just too inadequate and incompetent to do a single thing. The problem with defending your lethargy in this manner is that you may really start believing what you are telling yourself! If you say, "I can't," over and over often enough it becomes like a hypnotic suggestion, and after a while you become genuinely convinced you really are a paralytic invalid who can't do anything. Typical "can't" thoughts include: "I can't cook," "I can't function," "I can't work," "I can't concentrate," "I can't read," "I can't get out of bed," and "I can't clean my apartment."

Not only do such thoughts defeat you, they will sour your relationships with those you love because they will see all your "I can't" statements as annoying whining. They won't perceive that it *really looks and seems* impossible for you to do anything. They will nag you, and set up frustrating power struggles with you.

An extremely successful cognitive technique involves testing your negative predictions with actual experiments. Suppose, for example, you've been telling yourself: "I'm so upset I can't concentrate well enough to read anything at all." As a way of testing this hypothesis, sit down with today's newspaper and read one sentence, and then see if you can summarize the sentence out loud. You might then predict—"But I could never read and understand a whole paragraph." Again—put this to the test. Read a paragraph and summarize. Many severe, chronic depressions have been cracked open with this powerful method.

The "Can't Lose" System. You may feel hesitant to put your "can'ts" to the test because you don't want to run the risk of failure. If you don't run any risks, at least you can maintain the secret belief that you're basically a terrific person who's decided for the time being not to get involved. Behind your aloofness and lack of commitment lurks a powerful sense of inadequacy and the fear of failure.

The "Can't Lose" System will help you combat this fear. Make a list of the negative consequences you might have to deal with if you took a risk and actually did fail. Then expose the distortions in your fears, and show how you could cope productively even if you did experience a disappointment.

The venture that you have been avoiding may involve a financial, personal, or scholastic risk. Remember that even if you do fail, some good can come from it. After all, this is how you learned how to walk. You didn't just jump up from your crib one day and waltz gracefully across the room. You stumbled and fell on your face and got up and tried again. At what age are you suddenly expected to know everything and never make any more mistakes? If you can love and respect yourself in failure, worlds of adventure and new experiences will open up before you, and your fears will vanish. An example of a written "Can't Lose" System is shown in Figure 5–10.

Don't Put the Cart Before the Horse!

I'll bet you still may not know for sure where motivation comes from. What, in your opinion, comes first—motivation or action?

If you said motivation, you made an excellent, logical choice. Unfortunately, you're wrong. Motivation does *not* come first, *action* does! You have to prime the pump. Then you will begin to get motivated, and the fluids will flow spontaneously.

Individuals who procrastinate frequently confuse motivation and action. You foolishly wait until you feel in the mood to do something. Since you don't feel like doing it, you automatically put it off.

Your error is your belief that motivation comes first, and then leads to activation and success. But it is usually the other way around; action must come first, and the motivation comes later on.

Take this chapter, for example. The first draft of this chapter was overwritten, clumsy, and stale. It was so long and boring that a true procrastinator would never even have the fortitude to read it. The task of revising it seemed to me like trying to go swimming with concrete shoes. When the day I had scheduled for revising it came—I had to push myself to sit down and get started. My motivation was about 1 percent, and my urge to avoid the task was 99 percent. What a hideous chore!

Figure 5–10. The "Can't Lose" System. A housewife used this technique to overcome her fear of applying for a part-time job.

Negative Consequences of Being Turned Down for a Job	*Positive Thoughts and Coping Strategies*
1. This means I'll never get a job.	1. Overgeneralization. This is unlikely. I can test this by applying for a series of other jobs and putting my best foot forward to see what happens.
2. My husband will look down on me.	2. Fortune teller error. Ask him. Maybe he will be sympathetic.
3. But what if he's not sympathetic? He might say this shows I belong in the kitchen and don't have what it takes.	3. Point out to him I'm doing my best and that his rejecting attitude doesn't help. Tell him that I am disappointed, but that I credit myself for trying.
4. But we're nearly broke. We need the money.	4. We've survived so far and haven't missed a single meal.
5. If I don't get a job, I won't be able to afford some decent new school clothes for the kids. They'll look scraggly.	5. I can get some clothes later on. We'll have to learn to get along with what we have for a while. Happiness doesn't come from clothes but from our self-respect.
6. A lot of my friends have jobs. They'll see I can't cut the mustard in the business world.	6. They're not all employed, and even my friends who do have jobs can probably remember a time when they were out of work. They haven't done anything so far to indicate they look down on me.

After I got involved in the task, I became highly motivated, and the job seems easy now. Writing became fun after all! It works like this:

First:	Action
	↓
Second:	Motivation ←
	↓
Third:	More Action ↵ (back to Motivation)

If you are a procrastinator, you probably aren't aware of this. So you lie around in bed waiting for inspiration to strike. When someone suggests you do something, you whine, "I don't *feel* like it." Well, who said you were supposed to feel like it? If you wait until you're "in the mood," you may wait forever!

The following table will help you review the various activation techniques and select what's most helpful to you.

Table 5–1. Synopsis of Self-Activation Methods

Target Symptoms	*Self-Activation Techniques*	*Purpose of the Method*
1. You feel disorganized. You have nothing to do. You get lonely and bored on weekends.	1. Daily Activity Schedule	1. Plan things one hour at a time and record the amount of mastery and pleasure. Virtually any activity will make you feel better than lying in bed and will undercut your sense of inadequacy.
2. You procrastinate because tasks seem too difficult and unrewarding.	2. The Antiprocrastination Sheet	2. You put your negative predictions to the test.
3. You feel overwhelmed by the urge to do nothing.	3. Daily Record of Dysfunctional Thoughts	3. You expose the illogical thoughts that paralyze you. You learn that motivation follows action, not *vice versa*.
4. You feel there's no point in doing anything when you're alone.	4. Pleasure-Predicting Sheet	4. Schedule activities with the potential for personal growth or satisfaction, and predict how rewarding they will be. Compare the actual satisfaction you experience when you are alone and when you are with others.

5. You give yourself excuses for avoiding things.	5. But-Rebuttal	5. You get off your "but" by combatting your "buts" with realistic rebuttals.
6. You have the idea that whatever you do isn't worth much.	6. Self-Endorsement	6. Write down the self-downing thoughts and talk back to them. Look for distorted thought patterns, such as "all-or-nothing thinking." Make a list of things you do accomplish each day.
7. You think about a task in a self-defeating manner.	7. TIC-TOC Technique	7. You substitute task-oriented cognitions (TOCS) for task-interfering cognitions (TICS).
8. You feel overwhelmed by the magnitude of everything you have to do.	8. Little Steps for Little Feet	8. Break the task down into its tiny component parts, and do these one step at a time.
9. You feel guilty, oppressed, obliged, and duty-bound.	9. Motivation Without Coercion	9. a. You eliminate "shoulds," "musts," and "oughts" when you give yourself instructions. b. You list the advantages and disadvantages of any activity so you can begin to think in terms of what you *want* to do rather than what you *must* do.

Table 5–1. cont.

Target Symptoms	*Self-Activation Techniques*	*Purpose of the Method*
10. Someone else nags and harangues you. You feel pressured and resentful, so you refuse to do anything at all.	10. Disarming Technique	10. You assertively agree with them and remind them that you are capable of doing your own thinking.
11. You have difficulty modifying a habit such as smoking.	11. Visualize Success	11. You make a list of the positive benefits of having changed the habit. You visualize these after inducing a state of deep relaxation.
12. You feel unable to do anything on your own initiative because you see yourself as "a procrastinator."	12. Count What Counts	12. You count the things you do each day on your own initiative, using a wrist counter. This helps you overcome your bad habit of constantly dwelling on your inadequacies.
13. You feel inadequate and incompetent because you say, "I can't."	13. Test Your Can'ts	13. You set up an experiment in which you challenge and disprove your negative predictions.
14. You are afraid to fail, so you risk nothing.	14. "Can't Lose" System	14. Write down any negative consequences of failure and develop a coping strategy ahead of time.

Chapter 6

Verbal Judo: Learn to Talk Back When You're Under the Fire of Criticism

You are learning that the cause of your sense of worthlessness is your ongoing self-criticism. This takes the form of an upsetting *internal* conversation in which you constantly harangue and persecute yourself in a harsh, unrealistic manner. Frequently your inner criticism will be triggered by someone else's sharp remark. You may dread criticism simply because you have never learned effective techniques for handling it. Because it is relatively *easy* to do, I want to emphasize the importance of mastering the art of handling verbal abuse and disapproval nondefensively and without a loss of self-esteem.

Many depressive episodes are set in motion by external criticism. Even psychiatrists, who are supposedly professional abuse-takers, can react adversely to criticism. A psychiatric resident called Art received negative feedback intended to be helpful from his supervisor. A patient had complained that several comments Art made during a therapy session were abrasive. The resident reacted with a wave of panic and depression when he heard this, due to his thought, "Oh God! The truth is out about me. Even my patients can see what a worthless, insensitive person I am. They'll probably *kick* me out of the residency program and *drum* me out of the state."

Why is criticism so hurtful to some people, while others can remain unperturbed in the face of the most abusive attack? In this chapter you will learn the secrets of people who face disapproval fearlessly, and you will be shown specific, concrete steps to overcome and eliminate your own exquisite vulnerability to criticism. As you read the following sections, keep this in mind: Overcoming your fear of criticism will require a moderate amount of practice. But it is not difficult to develop and master this skill, and the positive impact on your self-esteem will be tremendous.

Before I show you the way out of the trap of crumbling inwardly when criticized, let me show you why criticism is more upsetting to some people than to others. In the first place, you must realize that it is *not* other people, or the critical comments they make, that upset you. To repeat, there has never been a single time in your life when the critical comments of some other person upset you—even to a small extent. No matter how vicious, heartless, or cruel these comments may be, they have *no* power to disturb you or to create even a *little bit* of discomfort.

After reading that paragraph you may get the impression that I am cracking up, mistaken, highly unrealistic, or some combination thereof. But I assure you I am not when I say: Only one person in this world has the power to *put you down*—and *you* are that person, no one else!

Here's how it works. When another person criticizes you, certain negative thoughts are automatically triggered in your head. Your emotional reaction will be created by these thoughts and not by what the other person says. The thoughts which upset you will invariably contain the same types of mental errors described in Chapter 3: overgeneralization, all-or-nothing thinking, the mental filter, labeling, etc.

For example, let's take a look at Art's thoughts. His panic was the result of his catastrophic interpretation: "This criticism shows how worthless I am." What mental errors is he making? In the first place, Art is jumping to conclusions when he arbitrarily concludes the patient's criticism is valid and reasonable. This may or may not be the case. Furthermore, he is *exaggerating* the importance of whatever he actually said to the patient that may have been undiplomatic (magnification), and he is *assuming* he could do nothing to correct any errors in his behavior (the fortune teller error). He unrealistically predicted he would be rejected and ruined professionally because he would repeat endlessly whatever error he made with this one patient (overgeneralization). He focused exclusively on his error (the mental filter) and overlooked his numerous other therapeutic successes (disqualifying or overlooking the positive). He identified with his erroneous behavior and concluded he was a "worthless and insensitive human being" (labeling).

The first step in overcoming your fear of criticism concerns your own mental processes: Learn to identify the negative thoughts you have when you are being criticized. It will be most helpful to write them down using the double-column technique described in the two previous chapters. This will enable you to analyze your thoughts and recognize where your thinking is illogical or wrong. Finally, write down rational responses that are more reasonable and less upsetting.

An excerpt from Art's written homework using the double-column technique is included (Figure 6–1). As he learned to *think* about the situation in a more realistic manner, he stopped wasting mental and emotional effort in catastrophizing, and was able to channel his energy into creative, goal-oriented problem solving. After evaluating precisely what he had said that was offensive or hurtful, he was able to take steps to modify his clinical style with patients so as to minimize future similar mistakes. As a result, he learned from the situation, and his clinical skills and maturity increased. This gave his self-confidence a boost and helped him overcome his fear of being imperfect.

To put it succinctly, if people criticize you the comments they make will be *right* or *wrong*. If the comments are

Figure 6–1. Excerpt from Art's written homework, using the double-column technique. He initially experienced a wave of panic when he received critical feedback from his supervisor about the way he handled a difficult patient. After writing down his negative thoughts, he realized they were quite unrealistic. Consequently, he felt substantial relief.

Automatic Thoughts (SELF-CRITICISM)	*Rational Responses* (SELF-DEFENSE)
1. Oh, God! The truth is out about me. Even the patients can see what a worthless, insensitive individual I am.	1. Just because one patient complains it doesn't mean that I am a "worthless, insensitive individual." The majority of my patients do, in fact, like me. Making a mistake doesn't reveal my "true essence." Everyone is entitled to make mistakes.
2. They'll probably kick me out of the residency program.	2. This is silly and rests on several erroneous assumptions: (a) all I do is bad things; (b) I have no capacity to grow. Since (a) and (b) are absurd, it is extremely unlikely my position here is threatened. I have on many occasions received praise from my supervisor.

wrong, there is really nothing for you to be upset about. Think about that for a minute! Many patients have come to me in tears, angry and upset because a loved one made a critical comment to them that was thoughtless and inaccurate. Such a reaction is unnecessary. Why should you be disturbed if someone else makes the mistake of criticizing you in an unjust manner? That's the other guy's error, not yours. Why upset yourself? Did you expect that other people would be perfect? On the other hand, if the criticism is *accurate*, there is still *no reason* for you to feel overwhelmed. You're not expected to be perfect. Just acknowledge your error and take whatever steps you can to correct it. It sounds *simple* (and it is!), but it may take some effort to transform this insight into an emotional reality.

Of course, you may fear criticism because you feel you need the love and approval of other people in order to be worthwhile and happy. The problem with this point of view is that you'll have to devote all your energies to trying to please people, and you won't have much left for creative, productive living. Paradoxically, many people may find you less interesting and desirable than your more self-assured friends.

Thus far, what I have told you is a review of the cognitive techniques introduced in the previous chapter. The crux of the matter is that only *your* thoughts can upset you and if you learn to *think* more realistically, you will *feel* less upset. Right now, write down the negative thoughts that ordinarily go through your head when someone criticizes you. Then identify the distortions and substitute more objective rational responses. This will help you feel less angry and threatened.

Now I would like to teach you some simple verbal techniques which may have considerable practical relevance. What can you say when someone is attacking you? How can you handle these difficult situations in a way that will enhance your sense of mastery and self-confidence?

Step One—Empathy. When someone is criticizing or attacking you, his (or her) motives may be to help you or to hurt you. What the critic says may be *right* or *wrong*, or *somewhere in between*. But it is not wise to focus on these issues initially. Instead, ask the person a series of specific questions designed to find out *exactly* what he or she means. Try to avoid being judgmental or defensive as you ask the questions. Constantly ask for more and more specific information. Attempt to see the world through the critic's eyes. If the person attacks you with vague, insulting labels, ask him or her to be more specific and to point out exactly what it is about you the person dislikes. This initial maneuver can itself go a long way to getting the critic off your back, and will help transform an attack-defense interaction into one of collaboration and mutual respect.

I often illustrate how to do this in a therapy session by role-playing an imaginary situation with the patient so that I can model this particular skill. I'll show you how to role-play; it's a useful skill to develop. In the dialogue that follows, I want you to imagine you are an angry critic. Say the most brutal and upsetting thing to me you can think of. What you say can be true, false, or partly both. I will respond to each of your assaults with the empathy technique.

YOU (playing the role of angry critic): Dr. Burns, you're a no-good shit.

DAVID: What about me is shitty?

YOU: Everything you say and do. You're insensitive, self-centered, and incompetent.

DAVID: Let's take each of these. I want you to try to be specific. Apparently I've done or said a number of things that upset you. Just *what* did I say that sounded insensitive? What gave you the impression I was self-centered? What did I *do* that seemed incompetent?

YOU: When I called to change my appointment the other day, you sounded rushed and irritable, as if you were in a big hurry and didn't give a damn about me.

DAVID: Okay, I came across in a rushed, uncaring way on the phone. What else have I done that irritated you?

YOU: You always seem to hurry me out at the end of the session—just like this was a big production line to make money.

DAVID: Okay, you feel I've been too rushed during sessions as well. I may have given you the impres-

> sion I'm more interested in your money than in you. What else have I done? Can you think of other ways I might have goofed up or offended you?

What I am doing is simple. By asking you specific questions I minimize the possibility that you will reject me completely. You—and I—become aware of some specific concrete problems that we can deal with. Furthermore, I am giving you your day in court by *listening* to you so as to understand the situation *as you see it*. This tends to defuse any anger and hostility and introduces a problem-solving orientation in the place of blame casting or debate. Remember the first rule—even if you feel the criticism is *totally* unjust, respond with empathy by asking specific questions. Find out precisely what your critic means. If the person is very hot under the collar, he or she may be hurling labels at you, perhaps even obscenities. Nevertheless, ask for more information. What do those words mean? Why does the person call you a "no-good shit"? *How* did you offend this individual? *What* did you do? *When* did you do it? *How often* have you done it? *What else* does the person dislike about you? Find out what your action means to him or her. Try to see the world through your critic's eyes. This approach will frequently calm the roaring lion and lay the groundwork for a more sensible discussion.

Step Two—Disarming the Critic. If someone is shooting at you, you have three choices: You can stand and shoot back—this usually leads to warfare and mutual destruction; you can run away or try to dodge the bullets—this often results in humiliation and a loss of self-esteem; or you can stay put and skillfully disarm your opponent. I have found that this third solution is by far the most satisfying. When you take the wind out of the other person's sails, you end up the winner, and your opponent more often than not will also feel like a winner.

How is this accomplished? It's simple: Whether your

critic is right or wrong, initially *find some way to agree with him or her*. Let me illustrate the easiest situation first. Let's assume the critic is primarily correct. In the previous example when you angrily accused me of sounding rushed and indifferent on several occasions, I might go on to say: "You're absolutely right. I was rushed when you called, and I probably *did* sound impersonal. Other people have also pointed this out to me at times. I want to emphasize that I didn't intend to hurt your feelings. You're also right that we *have* been rushed during several of our sessions. You might recall that sessions can be any length you like, as long as we decide this ahead of time so that the scheduling can be appropriately adjusted. Perhaps you'd like to schedule sessions that are fifteen or thirty minutes longer, and see if that's more comfortable."

Now, suppose the person who's attacking you is making criticisms you feel are unfair and not valid. What if it would be unrealistic for you to change? How can you agree with someone when you feel certain that what is being said is utter nonsense? It's easy—you can agree *in principle* with the criticism, or you can find some *grain* of truth in the statement and agree with that, or you can acknowledge that the person's upset is understandable because it is based on how he or she views the situation. I can best illustrate this by continuing the role-playing; you attack me, but this time say things that are primarily false. According to the rules of the game, I must (1) find some way to agree with *whatever* you say; (2) avoid sarcasm or defensiveness; (3) always speak the truth. Your statements can be as bizarre and as ruthless as you like, and I guarantee I will stick by these rules! Let's go!

YOU (continuing to play the role of angry critic): Dr. Burns, you're a shit.

DAVID: I feel that way at times. I often goof up at things.

YOU: This cognitive therapy is no damn good!

DAVID: There's certainly plenty of room for improvement.

YOU: And you're stupid.

DAVID: There are lots of people who are brighter than I am. I'm sure not the smartest person in the world.

YOU: You have no real feelings for your patients. Your approach to therapy is superficial and gimmicky.

DAVID: I'm not always as warm and open as I'd like to be. Some of my methods might seem gimmicky at first.

YOU: You're not a real psychiatrist. This book is pure trash. You're not trustworthy or competent to manage my case.

DAVID: I'm terribly sorry I seem incompetent to you. It must be quite disturbing to you. You seem to find it difficult to trust me, and you are genuinely skeptical about whether we can work together effectively. You're absolutely right—we can't work together successfully unless we have a sense of mutual respect and teamwork.

By this time (or sooner) the angry critic will usually lose steam. Because I do not fight back but instead find a way to agree with my opponent, the person quickly seems to run out of ammunition, having been successfully disarmed. You might think of this as winning by avoiding battle. As the critic begins to calm down, he or she will be in a better mood to communicate.

Once I have demonstrated these first two steps to a patient in my office, I usually propose we reverse roles to give the patient the chance to master the method. Let's do this. I will criticize and attack you, and you will practice the empathy and make up your own answers. Then see how closely they are accurate or nonsensical. To make the following dialogue a more useful exercise, cover up the responses called "You" and make up your own answers. Then see how closely they correspond with what I have written. Remember to ask questions using the empathy method and find valid ways to agree with me using the disarming technique.

DAVID (playing the role of angry critic): You're not here to get better. You're just looking for sympathy.

YOU (playing the role of the one under attack): What gives you the impression I'm just looking for sympathy?

DAVID: You don't do anything to help yourself between sessions. All you want to do is come here and complain.

YOU: It's true that I haven't been doing *some* of the written homework you suggested. Do you feel I shouldn't complain during sessions?

DAVID: You can do whatever you want. Just admit you don't give a damn.

YOU: You mean you think I don't want to get better, or what?

DAVID: You're no good! You're just a piece of garbage!

YOU: I've been feeling that way for years! Do you have some ideas about what I can do to feel differently?

DAVID: I give up. You win.

YOU: You're right. I *did* win!

I strongly suggest you practice this with a friend. The role-playing format will help you master the necessary skills needed when a real situation arises. If there is no one you feel comfortable with who could role-play with you effectively, a good alternative would be to write out imaginary dialogues between you and a hostile critic, similar to the ones you've been reading. After each harangue write down how you might answer using the empathy and disarming technique. It may seem difficult at first, but I think you'll catch on quite readily. It's really quite easy once you get the gist of it.

You will notice you have a profound, almost irresistible tendency to *defend* yourself when you are unjustly accused. This is a MAJOR mistake! If you give in to this tendency, you will find that the intensity of your opponent's attack *increases*! You will paradoxically be adding bullets to that person's arsenal every time you defend yourself. For example, you be the critic again, and this time I'll *defend* myself against your absurd accusations. You'll see how quickly our interaction will escalate to full-scale warfare.

YOU (in the role of critic again): Dr. Burns, you don't care about your patients.

DAVID (responding in a defensive manner): That's untrue and unfair. You don't know what you're talking about! My patients respect all the hard work I put in.

YOU: Well, here's one who doesn't! Good-bye! (You exit, having decided to fire me. My defensiveness leads to a total loss.)

In contrast, if I respond with empathy and disarm your hostility, more often than not you will feel I am *listening* to you and *respecting* you. As a result you lose your ardor to do battle and quiet down. This paves the way for step three—feedback and negotiation.

You may find initially that in spite of your determination to apply these techniques, when a real situation arises in which you are criticized, you will be caught up by your emotions and your old behavior patterns. You may find yourself sulking, arguing, defending yourself vehemently, etc. This is understandable. You're not expected to learn it all overnight, and you don't have to win every battle. It is important, however, to analyze your mistakes afterward so that you can review how you might have handled the situation differently along the lines suggested. It can be immensely helpful to find a friend to role-play the difficult situation with you afterward so that you can practice a variety of responses until you have mastered an approach you are comfortable with.

Step Three—Feedback and Negotiation. Once you have *listened* to your critic, using the empathy method, and *disarmed* him by finding some way to agree with him, you will then be in a position to explain your position and emotions *tactfully* but *assertively*, and to negotiate any real differences.

Let's assume that the critic is just plain wrong. How can you express this in a nondestructive manner? This is simple: You can express your point of view objectively with an acknowledgment you *might* be wrong. Make the conflict one based on fact rather than personality or pride. Avoid directing destructive labels at your critic. Remember, his error does *not* make him stupid, worthless, or inferior.

For example, a patient recently claimed that I sent a bill for a session for which she had already paid. She assaulted me with "Why don't you get your bookkeeping straight!" Knowing she was in error, I responded, "My records may indeed be wrong. I seem to recall that you forgot your checkbook that day, but I might be confused on this point. I hope you'll allow for the possibility that you or I *will* make errors at times. Then we can be more relaxed with each other. Why not see if you have a canceled check? That way we can find out the truth and make appropriate adjustments."

In this case my nonpolarizing response allowed her to save face and avoided a confrontation in which her self-respect was at risk. Although it turned out she was wrong, she later expressed relief that I acknowledged I do make mistakes. This helped her feel better about me, as she was afraid I would be as perfectionistic and demanding with her as she was with herself.

Sometimes you and the critic will differ not on a matter

of fact but of taste. Once again, you will be a winner if you present your point of view with diplomacy. For example, I have found that no matter how I dress, some patients respond favorably and some negatively. I feel most comfortable in a suit and tie, or in a sports coat and tie. Suppose a patient criticizes me because my clothes are too formal and this is iritating because it makes me appear to be part of the "Establishment." After eliciting further specific information about other things this person might dislike about me, I could then respond, "I can certainly agree with you that suits are a bit formal. You *would* be more comfortable with me if I dressed more casually. I'm sure you'll understand that after dressing in a variety of ways, I have found that a nice suit or sports coat is most acceptable to the majority of the people I work with, and that's why I've decided to stick with this style of dressing. I'm hopeful you won't let this interfere with our continued work together."

You have a number of options when you negotiate with the critic. If he or she continues to harangue you, making the same point again and again, you can simply repeat your assertive response politely but firmly over and over until the person tires out. For example, if my critic continued to insist I stop wearing suits, I might continue to say each time, "I understand your point entirely, and there *is* some truth to it. Nevertheless, I've decided to stick with more formal attire at this time."

Sometimes the solution will be in between. In this case negotiation and compromise are indicated. You may have to settle for *part* of what you want. But if you have conscientiously applied the *empathy* and *disarming techniques* first, you will probably get *more* of what you want.

In many cases you will be just plain wrong, and the critic will be right. In such a situation your critic's respect for you will probably increase by an orbital jump if you assertively *agree with the criticism*, thank the person for providing you with the information, and apologize for any hurt you might have caused. It sounds like old-fashioned common sense (and it is), but it can be amazingly effective.

By now you may be saying, "But don't I have a *right* to defend myself when someone criticizes me? Why should I always have to empathize with the other person? After all, *he* may be the ninny, not I. Isn't it *human* just to get angry and blow your stack? Why should I always have to *smooth* things out?"

Well, there is considerable truth in what you say. You *do* have the right to defend yourself vigorously from criticism and to get angry at anyone you choose whenever you like. And you are right on target when you point out that it is often your critic, and not you, whose thinking is fouled up. And there is more than a grain of truth behind the slogan "Better mad than sad." After all, if you're going to conclude that someone is "no damn good," why not let it be the other fellow? And furthermore, sometimes it *does* feel so much better to be mad at the other person.

Many psychotherapists would agree with you on this point. Freud felt that depression was "anger turned inward." In other words he believed depressed individuals direct their rage against themselves. In keeping with this view, many therapists urge their patients to get in touch with their anger and to express it more frequently to others. They might even say that some of the methods described in this section amount to a repressive cop-out.

This is a false issue. The crucial point is not whether or not you express your feelings, but the manner in which you do it. If your message is "I'm angry because you're criticizing me and you're no damn good," you will poison your relationship with that person. If you defend yourself from negative feedback in a defensive and vengeful way, you will reduce the prospect for productive interaction in the future. Thus, while your angry outburst momentarily *feels good*, you may defeat yourself in the long run by burning your bridges. You have polarized the situation prematurely and unnecessarily, and eliminated your chance to learn what the critic was trying to convey. And what is worse, you may experience a depressive backlash and punish yourself inordinately for your burst of temper.

Antiheckler Technique. A specialized application of the techniques discussed in this chapter might be particularly helpful for those of you who are involved in lecturing or teaching. I developed the "antiheckler technique" when I began lecturing to university and professional groups on current depression research. Although my lectures are usually well received, I occasionally find there is a single heckler in the audience. The heckler's comments usually have several characteristics: (1) They are intensely critical, but seem inaccurate or irrelevant to the material presented; (2) they often come from a person who is not well accepted or regarded among his or her local peers; and (3) they are expressed in a haranguing, abusive style.

I therefore had to develop an antiheckler technique which I could use to silence such a person in an inoffensive manner so that the rest of the audience could have an equal opportunity to ask questions. I find that the following method is highly effective: (1) I immediately *thank* the person for his or her comments; (2) acknowledge that the points brought up *are indeed* important; and (3) I emphasize that there is a *need for more knowledge* about the points raised, and I encourage my critic to pursue meaningful research and investigation of the topic. Finally, I invite the heckler to share his or her views with me further after the close of the session.

Although no verbal technique is guaranteed to bring a particular result, I have rarely failed to achieve a favorable effect when using this upbeat approach. In fact, these heckling individuals have frequently approached me after the lecture to compliment and thank me for my kind comments. It is sometimes the heckler who turns out to be most demonstrative and appreciative of my lecture!

Summary. The various cognitive and verbal principles for coping with criticism are summarized in the accompanying diagram (see Figure 6-2, page 146). As a general rule, when someone insults you, you will immediately go down one of three pathways—the *sad* route, the *mad* route, or the *glad* route. Whichever option you choose will be a

Figure 6–2. The three ways that you might react to criticism. Depending on how you think about the situation, you will feel sad, mad, or glad. Your behavior and the outcome will also be greatly influenced by your mental set.

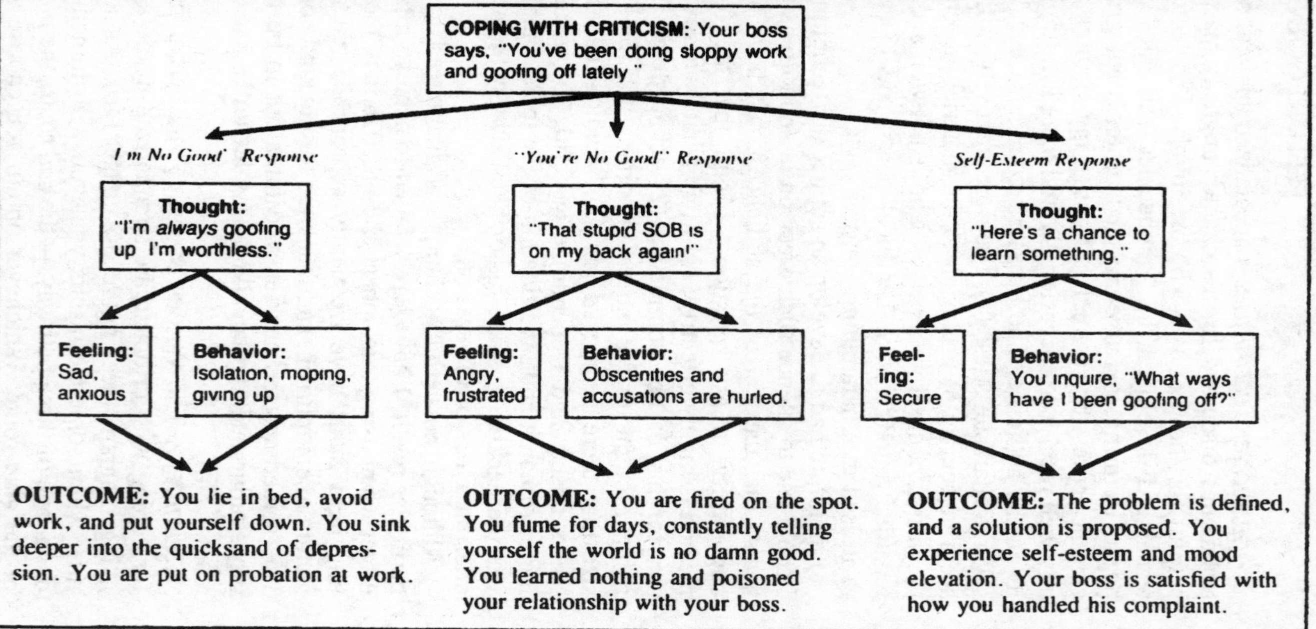
COPING WITH CRITICISM: Your boss says, "You've been doing sloppy work and goofing off lately."
"I'm No Good" Response
Thought: "I'm *always* goofing up. I'm worthless."
Feeling: Sad, anxious
Behavior: Isolation, moping, giving up
OUTCOME: You lie in bed, avoid work, and put yourself down. You sink deeper into the quicksand of depression. You are put on probation at work.
"You're No Good" Response
Thought: "That stupid SOB is on my back again!"
Feeling: Angry, frustrated
Behavior: Obscenities and accusations are hurled.
OUTCOME: You are fired on the spot. You fume for days, constantly telling yourself the world is no damn good. You learned nothing and poisoned your relationship with your boss.
Self-Esteem Response
Thought: "Here's a chance to learn something."
Feeling: Secure
Behavior: You inquire, "What ways have I been goofing off?"
OUTCOME: The problem is defined, and a solution is proposed. You experience self-esteem and mood elevation. Your boss is satisfied with how you handled his complaint.

total experience, and will involve your thinking, your feelings, your behavior, and even the way your body functions.

Most people with a tendency to depression choose the sad route. You *automatically* conclude the critic is right. Without any systematic investigation, you jump to the conclusion that you were in the wrong and made a mistake. You then magnify the importance of the criticism with a series of thinking errors. You might *overgeneralize* and wrongly conclude that your whole life consists of nothing but a string of errors. Or you might *label* yourself a "total goof-up." And because of your perfectionistic expectation that you are supposed to be flawless, you will probably feel convinced that your (presumed) error indicates you are worthless. As a result of these mental errors, you will experience depression and a loss of self-esteem. Your verbal responses will be ineffectual and passive, characterized by avoidance and withdrawal.

In contrast, you may choose the mad route. You will *defend* yourself from the horrors of being imperfect by trying to convince the critic that he or she is a monster. You will stubbornly refuse to admit any error because according to your perfectionistic standards, this would be tantamount to admitting you are a worthless worm. So you hurl accusations back on the assumption that the best defense is a good offense. Your heart beats rapidly, and hormones pour into your bloodstream as you prepare for battle. Every muscle tightens and your jaws are clenched. You may feel a temporary exhilaration as you tell your critic off in self-righteous indignation. You'll show him what a no-good piece of crap he is! Unfortunately, he doesn't agree, and in the long run your outburst is self-defeating because you've poisoned the relationship.

The third option requires that you either *have* self-esteem or at least act *as if you did*. It is based on the premise that you are a worthwhile human being and have no need to be perfect. When you are criticized, your initial response is *investigative*. Does the criticism contain a grain of truth?

Just what did you do that was objectionable? Did you in fact goof up?

Having defined the problem by asking a series of nonjudgmental questions, you are in a position to propose a solution. If a compromise is indicated, you can negotiate. If you were clearly in the wrong, you can admit it. If the critic was mistaken, you can point this out in a tactful manner. But whether your behavior was right or wrong, you will know that you are *right* as a human being, because you have finally perceived that your self-esteem was never at issue in the first place.

Chapter 7

Feeling Angry? What's Your IQ?

What's your IQ? I'm not interested in knowing how smart you are because your intelligence has little, if anything, to do with your capacity for happiness. What I want to know is what your *I*rritability *Q*uotient is. This refers to the amount of anger and annoyance you tend to absorb and harbor in your daily life. If you have a particularly high IQ, it puts you at a great disadvantage because you overreact to frustrations and disappointments by creating feelings of resentment that blacken your disposition and make your life a joyless hassle.

Here's how to measure your IQ. Read the list of twenty-five potentially upsetting situations described below. In the space provided after each incident, estimate the degree it would ordinarily anger or provoke you, using this simple rating scale:

0—You would feel very little or no annoyance.
1—You would feel a little irritated.
2—You would feel moderately upset.
3—You would feel quite angry.
4—You would feel very angry.

Mark your answer after each question as in this example:

You are driving to pick up a friend at the airport, and you are forced to wait for a long freight train. __2__

The individual who answered this question estimated his reaction as a two because he would feel moderately irritated, but this would quickly pass as soon as the train was gone. As you describe how you would ordinarily react to each of the following provocations, make your best general estimate even though many potentially important details are omitted (such as what kind of day you were having, who was involved in the situation, etc.).

Novaco Anger Scale*

1. You unpack an appliance you have just bought, plug it in, and discover that it doesn't work. ______
2. Being overcharged by a repairman who has you over a barrel. ______
3. Being singled out for correction, when the actions of others go unnoticed. ______
4. Getting your car stuck in the mud or snow. ______
5. You are talking to someone and they don't answer you. ______
6. Someone pretends to be something they are not. ______
7. While you are struggling to carry four cups of coffee to your table at a cafeteria, someone bumps into you, spilling the coffee. ______
8. You have hung up your clothes, but someone knocks them to the floor and fails to pick them up. ______

*This scale was developed by Dr. Raymond W. Novaco of the Program in Social Ecology at the University of California, Irvine, and part of it is reproduced here with his permission. The full scale contains eighty items.

9. You are hounded by a salesperson from the moment that you walk into a store. ______
10. You have made arrangements to go somewhere with a person who backs off at the last minute and leaves you hanging. ______
11. Being joked about or teased. ______
12. Your car is stalled at a traffic light, and the guy behind you keeps blowing his horn. ______
13. You accidentally make the wrong kind of turn in a parking lot. As you get out of your car someone yells at you, "Where did you learn to drive?" ______
14. Someone makes a mistake and blames it on you. ______
15. You are trying to concentrate, but a person near you is tapping their foot. ______
16. You lend someone an important book or tool, and they fail to return it. ______
17. You have had a busy day, and the person you live with starts to complain about how you forgot to do something that you agreed to do. ______
18. You are trying to discuss something important with your mate or partner who isn't giving you a chance to express your feelings. ______
19. You are in a discussion with someone who persists in arguing about a topic they know very little about. ______
20. Someone sticks his or her nose into an argument between you and someone else. ______
21. You need to get somewhere quickly, but the car in front of you is going 25 mph in a 40 mph zone, and you can't pass. ______
22. Stepping on a gob of chewing gum. ______
23. Being mocked by a small group of people as you pass them. ______

24. In a hurry to get somewhere, you tear a good pair of slacks on a sharp object. ______

25. You use your last dime to make a phone call, but you are disconnected before you finish dialing and the dime is lost. ______

Now that you have completed the Anger Inventory, you are in a position to calculate your IQ, your Irritability Quotient. Make sure that you have not skipped any items. Add up your score for each of the twenty-five incidents. The lowest possible total score on the test would be zero. This would mean you put down a zero on each item. This indicates you are either a liar or a guru! The highest score would be a hundred. This would mean you recorded a four on each of the twenty-five items, and you're constantly at or beyond the boiling point.

You can now interpret your total score according to the following scale:

0–45: The amount of anger and annoyance you generally experience is remarkably low. Only a few percent of the population will score this low on the test. You are one of the select few!

46–55: You are substantially more peaceful than the average person.

56–75: You respond to life's annoyances with an average amount of anger.

76–85: You frequently react in an angry way to life's many annoyances. You are substantially more irritable than the average person.

86–100: You are a true anger champion, and you are plagued by frequent intense furious reactions that do not quickly disappear. You probably harbor negative feelings long after the initial insult has passed. You may have the reputation of a firecracker or a hothead among

> people you know. You may experience frequent tension headaches and elevated blood pressure. Your anger may often get out of control and lead to impulsive hostile outbursts which at times get you into trouble. Only a few percent of the adult population react as intensely as you do.

Now that you know how much anger you have, let's see what you can do about it. Traditionally psychotherapists (*and* the general public) have conceptualized two primary ways to deal with anger: (a) anger turned "inward"; or (b) anger turned "outward." The former solution is felt to be the "sick" one—you internalize your aggression and absorb resentment like a sponge. Ultimately it corrodes you and leads to guilt and depression. Early psychoanalysts such as Freud felt that internalized anger was the cause of depression. Unfortunately, there is no convincing evidence in support of this notion.

The second solution is said to be the "healthy" one—you express your anger, and as you ventilate your feelings, you presumably feel better. The problem with this simplistic approach is that it doesn't work very well. If you go around ventilating all your anger, people will soon regard you as loony. And at the same time you aren't learning how to deal with people in society *without* getting angry.

The cognitive solution transcends both of these. You have a third option: *Stop creating* your anger. You don't have to choose between holding it in or letting it out because it won't exist.

In this chapter I provide guidelines to help you assess the pros and cons of experiencing anger in a variety of situations so you can decide when anger is and isn't in your best self-interest. If you choose, you can develop control over your feelings; you will gradually cease to be plagued by excessive irritability and frustration that sour your life for no good reason.

Just Who Is Making You Angry?

"People!
Shit!
I'm fed up with them!
I need a vacation from people."

The woman who recorded this thought at 2:00 A.M. couldn't sleep. How could the dogs and noisy neighbors in her apartment building be so thoughtless? Like her, I'll bet you're convinced it's other people's stupid, self-centered actions that make you angry.

It's natural to believe that external events upset you. When you're mad at someone, you automatically make them the cause of all your bad feelings. You say, "*You're* annoying me! *You're* getting on my nerves." When you think like this, you're actually fooling yourself because other people really cannot make you angry. Yes—you heard me right. A pushy teenager might crowd in front of you in line at the movie theater. A con artist might sell you a fake ancient coin at an antique shop. A "friend" might screw you out of your share of a profitable business deal. Your boyfriend might always show up late for dates in spite of his knowing how important promptness is to you. No matter how outrageous or unfair others might appear to you, *they* do not, never did, and never will upset you. The bitter truth is that you're the one who's *creating* every last ounce of the outrage you experience.

Does that sound like heresy or stupidity to you? If you think I'm contradicting the obvious, you may feel like burning this book or throwing it down in disgust. If so, I dare you to read on, because—

Anger, like all emotions, is created by your cognitions. The relationship between your thoughts and your anger is shown in Figure 7–1. As you will note, before you can feel irritated by any event you must first become aware of what is occurring and come to your own interpretation of it. Your feelings result from the meaning you give to the event, *not* from the event itself.

Figure 7–1. It is not negative events but your perceptions and thoughts about these events that create your emotional response.

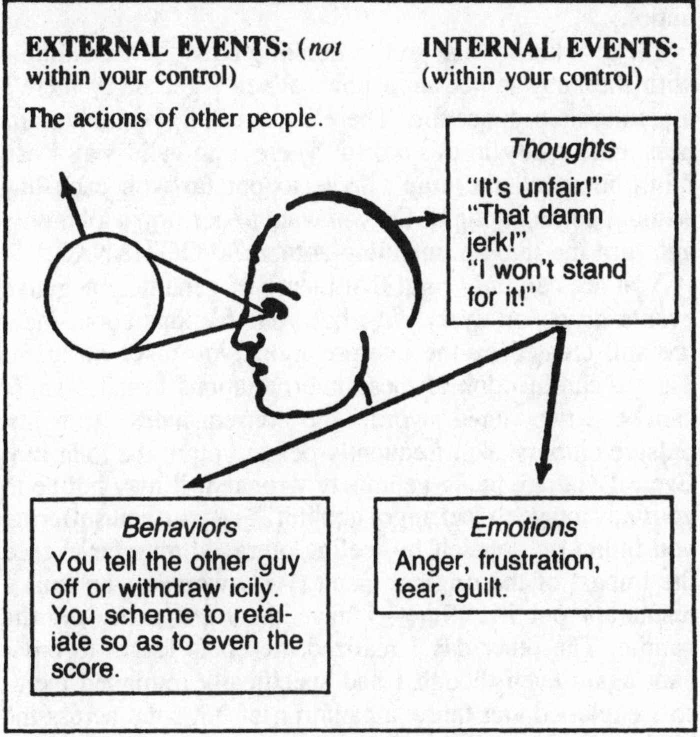
EXTERNAL EVENTS: (not within your control)
The actions of other people.
INTERNAL EVENTS: (within your control)
Thoughts
"It's unfair!"
"That damn jerk!"
"I won't stand for it!"
Behaviors
You tell the other guy off or withdraw icily. You scheme to retaliate so as to even the score.
Emotions
Anger, frustration, fear, guilt.

For example, suppose that after a hectic day you put your two-year-old child to sleep in his crib for the night. You close his bedroom door and sit down to relax and watch television. Twenty minutes later he suddenly opens the door to his room and walks out giggling. You might react to this in a variety of ways, depending on the meaning you attach to it. If you feel irritated, you're probably thinking, "Damn it! He's always a bother. Why can't he stay in bed and behave like he should? He never gives me a minute's rest!" On the other hand, you could be delighted to see him pop out of his room because you're thinking, "Great! He just crawled out of his crib on his own for the first time. He's growing up and getting more independent." The event is the same in both cases. Your emotional reaction is determined entirely by the way you are thinking about the situation.

I'll bet I know what you're thinking now: "That example with the baby is not applicable. When *I* get angry there's a justifiable provocation. There's plenty of *genuine* unfairness and cruelty in this world. There's no valid way I can think about all the crap I have to put up with each day without getting uptight. Do you want to perform a lobotomy and turn me into an unfeeling zombie? NO THANKS!"

You are certainly right that plenty of genuinely negative events *do* go on every day, but your feelings about them are still created by the interpretations you place on them. Take a careful look at these interpretations because anger can be a two-edged sword. The consequences of an impulsive outburst will frequently defeat you in the long run. Even if you are being genuinely wronged, it may not be to your advantage to feel angry about it. The pain and suffering you inflict on yourself by feeling outraged may far exceed the impact of the original insult. As a woman who runs a restaurant put it, "Sure—I have the *right* to fly off the handle. The other day I realized the chefs forgot to order ham again even though I had specifically reminded them, so I exploded and threw a cauldron of hot soup across the kitchen floor in disgust. Two minutes later I knew I'd acted like the biggest asshole in the world, but I didn't want to admit it, so I had to spend all my energy for the next forty-eight hours trying to convince myself I had the right to make a jackass of myself in front of twenty employees! It wasn't worth it!"

In many cases your anger is created by subtle cognitive distortions. As with depression, many of your perceptions are twisted, one-sided, or just plain wrong. As you learn to replace these distorted thoughts with others that are more realistic and functional, you will feel less irritable and gain greater self-control.

What kinds of distortion occur most often when you are

angry? One of the greatest offenders is *labeling*. When you describe the person you're mad at as "a jerk" or "a bum" or "a piece of shit," you see him in a totally negative way. You could call this extreme form of overgeneralization "globalizing" or "monsterizing." Someone may in fact have betrayed your trust, and it is absolutely right to resent what that person *did*. In contrast, when you label someone, you create the impression that he or she has a bad essence. You are directing your anger toward what that person "is."

When you write people off this way, you catalog in your mind's eye every single thing about them you don't like (the mental filter) and ignore or discount their good points (disqualifying the positive). This is how you set up a false target for your anger. In reality, every human being is a complex mix of positive, negative, and neutral attributes.

Labeling is a distorted thinking process that causes you to feel inappropriately indignant and morally superior. It's destructive to build your self-image this way: Your labeling will inevitably give way to your need to blame the other person. Your thirst for retaliation intensifies the conflict and brings out similar attitudes and feelings in the person you're mad at. Labeling inevitably functions as a self-fulfilling prophecy. You polarize the other person and bring about a state of interpersonal warfare.

What's the battle really all about? Often you're involved in a defense of your self-esteem. The other person may have threatened you by insulting or criticizing you, or by not loving or liking you, or by not agreeing with your ideas. Consequently, you may perceive yourself in a duel of honor to the death. The problem with this is that the other person is *not* a totally worthless shit, no matter how much you insist! And, furthermore, you cannot enhance your own esteem by denigrating someone else even if it does feel good temporarily. Ultimately only your own negative, distorted thoughts can take away your self-respect, as pointed out in Chapter 4. There is *one and only one* person in this world who has the power to threaten your self-esteem—and that is you. Your sense of worth can go down *only* if you put yourself down. The real solution is to put an end to your absurd inner harangue.

Another distortion characteristic of anger-generating thoughts is *mind reading*—you invent motives that explains to *your* satisfaction why the other person did what he or she did. These hypotheses are frequently erroneous because they will not describe the actual thoughts and perceptions that motivated the other person. Due to your indignation, it may not occur to you to check out what you are saying to yourself.

Common explanations you might offer for the other person's objectionable behavior would be "He has a mean streak"; "She's unfair"; "He's just like that"; "She's stupid"; "They're bad kids"; and so forth. The problem with these so-called explanations is that they are just additional labels that don't really provide any valid information. In fact, they are downright misleading.

Here's an example: Joan got hot under the collar when her husband told her he'd prefer to watch the Sunday football game on TV rather than go with her to a concert. She felt miffed because she told herself, "He doesn't love me! He always has to get his own way! It's unfair!"

The problem with Joan's interpretation is that it is not valid. He *does* love her, he doesn't always get his way, and he isn't intentionally being "unfair." On this particular Sunday the Dallas Cowboys are locking spurs with the Pittsburgh Steelers, and he *really* wants to see that game! There's no way he's going to want to get dressed and go to a concert.

When Joan thinks about her husband's motivations in such an illogical fashion, she creates two problems for the price of one. She has to put up with the self-created illusion that she's unloved in addition to missing out on his company at the concert.

The third form of distortion that leads to anger is *magnification*. If you exaggerate the importance of the negative event, the intensity and duration of your emotional reaction may get blown up out of all proportion. For example, if you are waiting for a late bus and you have an important appointment, you might tell yourself, "I can't take this!" Isn't that a slight exaggeration? Since you are taking it, you *can* take it, so why tell yourself you *can't*? The inconvenience of waiting for the bus is bad enough without creating additional discomfort and self-pity in this way. Do you really want to fume like that?

Inappropriate *should* and *shouldn't* statements represent the fourth type of distortion that feeds your anger. When you find that some people's actions are not to your liking, you tell yourself they "shouldn't" have done what they did, or they "should have" done something they failed to do. For example, suppose you register at a hotel and discover they lost the record of your reservation, and now there are no rooms available. You furiously insist, "This *shouldn't* have happened! Those stupid goddam clerks!"

Does the actual deprivation cause your anger? No. The deprivation can only create a sense of loss, disappointment, or inconvenience. Before you can feel anger, you must necessarily make the interpretation you are *entitled* to get what you want in this situation. Consequently, you see the goof-up on your reservation as an injustice. This perception leads to your feeling angry.

So what's wrong with that? When you say the clerks *shouldn't* have made a mistake, you are creating unnecessary frustration for yourself. It's unfortunate your reservation was lost, but it's highly unlikely anyone intended to treat you unjustly, or that the clerks are especially stupid. But they *did* make an error. When you insist on perfection from others, you will simply make *yourself* miserable and become immobilized. Here's the rub: Your anger probably won't cause a room to appear magically, and the inconvenience of going to another hotel will be far less than the misery you inflict on yourself by brooding for hours or days about the lost reservation.

Irrational should statements rest on your assumption that you are *entitled* to instant gratification at all times. So on those occasions when you don't get what you want, you go into panic or rage because of your attitude that unless you get X, you will either die or be tragically deprived of joy forever (X can represent love, affection, status, respect, promptness, perfection, niceness, etc.). This insistence that your wants be gratified at all times is the basis for much self-defeating anger. People who are anger-prone often formulate their desires in moralistic terms such as this: If I'm nice to someone, they *should* be appreciative.

Other people have free will, and often think and act in ways that aren't to your liking. All of your insistence that they must fall in line with your desires and wishes will not produce this result. The opposite is more often true. Your attempts to coerce and manipulate people with angry demands most often will alienate and polarize them and make them much less likely to want to please you. This is because other people don't like being controlled or dominated any more than you do. Your anger will simply limit the creative possibilities for problem solving.

The perception of unfairness or injustice is the ultimate cause of most, if not all, anger. In fact, we could define anger as the emotion which corresponds in a one-to-one manner to your belief that you are being treated unfairly.

Now we come to a truth you may see either as a bitter pill or an enlightening revelation. There is no such thing as a universally accepted concept of fairness and justice. There is an undeniable *relativity* of fairness, just as Einstein showed the relativity of time and space. Einstein postulated—and it has since been experimentally validated—there is no "absolute time" that is standard throughout the universe. Time can appear to "speed up" and "slow down," and is relative to the frame of reference of the observer. Similarly, "absolute fairness" does not exist. "Fairness" is relative to the observer, and what is fair to one person can appear quite unfair to another. Even social rules and moral strictures which are accepted within one culture can vary substantially in another. You can protest that this is not the case and insist that your own personal moral system is universal, but it just ain't so!

Here's proof: When a lion devours a sheep, is this unfair?

From the point of view of the sheep, it is *unfair*; he's being viciously and intentionally murdered with no provocation. From the point of view of the lion, it is *fair*. He's hungry, and this is the daily bread he feels entitled to. Who is "right"? There is *no ultimate or universal answer* to this question because there's no "absolute fairness" floating around to resolve the issue. In fact, fairness is simply a perceptual interpretation, an abstraction, a self-created concept. How about when *you* eat a hamburger? Is this "unfair"? To you, it's not. From the point of view of the cow, it certainly is (or was)! Who's "right"? There is no ultimate "true" answer.

In spite of the fact that "absolute fairness" does not exist, personal and social moral codes are important and useful. I am not recommending anarchy. I am saying that moral statements and judgments about fairness are stipulations, not objective facts. Social moral systems, such as the Ten Commandments, are essentially sets of rules that groups decide to abide by. One basis for such systems is the enlightened self-interest of each member of the group. If you fail to act in a manner that takes into account the feelings and interests of others you are likely to end up less happy because sooner or later they will retaliate when they notice you are taking advantage of them.

A system which defines "fairness" varies in its generality depending on how many people accept it. When a rule of behavior is unique to one person, other people may see it as eccentric. An example of this would be my patient who washes her hands ritualistically over fifty times a day to "set things right" and to avoid extreme feelings of guilt and anxiety. When a rule is nearly universally accepted it becomes part of a general moral code and may become a part of the body of law. The prohibition against murder is an example. Nevertheless, no amount of general acceptance can make such systems "absolute" or "ultimately valid" for everyone under all circumstances.

Much everyday anger results when we confuse our own personal wants with general moral codes. When you get mad at someone and you claim that they are acting "unfairly," more often than not what is really going on is that they are acting "fairly" relative to a set of standards and a frame of reference that differs from yours. Your assumption that they are "being unfair" implies that your way of looking at things is universally accepted. For this to be the case, everyone would have to be the same. But we aren't. We all think differently. When you overlook this and blame the other person for being "unfair" you are unnecessarily polarizing the interaction because the other person will feel insulted and defensive. Then the two of you will argue fruitlessly about who is "right." The whole dispute is based on the illusion of "absolute fairness."

Because of your relativity of fairness, there is a logical fallacy that is inherent in your anger. Although you feel convinced the other guy is acting *unfairly*, you must realize he is only acting unfairly relative to *your* value system. But he is operating from *his* value system, not yours. More often than not, his objectionable action will seem quite fair and reasonable to him. Therefore, from his point of view—which is his only possible basis for action—what he does is "fair." Do you want people to act fairly? Then you should *want* him to act as he does even though you *dislike* what he does, since he is acting fairly within his system! You can work to try to convince him to change his attitudes and ultimately modify his standards and his actions, and in the meantime you can take steps to make certain you won't suffer as a result of what he does. But when you tell yourself, "He's acting unfairly," you are fooling yourself and you are chasing a mirage!

Does this mean that all anger is inappropriate and that the concepts of "fairness" and "morality" are useless because they are relative? Some popular writers do give this impression. Dr. Wayne Dyer writes:

> We are conditioned to look for justice in life and when it doesn't appear, we tend to feel anger, anxiety or

> frustration. Actually, it would be equally productive to search for the fountain of youth, or some such myth. Justice does not exist. It never has, and it never will. The world is simply not put together that way. Robins eat worms. That's not fair to the worms. . . . You have only to look at nature to realize there is no justice in the world. Tornadoes, floods, tidal waves, droughts are all unfair.*

This position represents the opposite extreme, and is an example of all-or-nothing thinking. It's like saying—throw your watches and clocks away because Einstein showed that absolute time does not exist. The concepts of time and fairness are socially *useful* even though they do not exist in an absolute sense.

In addition to this contention that the concept of fairness is an illusion, Dr. Dyer seems to suggest that anger is useless:

> You may accept anger as a part of your life, but do you realize it serves no utilitarian purpose? . . . You do not have to possess it, and it serves no purpose that has anything to do with being a happy, fulfilled person. . . . The irony of anger is that it never works in changing others. . . . **

Again, his arguments seem to be based on cognitive distortion. To say anger serves *no* purpose is just more all-or-nothing thinking, and to say it never works is an overgeneralization. Actually, anger can be adaptive and productive in certain situations. So the real question is not "Should I or should I not feel anger?" but rather "Where will I draw the line?"

---

*Dr. Wayne W. Dyer, *Your Erroneous Zones* (New York: Avon Books, 1977), p. 173.

***Ibid.*, pp. 218–220.

The following two guidelines will help you to determine when your anger is productive and when it is not. These two criteria can help you synthesize what you are learning and to evolve a meaningful personal philosophy about anger:

1. Is my anger directed toward someone who has *knowingly, intentionally*, and *unnecessarily* acted in a hurtful manner?
2. Is my anger useful? Does it help me achieve a desired goal or does it simply defeat me?

Example: You are playing basketball, and a fellow on the other team elbows you in the stomach intentionally so as to upset you and get you off your game. You may be able to channel your anger productively so you will play harder and win. So far your anger is *adaptive*.* Once the game is over, you may no longer want that anger. Now it's *maladaptive*.

Suppose your three-year-old son runs mindlessly into the street and risks his life. In this case he is *not* intentionally inflicting harm. Nevertheless, the angry mode in which you express yourself may be adaptive. The emotional arousal in your voice conveys a message of alarm and importance that might not come across if you were to deal with him in a calm, totally objective manner. In both these examples, you *chose* to be angry, and the magnitude and expression of the emotion were under your control. The *adaptive* and *positive* effects of your anger differentiate it from hostility, which is impulsive and uncontrolled and leads to aggression.

Suppose you are enraged about some senseless violence you read about in the paper. Here the act seems clearly hurtful and immoral. Nevertheless, your anger may not be adaptive if—as is usually the case—there is nothing you plan to do about it. If, in contrast, you choose to help the

*Adaptive means useful and self-enhancing; maladaptive means useless and self-destructive.

victims or begin a campaign to fight crime in some way, your anger might again be adaptive.

Keeping these two criteria in mind, let me give you a series of methods you can use to reduce your anger in those situations where it is not in your best interest.

Develop the Desire. Anger can be the most difficult emotion to modify, because when you get mad you will be like a furious bulldog, and persuading you to stop sinking your teeth into the other person's leg can be extremely tough. You *won't really want* to rid yourself of those feelings because you will be consumed by the desire for revenge. After all, because anger is caused by what you perceive to be unfair, it is a *moral emotion*, and you will be extremely hesitant to let go of the righteous feeling. You will have the nearly irresistible urge to defend and justify your anger with *religious zeal*. Overcoming this will require an act of great willpower. So why bother?

The first step: Use the double-column technique to make a list of the advantages and disadvantages of feeling angry and acting in a retaliatory manner. Consider both the short- and long-term consequences of your anger. Then review the list and ask yourself which are greater, the costs or the benefits? This will help you determine if your resentment is really in your best self-interest. Since most of us ultimately want what's best for us, this can pave the way for a more peaceful and productive attitude.

Here's how it works. Sue is a thirty-one-year-old woman with two daughters from a previous marriage. Her husband, John, is a hard-working lawyer with one teenage daughter from his prior marriage. Because John's time is very limited, Sue often feels deprived and resentful. She told me she felt he wasn't giving her a fair shake in the marriage because he was not giving her enough of his time and attention. She listed the advantages and disadvantages of her irritability in Figure 7–2.

She also made a list of the positive consequences that

Figure 7–2. The Anger Cost-Benefit Analysis.

Advantages of My Anger	*Disadvantages of My Anger*
1. It feels good	1 I will be souring my relationship with John even more
2. John will understand that I strongly disapprove of him.	2. He will want to reject me.
3. I have the *right* to blow my stack if I want to.	3 I will often feel guilty and down on myself after I blow my stack
4. He'll know I'm not a doormat.	4 He will probably retaliate against me and get angry right back, since he doesn't like being taken advantage of either
5. I'll show him I won't stand for being taken advantage of	5 My anger inhibits both of us from correcting the problem that caused the anger in the first place. It prevents resolution and sidetracks us from dealing with the issues.
6. Even though I don't get what I want, I can at least have the satisfaction of getting revenge I can make him squirm and feel hurt like I do. Then he'll have to shape up.	6. One minute I'm up, one minute I'm down. My irritability makes John and the people around me never know what to expect. I get labeled as moody and cranky and spoiled and immature. They see me as a childish brat
	7. I might make neurotics out of my kids. As they grow up, they may resent my explosions and see me as someone to stay away from rather than to go to for help.
	8 John may leave me if he gets enough of my nagging and bitching.
	9 The unpleasant feelings I create make me feel miserable. Life becomes a sour and bitter experience, and I miss out on the joy and creativity I used to prize so highly.

might result from eliminating her anger: (1) People will like me better. They will want to be near me; (2) I will be more predictable; (3) I will be in better control of my emotions; (4) I will be more relaxed; (5) I will be more comfortable with myself; (6) I will be viewed as a positive, nonjudgmental, practical person; (7) I will behave more often as an adult than as a child who has to get what it wants; (8) I will influence people more effectively, and I'll get more of what I want through assertive, calm, rational negotiation than through tantrums and demands; and (9) my kids, husband, and parents will respect me more. As a result of this assessment, Sue told me she was convinced that the price of her anger substantially exceeded the benefits.

It is crucial that you perform this same type of analysis as a first step in coping with your anger. After you list the advantages and disadvantages of your anger, give yourself the same test. Ask yourself, if the upsetting situation that provokes me doesn't change immediately, would I be willing to cope with it instead of getting angry? If you can answer yes, then you are clearly motivated to change. You will probably succeed in gaining greater inner peace and self-esteem, and you will increase your effectiveness in life. This choice is up to you.

Cool Those Hot Thoughts. Once you've decided to cool down, an invaluable method that can help you is to write down the various "hot thoughts" that are going through your mind when you are upset. Then substitute less upsetting, more objective "cool thoughts," using the double-column method (Figure 7–3). Listen for those "hot thoughts" with your "third ear" so as to tune in to the antagonistic statements that go through your head. Record this private dialogue without any censorship. I'm sure you'll notice all kinds of highly colorful language and vengeful fantasies—write them all down. Then substitute "cool thoughts" that are more objective and less inflammatory. This will help you feel less aroused and overwhelmed.

Sue used this technique to deal with the frustration she felt when John's daughter, Sandy, acted manipulative and wrapped John around her finger. Sue kept telling him to be more assertive with Sandy and less of a soft touch, but he often reacted negatively to her suggestions. He felt Sue was nagging and making demands to get her way. This made him want to spend even *less* time with her, which contributed to a vicious cycle.

Figure 7–3. Sue wrote down her "Hot Thoughts" when her husband acted like a soft touch in response to his teenage daughter's selfish manipulations. When she substituted less upsetting "Cool Thoughts," her jealousy and resentment diminished.

Hot Thoughts	*Cool Thoughts*
1. How dare he not listen to me!	1. Easily. He's not obliged to do everything my way. Besides he *is* listening, but he's being defensive because I'm acting so pushy.
2. Sandy lies. She says she's working, but she's not. Then she expects John's help.	2. It's her nature to lie and to be lazy and to use people when it comes to work in school. She hates work. That's her problem.
3. John doesn't have much free time and if he spends it helping her, I will have to be alone and take care of my kids by myself.	3. So what. I like being alone. I'm capable of taking care of my kids by myself. I'm not helpless. I can do it. Maybe he'll want to be with me more if I learn not to get angry all the time.
4. Sandy's taking time away from me.	4. That's true. But I'm a big girl. I can tolerate some time alone. I wouldn't be so upset if he were working with my kids.
5. John's a schmuck. Sandy uses people.	5. He's a big boy. If he wants to help her he can. Stay out of it. It's not my business.
6. I can't stand it!	6. I can. It's only temporary. I've stood worse.
7. I'm a baby brat. I deserve to feel guilty.	7. I'm entitled to be immature at times. I'm not perfect and I don't need to be. It's not necessary to feel guilty. This won't help.

Sue wrote down the "hot thoughts" that made her feel jealous and guilty (see Figure 7–3). As she substituted "cool thoughts," she felt better, and this served as an antidote to her urge to try to control John. Although she still felt he was wrong in letting Sandy manipulate him, she decided he had the "right" to be "wrong." Consequently, Sue pushed John less, and he began to feel less pressured. Their relationship improved and ripened in a climate of mutual freedom and respect. Simply talking back to her "hot thoughts" was, of course, not the only ingredient that led to a successful second marriage for Sue and John, but it was a necessary and gigantic first step without which both of them could have easily ended up stalemated again!

You can also use the more elaborate chart, the "Daily Record of Dysfunctional Thoughts," to deal with your anger (see Figure 7–4, page 170). You can describe the provocative situation and assess how angry you feel before and after you do the exercise. Figure 7–4 shows how a young woman coped with her frustration when she was dealt with tersely by a prospective employer over the telephone. She reported that pinpointing her "hot thoughts" and putting the lie to them helped her nip an emotional explosion in the bud. This prevented the fretting and fuming that normally would have soured her entire day. She told me, "Before I did the exercise I thought my enemy was the man on the other end of the phone. But I learned that *I* was treating myself ten times worse than he was. Once I recognized this, it was relatively easy to substitute cooler thoughts, and I surprised myself by feeling a whole lot better right away!"

Imagining Techniques. Those negative "hot thoughts" that go through your mind when you are angry represent the script of a private movie (usually X-rated) that you project onto your mind. Have you ever noticed the picture on the screen? The images, daydreams, and fantasies of revenge and violence can be quite colorful indeed!

You may not be aware of these mental pictures unless

Figure 7–4. Daily Record of Dysfunctional Thoughts.

Provocative Situation	*Emotions*	*Hot Thoughts*	*Cool Thoughts*	*Outcome*
Called ad in paper for part-time medical transcriptionist. Ad said—needs "some experience." First, the man wouldn't even tell me what kind of company it was. Then he turned me down for job 'cause he didn't think I had enough experience!	Anger hatred frustration 98%	1. That jerk! *Who the hell does he think he is*! I have more than enough experience	1. Why am I getting so excited? I didn't like the tone of his voice anyway. So he didn't allow me to really explain my experience. I know I'm good. So it's not my fault I didn't get the job—it's his. Besides, would I want to work for someone like that?	Anger hatred frustration 15%
		2. That was the best ad in the paper, and I lost it.	2. I'm blowing things out of proportion. There are many other jobs I can get.	
		3. My parents will kill me.	3. Of course they won't. At least I'm trying.	
		4. I'm going to cry.	4. Now isn't that ridiculous? Why should someone make me cry? This isn't worth crying over. I *know* my worth—that's what counts.	

you look for them. Let me illustrate. Suppose I ask you to visualize a red apple in a brown basket right now. You can do this with your eyes open or closed. There! Do you see it now? That's what I'm referring to. Most of us have these visual images all day long. They are a part of normal consciousness, the pictorial illustrations of our thoughts. For example, memories sometimes occur to us as mental pictures. Conjure up an image now of some vivid past event—your high-school graduation, your first kiss (do you still remember it?), a long hike, etc. Do you see it now?

These images can affect you strongly, and their influence can be positively or negatively arousing, just like erotic dreams or nightmares. The exhilarating effect of a positive image can be intense. For instance, on your way to an amusement park you might have an image of that first dazzling descent down the roller coaster, and you may experience the excited rush in your belly. The daydream actually creates the pleasurable anticipation. Similarly, negative images play a powerful role in your level of emotional arousal. Visualize right now someone whom you've gotten good and mad at sometime in your life. What images come to mind? Do you imagine punching them in the nose or tossing them into a vat of boiling oil?

These daydreams actually keep your anger alive *long* after the initial insult has occurred. Your sense of rage may eat away at you for hours, days, months, or even years after the irritating event has long since passed. Your fantasies help keep the pain alive. Every time you fantasize about the occurrence you shoot new doses of arousal into your system. You become like a cow chewing on poison cud.

And who is creating this anger? You are because you chose to put those images in your mind! For all you know, the person you are mad at lives in Timbuktu, or maybe isn't even alive anymore, so he or she could hardly be the culprit! You are the director and producer of the film now, and, what's worse, you're the only one in the audience. Who has to watch and experience all the arousal? YOU DO! You're the one who's subjected to a continual clenching, a tightening of back muscles, and an outpouring of adrenal hormones into the bloodstream. You're the one whose blood pressure is going up. IN A NUTSHELL: *You're making yourself hurt*. Do you want to keep this up?

If not, you will want to do something to reduce the anger-generating images that you are projecting onto your mind. One helpful technique is to transform them in a creative way so they become less upsetting. Humor represents one powerful tool you can use. For example, instead of imagining wringing the neck of the person you are furious with, fantasize that he is walking around in diapers in a crowded department store. Visualize all the details: the potbelly, the diaper pins, the hairy legs. Now what's happening to your anger? Is that a broad smile spreading across your face?

A second method involves thought stoppage. As you notice the images crossing your mind each day, remind yourself that you have the right to turn the projector off. Think about something else. Find someone and engage him or her in conversation. Read a good book. Bake bread. Go jogging. When you don't reward the anger images with your arousal, they will recur less and less often. Instead of dwelling on them, think about an upcoming event that excites you, or switch to an erotic fantasy. If the upsetting memory is persistent, engage in vigorous physical exercise such as push-ups, rapid jogging, or swimming. These have the additional benefit of rechanneling your potentially hurtful arousal in a highly beneficial way.

Rewrite the Rules. You may frustrate and upset yourself needlessly because you have an unrealistic rule about personal relationships that causes you to be let down all the time. The key to Sue's anger was her belief she was *entitled* to John's love because of her rule "If I'm a good and faithful wife, I deserve to be loved."

As a result of this innocent-sounding assumption, Sue experienced a constant sense of danger in her marriage because anytime John wasn't giving her an appropriate helping of love and attention, she would experience it as a confirmation of her inadequacy. She would then manipulate and demand attention and respect in a constant battle to defend herself against a loss of self-esteem. Intimacy with him became like slipping slowly toward the edge of an icy cliff. No wonder she was desperately grabbing onto John, and no wonder she would explode when she sensed his indifference—didn't he realize her life was at stake?

In addition to the intense unpleasantness that her "love" rule created, it didn't work well in the long run. For a while Sue's manipulations did, in fact, get her some of the attention she craved. After all, she could *intimidate* John with her emotional explosions, she could *punish* him with her icy withdrawal, and she could *manipulate* him by arousing his guilt.

But the price Sue pays is that the love she receives isn't—and can't—be given freely and spontaneously. He will feel exhausted, trapped, and controlled. The resentment he's been storing up will press for release. When he stops buying into her belief that he *has* to give in to her demands, his desire for freedom will overpower him, and he will explode. The destructive effects of what passes for love never cease to amaze me!

If your relationships are characterized by this cyclic tension and tyranny, you may be better off rewriting the rules. If you adopt a more realistic attitude, you can end your frustration. It's much easier than trying to change the world. Sue decided to revise her "love" rule in the following way: "If I behave in a positive manner toward John, he will respond in a loving way a good bit of the time. I can still respect myself and function effectively when he doesn't." This formulation of her expectations was more realistic and didn't put her moods and self-esteem at the mercy of her husband.

The rules that get you into interpersonal difficulty often won't appear to be malignant. On the contrary, they often seem highly moral and humanistic. I recently treated a woman named Margaret who had the notion that "marriages *should* be fifty-fifty. Each partner *should* do for the other equally.'' She applied this rule to all human relationships. ''If I do nice things for people, they *should* reciprocate.''

So what's wrong with that? It certainly sounds ''reasonable'' and ''fair.'' It's kind of a spin-off from the Golden Rule. Here's what's wrong with it: It's an undeniable fact that human relationships, including marriages, are rarely spontaneously ''reciprocal'' because people are different. Reciprocity is a transient and inherently unstable ideal that can only be approximated through continued effort. This involves mutual consensus, communication, compromise, and growth. It requires negotiation and hard work.

Margaret's problem was that she didn't recognize this. She lived in a fairytale world where reciprocity existed as an assumed reality. She went around always doing good things for her husband and others and then waited for their reciprocity. Unfortunately, these unilateral contracts fell apart because other people usually weren't aware that she expected to be repaid.

For example, a local charity organization advertised for a salaried assistant director to start in several months. Margaret was quite interested in this position and submitted her application. She then gave large amounts of her time doing volunteer work for the organization and assumed that the other employees would ''reciprocate'' by liking and respecting her, and that the director would ''reciprocate'' by giving her the job. In reality, the other employees did not respond to her warmly. Perhaps they sensed and resented her attempt to control them with her ''niceness'' and virtue. When the director chose another candidate for the position, she hit the roof and felt bitter and disillusioned because her ''reciprocity'' rule had been violated!

Since her rule caused her so much trouble and disappointment she opted to rewrite it, and to view reciprocity not as a *given* but as a goal she could work toward by pursuing her own self-interest. At the same time she relinquished her demand that others read her mind and respond as she wanted. Paradoxically, as she learned to *expect* less, she *got* more!

Figure 7–5. Revising "Should Rules."

Self-Defeating Should Rule	*Revised Version*
1. If I'm nice to someone, they should be appreciative.	1. It would be nice if people were always appreciative, but this isn't realistic. They will often be appreciative, but sometimes they won't be.
2. Strangers should treat me courteously.	2. Most strangers will treat me courteously if I don't act like I have a chip on my shoulder. Occasionally some sourpuss will act obnoxious. Why let this bother me? Life is too short to waste time concentrating on negative details.
3. If I work hard for something, I should get it.	3. This is ridiculous. I have no guarantee I'll *always* be successful in everything. I'm not perfect and I don't have to be.
4. If someone treats me unfairly, I should get mad because I have the right to get mad and because it makes me more human.	4. All human beings have the right to get mad whether or not they're treated unfairly. The real issue is—is it to my advantage to get mad? Do I want to feel angry? What are the costs and benefits?
5. People shouldn't treat me in ways I wouldn't treat them.	5. Hogwash. Everyone doesn't live by my rules, so why expect they will? People will *often* treat me as well as I treat them, but not *always*.

If you have a "should" or "shouldn't" rule that has been causing you disappointment and frustration, rewrite it in more realistic terms. A number of examples to help you do this are shown in Figure 7–5. You will notice that the substitution of one word—"it would be nice *if*" in place of "should"—can be a useful first step.

Learn to Expect Craziness. As the anger in Sue's relationship with John cooled down, they became closer and more loving. However, John's daughter, Sandy, responded to his increased intimacy by even greater manipulations. She began to lie, borrowed money without returning it; she sneaked into Sue's bedroom, went through drawers, and stole Sue's personal items; she left the kitchen messy, etc. All these actions effectively got Sue's goat because she told herself, "Sandy shouldn't act so sneaky. She's crazy! It's unfair!" Sue's sense of frustration was the product of two necessary ingredients:

1. Sandy's obnoxious behavior;
2. Sue's expectation that she should act in a more mature way.

Since the evidence suggested that Sandy *wasn't* about to change, Sue had only one alternative: She could discard her unrealistic expectation that Sandy behave in an adult, ladylike fashion! She decided to write the following memo to herself entitled:

Why Sandy Should Act Obnoxiously

It is Sandy's nature to be manipulative because she believes that she's entitled to love and attention. She believes that getting love and attention is a matter of life and death. She thinks she needs to be the center of attention in order to survive. Therefore, she will see any lack of love as unfair and a great danger to her sense of self-esteem.

Because she feels she has to manipulate in order to get attention, she *should* act in a manipulative way. Therefore, I can expect and predict that she will continue to act this way until she changes. Since it is unlikely that she will change in the near future, I can expect her to continue to behave this way for a period of time. Therefore, I will have no reason to feel frustrated or surprised because she will be acting the way she *should* act.

Furthermore, I want all humans including Sandy to act in a manner that they believe to be fair. Sandy feels she's entitled to more attention. Since her obnoxious behavior is based on her sense of entitlement, I can remind myself that what she does is fair from her point of view.

Finally, I want my moods to be under my control, not hers. Do I want to make myself feel upset and angry at her "fair, obnoxious" behavior? No! Therefore, I can begin to change the way I react to her:

1. I can thank her for stealing since this is what she "should" do!
2. I can laugh to myself about her manipulations since they are childish.
3. I can choose not to be angry unless it is my decision to use the anger to accomplish a specific goal.
4. If I feel a loss of self-esteem due to Sandy's manipulations, I can ask myself, Do I want to give a child such power over me?

What is the desired effect of such a memorandum? Sandy's provocative actions are probably knowingly malicious. Sandy consciously targets Sue because of the resentment and helpless frustration she feels. When Sue gets upset, she paradoxically gives Sandy exactly what she wants! She can greatly reduce her frustration as she changes her expectations.

Enlightened Manipulation. You may fear that you will be a pushover if you change your expectations and give up your anger. You might sense that other people would take advantage of you. This apprehension reflects your sense of inadequacy as well as the fact that you probably have not been trained in more enlightened methods of going after what you want. You probably believe that if you didn't make demands on people you'd end up empty-handed.

So what's the alternative? Well, as a starting point let's review the work of Dr. Mark K. Goldstein, a psychologist who has done some brilliant and creative clinical research on the behavioral conditioning of husbands by wives. In his work with neglected and angry wives, he became aware of the self-defeating methods they used to get what they wanted from their husbands. He asked himself: What have we learned in the laboratory about the most effective scientific methods for influencing *all* living organisms, including bacteria, plants, and rats? Can we apply these principles to wayward and sometimes brutal husbands?

The answer to these questions was straightforward—*reward* the desired behavior instead of *punishing* the undesired behavior. Punishment causes aversion and resentment and brings about alienation and avoidance. Most of the deprived and abandoned wives he treated were misguidedly trying to punish their husbands into doing what they wanted. By switching them to a reward model in which the desired behavior got copious attention, he observed some dramatic turnabouts.

The wives Dr. Goldstein treated were not unique. They were ensnarled in the ordinary marital conflicts that most of us confront. These women had a long history of giving their spouses attention either indiscriminately or, in some cases, primarily in response to undesirable behavior. A major shift had to occur in order for them to elicit the kind of response they desired from their husbands but were not getting. By keeping meticulous scientific records of their interactions with their husbands, the women were able to achieve control over how they responded.

Here's how it worked for one of Dr. Goldstein's patients. After years of fighting, wife X reported she lost her husband. He abandoned her and moved in with his girl friend. His primary interactions with wife X had centered around abuse and indifference. It appeared on the surface as if he didn't care much about her. Nevertheless, he did call her occasionally, indicating he might have some interest in her. She had the choice of cultivating this attention or crushing it further by continued inappropriate responses.

Wife X defined her goals. She would experiment to see if she *could* in fact get her husband back. The first milestone would be to determine if she could effectively increase his rate of contact with her. She measured meticulously the frequency and duration of his every telephone call and visit home, recording this information on a piece of graph paper taped to the refrigerator door. She carefully assessed the crucial relationship between her behavior (the stimulus) and the frequency of his contacts (the response).

She initiated no contacts with him at all on her own, but instead responded positively and affectionately to his calls. Her strategy was straightforward. Rather than noticing and reacting to all the things about him that she didn't like, she began to reinforce systematically those that she did like. The rewards she used were all the things that turned him on—praise, food, sex, affection, etc.

She began by responding to his rare calls in an upbeat, positive, complimentary manner. She flattered and encouraged him. She avoided any criticism, argument, demands, or hostility, and found a way to *agree* with everything he said, using the disarming technique described in Chapter 7. Initially she terminated all these calls after five to ten minutes to ensure the likelihood the conversations would not deteriorate into an argument or become boring to him. This guaranteed that her feedback would be pleasant to him, and that his response to it would not be suppressed or eliminated.

After she did this a few times, she noticed her husband began to call more and more frequently because the calls were positive, rewarding experiences for him. She noted this increased rate of telephoning on her graph paper just as a scientist observes and documents the actions of an experimental rat. As his phone calls increased, she began to feel encouraged, and some of her irritation and resentment melted away.

One day he appeared at the house and according to her plan, she announced, "I'm so happy you dropped by because I just happen to have a fresh, fancy imported Cuban cigar in the freezer for you. It's the expensive type you really like." She actually had a whole box of them waiting so she was able to repeat this each time he visited—regardless of why or when he came. She noticed the frequency of his visits substantially increased.

In a similar manner, she continued to "shape" his behavior using *rewards* rather than coercion. She realized how successful she had been when her husband decided to leave his girl friend and asked if he could move back in with her.

Am I saying that is the *only* way to relate and to influence people? No—that would be absurd. It's just a pleasant spice, not the whole banquet or even the main course. But it's a frequently overlooked delicacy that few appetites can resist. There's no *guarantee* it will work—some situations may be irreversible, and you can't always get what you want.

At any rate, *try* the upbeat reward system. You may be pleasantly surprised at the remarkable effectiveness of your secret strategy. In addition to motivating the people you care about to want to be around you, it will improve your mood because you learn to notice and focus on the positive things that others do rather than dwell on their negatives.

"Should" Reduction. Because many of the thoughts which generate your anger involve moralistic "should" statements, it will help you to master some "should" removal methods. One way is to make a list, using the double-column method, of all the reasons why you believe the other person "shouldn't" have acted as he did. Then challenge these reasons until you can see why they are unrealistic and don't actually make good sense.

Example: Suppose the carpenter on your new house did a sloppy job on the kitchen cabinets. The doors are poorly aligned and don't close properly. You feel irate because you see this as "unfair." After all, you paid full union wages, so you feel entitled to excellent workmanship from

Figure 7–6.

Reasons He Should Have Taken More Pride in His Work	*Rebuttals*
1. Because I paid top dollar.	1. He gets paid the same wage whether or not he takes extra pride in his work.
2. Because it's only decent to do a good job.	2. He probably felt he did an adequate job. And the paneling he did actually looks quite decent.
3. Because he should make sure he gets it done *right*.	3. Why should he?
4. Because *I* would if I were a carpenter.	4. But he's not me—he's not trying to meet my standards.
5. Because he should care more about his product.	5. There's no reason for him to care more. Some carpenters care a lot about their work, and for others it's just a job.
6. So why must *I* get the one who does sloppy work?	6. All the people who worked on your house didn't do sloppy work. You can't expect to get 100 percent top-notch people. That would just be unrealistic.

a top craftsman. You fume as you tell yourself, "The lazy bastard should take some pride in his work. What's the world coming to?" You list the reasons and rebuttals detailed in Figure 7–6.

The rationale for eliminating your "should" statement is simple: It's not true that you are entitled to get what you want just because you want it. You'll have to negotiate. Call the carpenter, complain, and insist the job be corrected.

But don't double your trouble by making yourself excessively hot and bothered. The carpenter probably wasn't *trying* to hurt you, and your anger might simply polarize him and put him on the defensive. After all, half of all the carpenters (and psychiatrists, secretaries, writers, and dentists, etc.) throughout human history have been below average. Do you believe that? It's true by definition because "average" is *defined* as the halfway point! It's ludicrous to fume and complain that this particular carpenter's average talent is "unfair," or that he "should" be other than he is.

Negotiating Strategies. At this point you may be bristling because you are thinking, "Well! That's a fine kettle of fish! Dr. Burns seems to be telling me I can find happiness by believing that lazy, incompetent carpenters *should* do mediocre work. After all, it's their nature, the good doctor claims! What weak-spined hogwash! I'm not going to be stripped of my human dignity and let people walk all over me and get away with second-rate crappy work I'm paying a fortune for."

Cool down! Nobody's asking you to let the carpenter pull the wool over your eyes. If you want to exert your influence in an effective way instead of moping angrily and creating inner turmoil, a calm, firm, assertive approach will usually be the most successful. Moralistic "shoulding," in contrast, will simply aggravate you and polarize him, and cause him to feel defensive and to counterattack. Remember—fighting is a form of intimacy. Do you really want to be so *intimate* with this carpenter? Wouldn't you prefer to get what you want instead?

As you stop consuming your energy in anger, you can focus your efforts on getting what you want. The following negotiating principles can work effectively in such a situation:

1. Instead of telling him off, *compliment* him on what he did right. It's an undeniable fact of human nature that few people can resist flattery even if it's blatantly

insincere. However, since you can find *something* good about him or his work, you can make your compliment honest. Then mention the problem with the cupboard doors tactfully, and calmly explain why you want him to come back and correct the alignment.

2. *Disarm him* if he argues by finding a way to agree with him regardless of how absurd his statements are. This will shut him up and take the wind out of his sails. Then immediately—
3. *Clarify* your point of view again calmly and firmly.

Repeat the above three techniques over and over in varying combinations until the carpenter finally gives in or an acceptable compromise is reached. Use ultimatums and intimidating threats only as a last resort, and make sure you are ready and willing to follow through when you do. As a general principle, use diplomacy in expressing your dissatisfaction with his work. Avoid labeling him in an insulting way or implying he is bad, evil, malignant, etc. If you decide to tell him about your negative feelings, do so objectively without magnification or an excess of inflammatory language. For example, "I resent shoddy work when I feel you have the ability to do a good professional job" is far preferable to "You mother——! Your——work is an outrage."

In the following dialogue I will identify each of these techniques.

YOU: I was pleased with how some of the work came out, and I'm hopeful I'll be able to tell other people I was happy with the whole job. The paneling was especially well done. I'm a little concerned about the kitchen cabinets, however. (Compliment)

CARPENTER: What seems to be the trouble?

YOU: The doors aren't lined up, and many of the handles are on crooked.

CARPENTER: Well, that's about the best I can do on those kinds of cabinets. They're mass-produced, and they just aren't made the best.

YOU: Well, that's true. They aren't as well made as a more expensive type might be. (Disarming technique) Nevertheless, they aren't acceptable this way, and I'd appreciate it if you'd do something to make them more presentable. (Clarification; tact)

CARPENTER: You'll have to talk to the manufacturer or the builder. There's nothing I can do about it.

YOU: I can understand your frustration (Disarming technique), but it's your responsibility to complete these cabinets to our satisfaction. They're simply not acceptable. They look shoddy, and they don't close properly. I know it's an inconvenience, but my position is that the job can't be considered complete and the bill won't be paid until you've corrected it. (Ultimatum) I can see from your other work that you have the skill to make them look right in spite of the extra time it will take. That way we'll be completely satisfied with your work, and we can give you a good recommendation. (Compliment)

Try these negotiating techniques when you are at loggerheads with someone. I think you'll find they work more effectively than blowing your stack, and you'll feel better because you'll usually end up getting more of what you want.

Accurate Empathy. Empathy is the ultimate anger antidote. It's the highest form of magic described in this book, and its spectacular effects are firmly entrenched in *reality*. No trick mirrors are needed.

Let's define the word. By empathy, I do *not* mean the capacity to feel the same way someone else feels. This is sympathy. Sympathy is highly touted but is, in my opinion, somewhat overrated. By empathy, I do *not* mean acting in a tender, understanding manner. This is support. Support is also highly valued and overrated.

So what is empathy? Empathy is the ability to comprehend with accuracy the precise thoughts and motivations of other people in such a way that they would say, "Yes, that *is exactly* where I'm coming from!" When you have this extraordinary knowledge, you will understand and accept without anger why others act as they do even though their actions might not be to your liking.

Remember, it is actually *your* thoughts that create your anger and not the other person's behavior. The amazing thing is that the moment you grasp why the other person is acting that way, this knowledge tends to put the lie to your anger-producing thoughts.

You might ask, If it's so easy to eliminate anger through empathy, why do people get so damn mad at each other every day? The answer is that empathy is difficult to acquire. As humans we are trapped in our own perceptions, and we react automatically to the meanings we attach to what people do. Getting inside the other person's skull requires hard work, and most people don't even know how to do this. Do you? You will learn how in the next few pages.

Let's start with an example. A businessman recently sought help because of his frequent episodes of angry outbursts and abusive behavior. When his family or employees didn't do what he wanted, he'd bite their heads off. He usually succeeded in intimidating people, and he enjoyed dominating and humiliating them. But he sensed that his impulsive explosions ultimately caused problems for him because of his reputation as a sadistic hothead.

He described a dinner party he attended where the waiter

forgot to fill his wineglass. He felt a surge of rage due to his thought, "The waiter thinks I'm unimportant. Who the hell does he think he is anyway? I'd like to wring the mother——'s neck."

I used the empathy method to demonstrate to him how illogical and unrealistic his angry thoughts were. I suggested that we do some role-playing. He was to play the waiter, and I would act the part of a friend. He was to try to answer my questions as truthfully as possible. The following dialogue evolved:

DAVID (playing the role of the waiter's friend): I noticed that you didn't fill the wineglass of that businessman there.

PATIENT (playing the role of waiter): Oh, I see that I didn't fill his glass.

DAVID: Why didn't you fill his glass? Do you think he is an unimportant person?

PATIENT (after a pause): Well, no, it wasn't that. I actually don't know much about him.

DAVID: But didn't you decide that he was an unimportant person and refuse to give him any wine because of that?

PATIENT (laughing): No, *that* isn't why I didn't give him any wine.

DAVID: Then why didn't you give him wine?

PATIENT (after thinking): Well, I was daydreaming about my date for tonight. Furthermore, I was looking at that pretty girl across the table. I was distracted by her low-cut dress, and I just overlooked his wineglass.

This role-playing episode created great relief for the patient because by placing himself in the waiter's shoes he was able to see how unrealistic his interpretation had been. His cognitive distortion was jumping to conclusions (mind reading). He automatically concluded the waiter was being *unfair*, which made him feel he had to retaliate to maintain his self-pride. Once he acquired some empathy, he was able to see that his righteous indignation was caused entirely and exclusively by his own distorted thoughts and *not* the waiter's actions. It is often extremely difficult for angry-prone individuals to accept this at first because they have a nearly irresistible urge to blame others and to retaliate. How about you? Does the idea that many of your angry thoughts are invalid seem abhorrent and unacceptable?

The empathy technique can also be quite useful when the other person's actions appear more obviously and intentionally hurtful. A twenty-eight-year-old woman named Melissa sought counseling around the time she was separating from her husband, Howard. Five years earlier Melissa discovered that Howard was having an affair with Ann, an attractive secretary who worked in his building. This revelation was a heavy blow to Melissa, but to make matters even worse, Howard was hesitant to make a clean break with Ann, and so the affair dragged on for eight additional months. The humiliation and rage Melissa felt during this period was a major factor that led to her ultimate decision to leave him. Her thoughts ran along these lines: (1) He had no right to act like that. (2) He was self-centered. (3) It was unfair. (4) He was a bad, rotten person. (5) I must have failed.

In the course of a therapy session, I asked Melissa to play Howard's role, and then I cross-examined her to see if she could explain precisely why he had had the affair with Ann and acted as he did. She reported that as the role-playing evolved, she suddenly saw where Howard had been coming from, and at that moment her anger toward him completely vanished. After the session she wrote a description of the dramatic disappearance of the anger she had harbored for years:

> After Howard's affair with Ann presumably ended, he insisted on continuing to see her and was still very

much bound up with her. This was painful to me. It made me feel that Howard really didn't respect me and considered himself more important than I was. I felt that if he really did love me he wouldn't put me through this. How could he continue to see Ann when he knew how miserable it made me feel? I felt really angry at Howard and down on myself. When I tried the empathy approach and played the role of Howard, I saw the "whole." I suddenly saw things differently. When I imagined I was Howard, I could see where he was coming from. Putting myself in his place, I saw the problem of loving Melissa my wife, as well as Ann my lover. It dawned on me that Howard was really trapped in a "can't-win" system created by his thoughts and feelings. He loved me but was desperately attracted to Ann. As much as he wanted to he couldn't stop seeing her. He felt very guilty and couldn't stop himself. He felt he would lose if he left Ann, and he would lose if he left me. He was unwilling and unable to come to terms with either form of loss, and it was *his indecisiveness rather than any inadequacy on my part* which caused him to be slow in making up his mind.

The experience was a revelation for me. I really saw what had happened for the first time. I knew Howard had not done anything deliberately to hurt me, but had been incapable of doing anything other than what he did. I felt good being able to see and understand this.

I told Howard when I spoke to him next. We both felt a lot better about this. I also got a really good feeling from the experience with the empathy technique. It was very exciting. More real than what I had seen before.

The key to Melissa's anger was her fear of losing self-esteem. Although Howard had indeed acted in a genuinely negative manner, it was the *meaning* she attached to the experience that caused her sense of grief and rage. She assumed that as a "good wife" she was entitled to a "good marriage." This is the logic that got her into emotional trouble:

Premise: If I am a good and adequate wife, my husband is bound to love me and be faithful to me.

Observation: My husband is not acting in a loving, faithful way.

Conclusion: Therefore, either I am not a good and adequate wife, or else Howard is a bad, immoral person because he is breaking my "rule."

Thus, Melissa's anger represented a feeble attempt to save the day because within her system of assumptions, this was actually the *only* alternative to suffering a loss of self-esteem. The only problems with her solution were (a) she wasn't *really* convinced he was "no good"; (b) she didn't really *want* to write him off since she loved him; and (c) her chronic sour anger didn't *feel* good, it didn't *look* good, and it drove him farther away.

Her premise that he would love her as long as she was good was a fairy tale she had never thought to question. The empathy method transformed her thinking in a highly beneficial way by allowing her to relinquish the *grandiosity* inherent in her premise. His misbehavior was caused by *his* distorted cognitions, not her inadequacy. Thus, *he* was responsible for the jam he was in, not she!

This sudden insight struck her like a lightning bolt. The moment she saw the world through *his* eyes, her anger vanished. She became a much *smaller* person in the sense that she no longer saw herself as responsible for the actions of her husband and the people around her. But at the same time she experienced a sudden increase in self-esteem.

In the next session I decided to put her new insight to the acid test. I confronted her with the negative thoughts that had originally upset her to see if she could answer them effectively:

DAVID: Howard could have stopped seeing her sooner. He made a fool out of you.

MELISSA: No—he couldn't stop because he was trapped. He felt a tremendous obsession, and he was attracted to Ann.

DAVID: But then he *should've* gone off with her and broken up with you so he could stop torturing you. That would've been the *only decent thing* to do!

MELISSA: He felt he couldn't break off with me either because he loved me and was committed to me and to our children.

DAVID: But that was unfair, to keep you dangling so long.

MELISSA: He didn't mean to be unfair. It just happened.

DAVID: It just happened! What Pollyanna nonsense! The fact is, *he shouldn't have* gotten into such a situation in the first place.

MELISSA: But that's where he was at. Ann represented excitement, and he felt bored and overwhelmed by life at the time. Eventually one day he just couldn't resist her flirting any more. He took one small step over the line in a moment of weakness, and then the affair was off and running.

DAVID: Well, you are less of a person because he wasn't faithful to you. This makes you inferior.

MELISSA: It has nothing to do with being less of a person. I don't have to get what I want all the time to be worthwhile.

DAVID: But he never would have sought excitement elsewhere if you were an adequate wife. You're undesirable and unlovable. You're second-rate, and that's why your husband had an affair.

MELISSA: The fact is, he ultimately chose me over Ann, but that doesn't make me any better than Ann, does it? Similarly, the fact that he chose to deal with his problems by escaping doesn't mean that I'm unlovable or less desirable.

I could see that Melissa was clearly unruffled by my vigorous attempts to get her goat, and this proved she had transcended this painful period of her life. She traded in her anger for joy and self-esteem. Empathy was the key that freed her from being trapped in hostility, self-doubt, and despair.

Putting It All Together: Cognitive Rehearsal. When you get angry, you may feel you react too rapidly to be able to sit down and assess the situation objectively and apply the various techniques described in this chapter. This is one of the characteristics of anger. Unlike depression, which tends to be steady and chronic, anger is much more eruptive and episodic. By the time you are aware you are upset you may already feel out of control.

"Cognitive rehearsal" is an effective method for solving this problem and for synthesizing and using the tools you have learned thus far. This technique will help you learn to overcome your anger ahead of time without actually experiencing the situation. Then when the real thing happens, you'll be prepared to handle it.

Begin by listing an "anger hierarchy" of the situations that most commonly trigger you off and rank these from +1 (the least upsetting) to +10 (the most infuriating), as shown in Figure 7–7. The provocations should be ones that you'd like to handle more effectively because your anger is maladaptive and undesirable.

Start with the first item on the hierarchy list that is the least upsetting to you, and fantasize as vividly as you can that you are *in* that situation. Then verbalize your "hot thoughts" and write them down. In the example given in Figure 7–7, you're feeling annoyed because you're telling

Figure 7–7. The Anger Hierarchy.

+1–I sit in a restaurant for fifteen minutes, and the waiter doesn't come.

+2–I call a friend who doesn't return the call.

+3–A client cancels an appointment at the last minute without explanation.

+4–A client fails to show up for an appointment without informing me.

+5–Someone criticizes me nastily.

+6–An obnoxious group of juveniles crowd in front of me in line at a theater.

+7–I read in the paper about senseless violence, such as rape.

+8–A customer refuses to pay a bill for goods I've delivered and skips town so that I can't collect.

+9–Local delinquents repeatedly knock down my mailbox in the middle of the night over a several-month period. There's nothing I can do to catch them or stop them.

+10–I see a television report that someone—presumably a group of teenagers—have broken into the zoo at night, and stoned a number of small birds and animals to death and mutilated others.

yourself, "The goddamn mother——ing waiters don't know what the——they're doing! Why don't the lazy bastards get off their butts and move? Who the hell do they think they are? Am I supposed to starve to death before they'll give me a menu and a glass of water?"

Next fantasize flying off the handle, telling off the maître d', and storming out and slamming the restaurant door. Now record how upset you feel between 0 and 100 percent.

Then go through the same mental scenario, but substitute more appropriate "cool thoughts" and fantasize that you feel *relaxed* and unperturbed; imagine that you handle the situation tactfully, assertively, and effectively. For example, you might tell yourself, "The waiters don't seem to be noticing me. Perhaps they're busy and overlooked the fact that I haven't gotten a menu yet. No point in getting hot under the collar about this."

Then instruct yourself to approach the headwaiter and explain the situation assertively, following these principles: Point out tactfully that you've been waiting; if he explains they are busy, disarm him by *agreeing* with him; compliment him on the good business they are doing; and repeat your request for better service in a firm but friendly way. Finally, imagine that he responds by sending a waiter who apologizes and gives you top-notch VIP service. You feel good and enjoy the meal.

Now practice going through this version of the scenario each night until you have mastered it and can fantasize handling the situation effectively and calmly in this manner. This cognitive rehearsal will enable you to program yourself to respond in a more assertive and relaxed way when the actual situation confronts you again.

You might have one objection to this procedure: You may feel it is unrealistic to fantasize a positive outcome in the restaurant since there is no guarantee the staff will in reality respond in a friendly way and give you what you want. The answer to this objection is simple. There's no guarantee they'll respond abrasively either, but if you *expect* a negative response, you'll enhance the probability of getting one because your anger will have an enormous capacity to act as a self-fulfilling prophecy. In contrast, if you expect and fantasize a positive outcome and apply an upbeat approach, it will be much more likely to occur.

You can, of course, also prepare for a negative outcome in a similar way, using the cognitive rehearsal method. Imagine you *do* approach the waiter, and he acts snotty and superior and gives you poor service. Now record your hot thoughts, then substitute cool thoughts and develop a new coping strategy as you did before.

You can continue to work your way up your hierarchy

list in this way until you have learned to think, feel, and act more peacefully and effectively in the majority of the provocative situations you encounter. Your approach to these situations will have to be flexible, and different coping techniques may be required for the different types of provocations listed. Empathy might be the answer in one situation, verbal assertiveness could be the key to another, and changing your expectations might be the most useful approach to a third.

It will be crucial not to evaluate your progress in your anger-reduction program in an all-or-nothing way because emotional growth takes some time, especially when it comes to anger. If you ordinarily react to a particular provocation with 99 percent anger and then find you become 70 percent upset next time, you could view this as a successful first try. Now keep working at it, using your cognitive rehearsal method, and see if you can reduce it to 50 percent and then to 30 percent. Eventually you will make it vanish altogether, or at least you will have brought it down to an acceptable, irreducible minimum.

Remember that the wisdom of friends and associates can be a potential gold mine you can utilize when you're stuck. They may see clearly in any area where you have a blind spot. Ask them how *they* think and behave in a particular situation that makes you feel frustrated, helpless, and enraged. What would they tell themselves? What would they actually do? You can learn a surprising amount rapidly if you are willing to ask.

Ten Things You Should Know About Your Anger

1. The events of this world don't make you angry. Your "hot thoughts" create your anger. Even when a genuinely negative event occurs, it is the meaning you attach to it that determines your emotional response.

   The idea that you are responsible for your anger is

ultimately to your advantage because it gives you the opportunity to achieve control and make a free choice about how you want to feel. If it weren't for this, you would be helpless to control your emotions; they would be irreversibly bound up with every external event of this world, most of which are ultimately out of your control.

2. Most of the time your anger will not help you. It will immobilize you, and you will become frozen in your hostility to no productive purpose. You will feel better if you place your emphasis on the active search for creative solutions. What can you do to correct the difficulty or at least reduce the chance that you'll get burned in the same way in the future? This attitude will eliminate to a certain extent the helplessness and frustration that eat you up when you feel you can't deal with a situation effectively.

   If no solution is possible because the provocation is totally beyond your control, you will only make yourself miserable with your resentment, so why not get rid of it? It's difficult if not impossible to feel anger and joy simultaneously. If you think your angry feelings are especially precious and important, then think about one of the happiest moments of your life. Now ask yourself, How many minutes of that period of peace or jubilation would I be willing to trade in for feeling frustration and irritation instead?
3. The thoughts that generate anger more often than not will contain distortions. Correcting these distortions will reduce your anger.
4. Ultimately your anger is caused by your belief that someone is acting unfairly or some event is unjust. The intensity of the anger will increase in proportion to the severity of the maliciousness perceived and if the act is seen as intentional.
5. If you learn to see the world through other people's eyes, you will often be surprised to realize their actions

are *not* unfair from their point of view. The unfairness in these cases turns out to be an illusion that exists *only in your mind*! If you are willing to let go of the unrealistic notion that your concepts of truth, justice, and fairness are shared by everyone, much of your resentment and frustration will vanish.

6. Other people usually do not feel they deserve your punishment. Therefore, your retaliation is unlikely to help you achieve any positive goals in your interactions with them. Your rage will often just cause further deterioration and polarization, and will function as a self-fulfilling prophecy. Even if you temporarily get what you want, any short-term gains from such hostile manipulation will often be more than counterbalanced by a long-term resentment and retaliation from the people you are coercing. No one likes to be controlled or forced. This is why a positive reward system works better.
7. A great deal of your anger involves your defense against loss of self-esteem when people criticize you, disagree with you, or fail to behave as you want them to. Such anger is *always* inappropriate because only your own negative distorted thoughts can cause you to lose self-esteem. When you blame the other guy for your feelings of worthlessness, you are always fooling yourself.
8. Frustration results from unmet expectations. Since the event that disappointed you was a part of "reality," it was "realistic." Thus, your frustration always results from your *unrealistic* expectation. You have the right to try to influence reality to bring it more in line with your expectations, but this is not always practical, especially when these expectations represent ideals that don't correspond to everyone else's concept of human nature. The simplest solution would be to

change your expectations. For example, some unrealistic expectations that lead to frustration include:

a. If I want something (love, happiness, a promotion, etc.), I deserve it.
b. If I work hard at something, I *should* be successful.
c. Other people *should* try to measure up to my standards and believe in my concept of "fairness."
d. I *should* be able to solve any problems quickly and easily.
e. If I'm a good wife, my husband is *bound* to love me.
f. People *should* think and act the way I do.
g. If I'm nice to someone, they *should* reciprocate.

9. It is just childish pouting to insist you have the *right* to be angry. Of course you do! Anger is legally permitted in the United States. The crucial issue is—is it to your advantage to feel angry? Will you or the world really benefit from your rage?
10. You rarely need your anger in order to be human. It is not true that you will be an unfeeling robot without it. In fact, when you rid yourself of that sour irritability, you will feel greater zest, joy, peace, and productivity. You will experience liberation and enlightenment.

Chapter 8

Ways of Defeating Guilt

No book on depression would be complete without a chapter on guilt. What is the function of guilt? Writers, spiritual leaders, psychologists, and philosophers have grappled forever with this question. What is the basis of guilt? Does it evolve from the concept of "original sin"? Or from Oedipal incestuous fantasies and the other taboos that Freud postulated? Is it a realistic and helpful component of human experience? Or is it a "useless emotion" that mankind would be better off without, as suggested by some recent pop psychology writers?

When the mathematics of calculus was developed, scientists found they could readily solve complex problems of motion and acceleration that were extremely difficult to handle using older methods. The cognitive theory has similarly provided us with a kind of "emotional calculus" that makes certain thorny philosophical and psychological questions much easier to resolve.

Let's see what we can learn from a cognitive approach. Guilt is the emotion you will experience when you have the following thoughts:

1. I have done something I shouldn't have (or I have failed to do something that I should have) because my

actions fall short of my moral standards and violate my concept of fairness.

2. This "bad behavior" shows that I am a bad person (or that I have an evil streak, or a tainted character, or a rotten core, etc.).

This concept of the "badness" of self is central to guilt. In its absence, your hurtful action might lead to a healthy feeling of remorse but not guilt. Remorse stems from the *un*distorted awareness that you have willfully and unnecessarily acted in a hurtful manner toward yourself or another person that violates your personal ethical standards. Remorse differs from guilt because there is no implication your transgression indicates you are inherently bad, evil, or immoral. To put it in a nutshell, remorse or regret are aimed at behavior, whereas guilt is targeted toward the "self."

If in addition to your guilt you feel depression, shame, or anxiety, you are probably making one of the following assumptions:

1. Because of my "bad behavior," I am inferior or worthless (this interpretation leads to depression).
2. If others found out what I did, they would look down on me (this cognition leads to shame).
3. I'm in danger of retaliation or punishment (this thought provokes anxiety).

The simplest way to assess whether the feelings created by such thoughts are useful or destructive is to determine if they contain any of the ten cognitive distortions described in Chapter 3. To the extent that these thinking errors are present, your guilt, anxiety, depression, or shame certainly cannot be valid or realistic. I suspect you will find that a great many of your negative feelings are in fact based on such thinking errors.

The first potential distortion when you are feeling guilty

is your assumption you have done something wrong. This may or may not actually be the case. Is the behavior you condemn in yourself in reality so terrible, immoral, or wrong? Or are you *magnifying* things out of proportion? A charming medical technologist recently brought me a sealed envelope containing a piece of paper on which she had written something about herself which was so terrible she couldn't bear to say it out loud. As she trembling handed the envelope to me, she made me promise not to read it out loud or laugh at her. The message inside was—"I pick my nose and eat it!" The apprehension and horror on her face in contrast to the triviality of what she had written struck me as so funny I lost all professional composure and burst into laughter. Fortunately, she too broke into a belly laugh and expressed a sense of relief.

Am I claiming that you *never* behave badly? No. That position would be extreme and unrealistic. I am simply insisting that to the extent your perception of goofing up is unrealistically magnified, your anguish and self-persecution are inappropriate and unnecessary.

A second key distortion that leads to guilt is when you *label* yourself a "bad person" because of what you did. This is actually the kind of superstitious destructive thinking that led to the medieval witch hunts! You may have engaged in a bad, angry, hurtful action, but it is counterproductive to label yourself a "bad" or "rotten" person because your energy gets channeled into rumination and self-persecution instead of creative problem-solving strategies.

Another common guilt-provoking distortion is *personalization*. You inappropriately assume responsibility for an event you did not cause. Suppose you offer a constructive criticism to your boyfriend, who reacts in a defensive and hurt manner. You may blame yourself for his emotional upset and arbitrarily conclude that your comment was inappropriate. In fact, his negative *thoughts* upset him, not your comment. Furthermore, these thoughts are probably distorted. He might be thinking that your criticism means he's no good and conclude that you don't respect him.

Now—did *you* put that illogical thought into his head? Obviously not. *He* did it, so you can't assume responsibility for his reaction.

Because cognitive therapy asserts that only your thoughts create your feelings, you might come to the nihilistic belief that you cannot hurt anybody no matter what you do, and hence you have license to do *anything*. After all, why not run out on your family, cheat on your wife, and screw your partner financially? If they're upset, it's their problem because it's their thoughts, right?

Wrong! Here we come again to the importance of the concept of cognitive distortion. To the extent that a person's emotional upset is caused by his distorted thoughts, then you can say he is responsible for his suffering. If you blame yourself for that individual's pain, it is a personalization error. In contrast, if a person's suffering is caused by valid, undistorted thoughts, then the suffering is real and may in fact have an external cause. For example, you might kick me in the stomach, and I could have the thoughts, "I've been kicked! It hurts! — — —!" In this case the responsibility for my pain rests with *you*, and your perception that you have hurt me is not distorted in any way. Your remorse and my discomfort are real and valid.

Inappropriate "should" statements represent the "final common pathway" to your guilt. Irrational should statements imply you are expected to be perfect, all-knowing, or all-powerful. Perfectionistic shoulds include rules for living that defeat you by creating impossible expectations and rigidity. One example of this would be, "I *should* be happy at all times." The consequence of this rule is that you will feel like a failure every time you are upset. Since it is obviously unrealistic for any human being to achieve the goal of perpetual happiness, the rule is self-defeating and irresponsible.

A should statement that is based on the premise you are all-knowing assumes you have all the knowledge in the universe and that you can predict the future with absolute certainty. For example, you might think, "I shouldn't have gone to the beach this weekend because I was coming down with the flu. What a jerk I am! Now I'm so sick I'll be in bed for a week." Berating yourself this way is unrealistic because you didn't know for certain that going to the beach would make you so ill. If you *had* known this, you would have acted differently. Being human, you made a decision, and your hunch turned out to be wrong.

Should statements based on the premise you are all-powerful assume that, like God, you are omnipotent and have the ability to control yourself and other people so as to achieve each and every goal. You miss your tennis serve and wince, exclaiming, "I *shouldn't* have missed that serve!" Why shouldn't you? Is your tennis so superb that you can't possibly miss a serve?

It is clear that these three categories of should statements create an inappropriate sense of guilt because they do not represent sensible moral standards.

In addition to distortion, several other criteria can be helpful in distinguishing abnormal guilt from a healthy sense of remorse or regret. These include the *intensity*, *duration*, and *consequences* of your negative emotion. Let's use these criteria to evaluate the incapacitating guilt of a married fifty-two-year-old grammar-school teacher named Janice. Janice had been severely depressed for many years. Her problem was that she continually obsessed about two episodes of shoplifting that had occurred when she was fifteen. Although she had led a scrupulously honest life since that time, she could not shake the memory of those two incidents. Guilt-provoking thoughts constantly plagued her: "I'm a thief. I'm a liar. I'm no good. I'm a fake." The agony of her guilt was so enormous that every night she prayed that God would let her die in her sleep. Every morning when she woke up still alive, she was bitterly disappointed and told herself, "I'm such a bad person even God doesn't want me." In frustration she finally loaded her husband's pistol, aimed it at her heart, and pulled the trigger. The gun misfired and did not go off. She had not cocked it properly. She felt the ultimate defeat: She couldn't even kill herself! She put the gun down and wept in despair.

Janice's guilt is inappropriate not only because of the obvious distortions, but also because of the *intensity, duration*, and *consequences* of what she was feeling and telling herself. What she feels cannot be described as a healthy remorse or regret about the actual shoplifting, but an irresponsible degradation of her self-esteem that blinds her to living in the here and now, and is far out of proportion to any actual transgression. The consequences of her guilt created the ultimate irony—her belief that she was a bad person caused her to attempt to murder herself, a most destructive and pointless act.

The Guilt Cycle

Even if your guilt is unhealthy and based on distortion, once you begin to feel guilty, you may become trapped in an illusion that makes the guilt appear valid. Such illusions can be powerful and convincing. You reason:

1. I feel guilty and worthy of condemnation. This means I've been bad.
2. Since I'm bad, I deserve to suffer.

Thus, your guilt convinces you of your badness and leads to further guilt. This cognitive-emotional connection locks your thoughts and feelings into each other. You end up trapped in a circular system which I call the "guilt cycle."

Emotional reasoning fuels this cycle. You automatically assume that because you're feeling guilty, you *must* have fallen short in some way and that you deserve to suffer. You reason, "I *feel* bad, therefore I must *be* bad." This is irrational because your self-loathing does not necessarily prove that you did anything wrong. Your guilt just reflects the fact that you *believe* you behaved badly. This *might* be the case, but it often is not. For example, children are frequently punished inappropriately when parents are feeling tired and irritable and misinterpret their behavior. Under these conditions, the poor child's guilt obviously does not prove he or she did anything wrong.

Your self-punishing behavior patterns intensify the guilt cycle. Your guilt-provoking thoughts lead to unproductive actions that reinforce your belief in your badness. For example, a guilt-prone neurologist was trying to prepare for her medical-board certification examination. She had difficulty studying for the test, and felt guilty about the fact that she wasn't studying. So she wasted time each night watching television while the following thoughts raced through her mind: "I *shouldn't* be watching TV. I *should* be preparing for my boards. I'm lazy. I don't deserve to be a doctor. I'm too self-centered. I ought to be punished." These thoughts made her feel intensely guilty. She then reasoned, "This guilt proves what a lazy no-good person I am." Thus, her self-punishing thoughts and her guilty feelings reinforced each other.

Like many guilt-prone people, she had the idea that if she punished herself enough she would eventually get moving. Unfortunately, quite the opposite was true. Her guilt simply drained her energy and reinforced her belief that she was lazy and inadequate. The only actions that resulted from her self-loathing were the nightly compulsive trips to the refrigerator to "pig out" on ice cream or peanut butter.

The vicious cycle that she trapped herself in is shown in Figure 8–1. Her negative thoughts, feelings, and behaviors all interacted in the creation of the self-defeating, cruel illusion that she was "bad" and uncontrollable.

The Irresponsibility of Guilt. If you have actually done something inappropriate or hurtful, does it follow that you deserve to suffer? If you feel the answer to this question is yes, then ask yourself, "How long must I suffer? One day? A year? For the rest of my life?" What sentence will you

Figure 8–1. A neurologist's self-critical thoughts caused her to feel so guilty that she had difficulty preparing for her certification examination. Her procrastination strengthened her conviction that she was bad and deserved punishment. This further undermined her motivation to solve the problem.

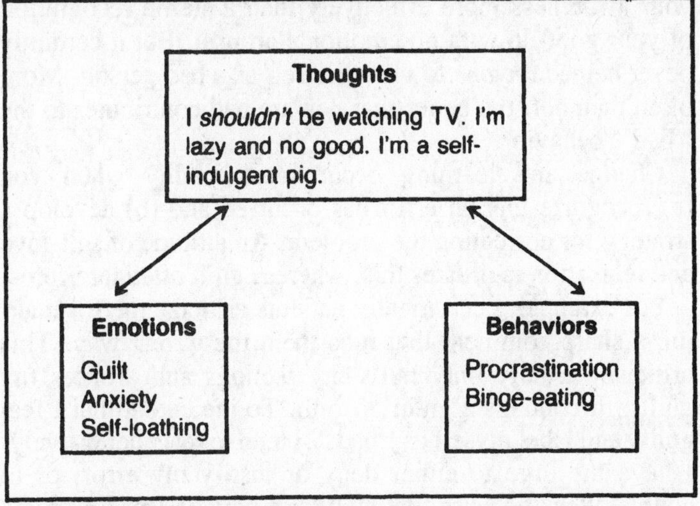
Thoughts
I *shouldn't* be watching TV. I'm lazy and no good. I'm a self-indulgent pig.
Emotions
Guilt
Anxiety
Self-loathing
Behaviors
Procrastination
Binge-eating

choose to impose on yourself? Are you willing to stop suffering and making yourself miserable when your sentence has expired? This would at least be a *responsible* way to punish yourself because it would be time-limited. But what is the point of abusing yourself with guilt in the first place? If you did make a mistake and act in a hurtful way, your guilt won't reverse your blunder in some magical manner. It won't speed your learning processes so as to reduce the chance you'll make the same mistake in the future. Other people won't love and respect you more because you are feeling guilty and putting yourself down in this manner. Nor will your guilt lead to productive living. So what's the point?

Many people ask, "But how could I behave morally and control my impulses if I don't feel guilt?" This is the probation-officer approach to living. Apparently you view yourself as so willful and uncontrollable that you must constantly castigate yourself in order to keep from going wild. Certainly, if your behavior has a needlessly hurtful impact on others, a small amount of painful remorse will add to your awareness more effectively than a sterile recognition of your goof-up with no emotional arousal. But it certainly never helped *anyone* to view himself as a bad person. More often than not, the belief that you are bad contributes to the "bad" behavior.

Change and learning occur most readily when you (a) recognize that an error has occurred and (b) develop a strategy for correcting the problem. An attitude of self-love and relaxation facilitates this, whereas guilt often interferes.

For example, occasionally patients criticize me for making a sharp comment that rubs them the wrong way. This criticism usually only hurts my feelings and arouses my guilt if it contains a grain of truth. To the extent that I feel guilty and label myself as "bad," I tend to react defensively. I have the urge to either deny or justify my error, or to counterattack because that feeling of being a "bad person" is so odious. This makes it much more difficult for me to admit and correct the error. If, in contrast, I do not harangue myself or experience any loss of self-respect, it is easy to admit my mistake. Then I can readily correct the problem and learn from it. The less guilt I have, the more effectively I can do this.

Thus, what is called for when you do goof up is a process of recognition, learning, and change. Does guilt help you with any of these? I don't believe it does. Rather than facilitating your recognition of your error, guilt engages you in a coverup operation. You want to close your ears to any criticism. You can't bear to be in the wrong because it feels so terrible. This is why guilt is counterproductive.

You may protest, "How can I know I've done something wrong if I don't feel guilty? Wouldn't I just indulge in a blind rampage of uncontrolled, destructive selfishness if it weren't for my guilt?"

Anything is possible, but I honestly doubt this would happen. You can replace your guilt with a more enlightened basis for moral behavior—empathy. Empathy is the ability to visualize the consequences, good and bad, of your behavior. Empathy is the capacity to conceptualize the impact of what you do on yourself and on the other person, and to feel appropriate and genuine sorrow and regret without labeling yourself as inherently bad. Empathy gives you the necessary mental and emotional climate to guide your behavior in a moral and self-enhancing manner in the absence of the whip of guilt.

Using these criteria, you can now readily determine whether your feelings represent a normal and healthy sense of remorse or a self-defeating, distorted sense of guilt. Ask yourself:

1. Did I consciously and willfully do something "bad," "unfair," or needlessly hurtful that I shouldn't have? Or am I irrationally expecting myself to be perfect, all-knowing, or all-powerful?
2. Am I labeling myself a *bad* or *tainted person* because of this action? Do my thoughts contain other cognitive distortions, such as magnification, overgeneralization, etc.?
3. Am I feeling a realistic regret or remorse, which results from an empathic awareness of the negative impact of my action? Are the intensity and duration of my painful emotional response appropriate to what I actually did?
4. Am I learning from my error and developing a strategy for change, or am I moping and ruminating nonproductively or even punishing myself in a destructive manner?

Now, let's review some methods that will allow you to rid yourself of inappropriate guilty feelings and maximize your self-respect.

1. Daily Record of Dysfunctional Thoughts. In earlier chapters you were introduced to a Daily Record of Dysfunctional Thoughts for overcoming low self-esteem and inadequacy. This method works handsomely for a variety of unwanted emotions, including guilt. Record the activating event that leads to your guilt in the column labeled "Situation." You may write, "I spoke sharply to an associate," or "Instead of contributing ten dollars, I threw my alumni fund-raising appeal in the wastebasket." Then "tune in" to that tyrannical loudspeaker in your head and identify the accusations that create your guilt. Finally, identify the distortions and write down more objective thoughts. This leads to relief.

An example of this is demonstrated in Figure 8–2. Shirley was a high-strung young woman who decided to move to New York to pursue her acting career. After she and her mother had spent a long and tiring day looking for apartments, they took a train back to Philadelphia. After boarding, they discovered they had mistakenly taken a train without food service or a lounge car. Shirley's mother began to complain about the lack of cocktail service, and Shirley felt flooded with guilt and self-criticism. As she recorded and talked back to her guilt-provoking thoughts, she felt substantial relief. She told me that by overcoming her guilt, she avoided the temper tantrum she would normally have thrown in such a frustrating situation (see Figure 8–2, page 209).

2. Should Removal Techniques. Here are some methods for reducing all those irrational "should" statements you've been hitting yourself with. The first is to ask yourself, "Who says I should? Where is it written that I should?" The point of this is to make you aware that you are being critical of yourself unnecessarily. Since you are ultimately making your own rules, once you decide that a rule is not useful you can revise it or get rid of it. Suppose you are telling yourself that you should be able to make your spouse happy all the time. If your experience teaches you that this is

Figure 8–2.

Situation	*Emotions*	*Guilt-provoking Thoughts*	*Cognitive Distortions*	*Rational Responses*	*Outcome*
My mother is very tired and due to her lack of understanding of the train schedule, we take a train without comforts.	Extreme guilt; frustration, anger; self-pity	1. Gee. Mom walked all over New York with me today, and now she can't even get a drink *because I really didn't explain the schedule properly*. I should have explained that "no food" did not mean snacks.	1. Personalization; mental filter; should statement.	1. I feel bad for Mom—but the train ride is only 1½ hours. I thought I explained everything. I guess we all make mistakes sometimes.	Substantial relief
		2. Now I feel terrible—I'm so selfish.	2. Emotional reasoning	2 I am more upset than Mom. What's done is done—don't cry over spilt milk.	
		3. Why do I always foul up everything?	3. Overgeneralization, personalization.	3. I don't foul up everything It's not my fault she misunderstood.	
		4. She's so good to me, and I'm a louse.	4. Labeling; all-or-nothing thinking.	4. One incident does not make a louse.	

neither realistic nor helpful, you can rewrite the rule to make it more valid. You might say, "I can make my spouse happy some of the time, but I certainly can't at all times. Ultimately, happiness is up to him or her. And I'm not perfect any more than he or she is. Therefore, I will not anticipate that what I do will always be appreciated."

In deciding about the usefulness of a particular rule, it can be helpful to ask yourself, "What are the advantages and disadvantages of having that rule for myself?" "How will it help me believe I *should always* be able to make my spouse happy, and what will the price be for believing this?" You can assess the costs and benefits, using the double-column method shown in Figure 8–3.

Another simple but effective way to rid yourself of should statements involves substituting other words for "should," using the double-column technique. The terms "It would be nice if" or "I wish I could" work well, and often sound more realistic and less upsetting. For example, instead of

Figure 8–3. The advantages and disadvantages of believing "I should be able to make my wife happy all the time."

Advantages	*Disadvantages*
1. When she is happy, I will feel I'm doing what I'm supposed to.	1. When she's unhappy, I'll feel guilty and I'll blame myself.
2. I'll work very hard to be a good husband.	2. She'll be able to manipulate me with my guilt. Anytime she wants her way she can act unhappy, and then I'll feel so bad I'll have to back down.
	3. Since she is unhappy a good bit of the time, I'll often feel like a failure. Since her unhappiness often has nothing to do with me, this will be a waste of energy.
	4. I'll end up feeling resentful that I'm paradoxically giving *her* so much power over *my* moods!

saying, "I *should* be able to make my wife happy," you could substitute "It *would be nice if* I could make my wife happy now because she seems upset. I can ask what she's upset about and see if there might be a way I could help." Or instead of "I *shouldn't* have eaten the ice cream," you can say, "It would have been better if I hadn't eaten the ice cream, but it's not the end of the world that I did."

Another anti-should method involves showing yourself that a should statement doesn't fit reality. For example, when you say, "I shouldn't have done X," you assume (1) it is a fact that you *shouldn't have*, and (2) it is going to help you to say this. The "reality method" reveals—to your surprise—that the truth is usually just the opposite: (a) In point of fact, you *should have* done what you did; and (b) it is going to hurt you to say you *shouldn't have*.

Incredulous? Let me demonstrate. Assume you've been trying to diet and you ate some ice cream. So you have the thought, "I *shouldn't* have eaten this ice cream." In our dialogue I want you to argue that it's *really true* that you *shouldn't have* eaten the ice cream, and I will try to put the lie to your arguments. The following is modeled after an actual conversation, which I hope you find as delightful and helpful as I did:

DAVID: I understand you're on a diet, and you ate some ice cream. I believe you *should have* eaten the ice cream.

YOU: Oh, no. That's impossible. I *shouldn't have* eaten it because I'm on a diet. You see, I'm trying to lose weight.

DAVID: Well, I believe you *should have* eaten the ice cream.

YOU: Burns, are you dense? I *shouldn't have* because I'm trying to lose weight. That's what I'm trying to tell you. How can I lose weight if I'm eating ice cream?

DAVID: But in point of fact you did eat it.

YOU: Yeah. That's the problem. I *shouldn't have* done that. Now do you see the light?

DAVID: And apparently you're claiming that "things should have been different" than they were. But things were the way they were. And things usually are the way they are for a good reason. Why do you think you did what you did? What's the reason you ate the ice cream?

YOU: Well, I was upset and I was nervous and I'm basically a pig.

DAVID: Okay, you were upset and you were nervous. Have you had a pattern in your life of eating when you've been upset and nervous?

YOU: Yeah. Right. I've never had any self-control.

DAVID: So, wouldn't it be natural to expect then that last week when you were nervous you would do what you have habitually done?

YOU: Yeah.

DAVID: So, wouldn't it be sensible therefore to conclude that you *should* have done that because you had a very long-standing habit of doing it?

YOU: I feel like you're telling me that I *should* just keep eating ice cream and end up like a fat pig or something.

DAVID: Most of my clients aren't as difficult as you! At any rate, I'm not telling you to act like a pig, and I'm not recommending you continue this bad habit of eating when you're upset. What I'm saying is that you're giving yourself two problems for the price of one. One is that you did in fact break your diet. If you're going to lose weight, this will slow you down. And the second problem is that you're being hard on yourself about having done that. The second headache you don't need.

YOU: So you're saying that because I have a habit of eating when I'm nervous it's predictable that until I learn some methods for changing the habit, I'll continue to do it.

DAVID: I wish I'd said it that well myself!

YOU: Therefore, I *should have* eaten the ice cream because I haven't changed the habit yet. As long as the habit continues, I *will* and *should* keep overeating when I'm nervous. I see what you mean. I feel a whole lot better, Doctor, except for one thing. How can I learn to stop doing this? How can I develop some strategies for modifying my behavior in a more productive way?

DAVID: You can motivate yourself with a whip or a carrot. When you tell yourself, "I *should* do this" or "I *shouldn't* do that" all day long, you get bogged down with a shouldy approach to life. And you already know what you end up with—emotional constipation. If you'd rather get things moving instead, I suggest you try to motivate yourself through rewards rather than punishment. You might find that these work more effectively.

In my case I used the "Dots and doughnuts" diet. Mason Dots (a gum candy) and glazed doughnuts are two of my favorite sweets. I found that the most difficult time to control my eating was in the evening when I was studying or watching TV. I'd have an urge to eat ice cream. So, I told myself that if I controlled this urge, I could reward myself with a big, fresh, glazed doughnut in the morning and a box of Mason Dots in the evening. Then I'd concentrate on how good they'd taste, and this helped me forget the ice cream. Incidentally, I also had the rule that if I *did* goof up and eat the ice cream, I could *still* have the Dots and the doughnut as a reward for trying or as a commiseration for slipping back. Either way it helped me, and I lost over fifty pounds this way.

I also made up the following syllogism:

(A) Human beings on diets goof up from time to time.
(B) I'm a human being.
(C) Therefore, I *should* goof up from time to time.

This helped me greatly too, and it enabled me to binge on weekends and feel good about it. I usually lost more during the week than I gained on weekends; so, overall I lost weight and enjoyed myself. Every time I goofed up in my diet I didn't allow myself to criticize the lapse or feel guilty. I began to think about it as the "Binge-on-whatever-you want-whenever-you-want-to-without-guilt-and-enjoy-it diet," and it was so much fun it was a mild disappointment when I finally achieved my aimed-for weight. I actually lost over ten more pounds at that point because the diet was so enjoyable. I believe that the proper *attitude* and *feelings* are the key. With them you can move mountains—even mountains of flesh.

The major thing that holds you back when you're trying to change a bad habit like eating, smoking, or drinking too much is your belief you are out of control. The cause of this lack of control is those should statements. They defeat you. Suppose, for example, you are trying to avoid eating ice cream. There you are watching TV, saying, "Oh, I really *should* study and I *shouldn't* eat any ice cream." Now ask yourself, "How do I feel when I say these things to myself?" I think you know the answer: You feel guilty and nervous. Then what do you do? You go and eat! That is the point. The reason you're eating is that you're telling yourself you shouldn't! Then you try to bury your guilt and anxiety under more piles of food.

Another simple should removal technique involves your wrist counter. Once you become convinced that the shoulds are not to your advantage, you can count them. Every time you make a should statement, click the counter. If you do this, be sure to set up a reward system based on the daily total. The more shoulds you spot this way, the greater the reward you deserve. Over a period of several weeks, your daily total of should statements will begin to go down, and you'll notice you're feeling less guilty.

Another should removal technique zeroes in on the fact that you don't really trust yourself. You may believe that without all these should statements you would just turn wild and go on a rampage of destruction or murder, or even ice-cream eating. A way to evaluate this is to ask yourself if there was any period in your life when you were particularly happy and felt reasonably fulfilled, productive, and under control. Think it over for a moment before you read on, and make sure you have a mental picture of this time. Now ask yourself, "During that period in my life, was I whipping myself with a lot of should statements?" I believe your answer will be no. Now tell me—were you doing all these wild, terrible things then? I think you'll realize you were "should-free" and under control. This is proof that you can lead a productive, happy life without all those shoulds.

You can test this hypothesis with an experiment in the next couple of weeks. Try reducing your should statements using these various techniques, and then see what happens to your mood and self-control. I think you'll be pleased.

Another method that you can fall back on is the obsessional-filibuster technique described in Chapter 4. Schedule two minutes three times a day to recite all your should statements and self-persecutions out loud: "I *should* have gone to the market before it closed," and "I *shouldn't* have picked my nose at the country club," and "I'm such a rotten bum," etc. Just rattle off all the most abusive self-criticisms you can think of. It might be especially helpful to write them down or dictate them into a tape recorder. Then read them later out loud, or listen to the tape. I think this will help you see how ludicrous these statements are. Try to limit your shoulds to these scheduled periods so you won't be bothered by them at other times.

Another technique to combat should statements involves getting in touch with the limits of your knowledge. When

I was growing up, I often heard people say, "Learn to accept your limits and you'll become a happier person," but no one ever bothered to explain what this meant or how to go about doing it. Furthermore, it always sounded like a bit of a put-down, as if they were saying, "Learn what a second-rate dud you actually are."

In reality, it's not as bad as all that. Suppose you frequently look into the past and mope about your mistakes. For example, as you review the financial section of the paper, you tell yourself, "I shouldn't have bought that stock. It's gone down two points." As a way out of this trap, ask yourself, "Now, at the time I bought the stock, did I know it was going to go down in value?" I suspect you'll say no. Now ask, "If I'd known it was going down, would I have bought it?" Again you'll answer no. So what you're really saying is that if you'd known this at the time, you'd have acted differently. To do this you would have to be able to predict the future with absolute certainty. Can you predict the future with absolute certainty? Again your answer must be no. You have two options: You can either decide to accept yourself as an imperfect human being with limited knowledge and realize that you will at times make mistakes, or you can hate yourself for it.

Another effective way to combat shoulds is to ask, "Why should I?" Then you can challenge the evidence you come up with so as to expose the faulty logic. In this way you can reduce your should statement to the level of absurdity. Suppose, for example, you hire someone to do some work for you. It could be lawn work, or a painting job, or anything. When he submits his bill, it seems higher than you understood it would be, but he gives you some fast talk, so you give in and end up paying his price. You feel taken advantage of. You begin to berate yourself for not acting more firmly. Let's do some role-playing, and you can pretend that you're the poor sucker who paid too much.

YOU: Yesterday I *should* have told that guy that his bill was too high.

DAVID: You should have told him that he gave you a lower estimate?

YOU: Yeah. I *should have* been more assertive.

DAVID: Why *should you* have? I agree that it would have been to your advantage to speak up for yourself. You can work on developing your assertive skills so that in the future you'll do better in situations like that. But the point is: Why *should you* have been more effective yesterday?

YOU: Well, because I'm always letting people take advantage of me.

DAVID: Okay, let's think about your line of reasoning. "Because I'm always letting people take advantage of me, I *should have* been more assertive yesterday." Now—what is the rational response to this? Is there anything about your statement that seems a little bit illogical? Is there anything fishy about your reasoning?

YOU: Mmmm . . . let me think. Well, in the first place, it's not exactly true that I'm *always* letting people take advantage of me. That would be an overgeneralization. I sometimes do get my way. In fact, I can be quite demanding at times. Furthermore, if it *were* true that I was *always* getting taken advantage of in certain situations, then it would follow that I *should have* behaved exactly as I did since this is my habit. Until I've mastered some new ways to deal with people, I'll probably continue to have this problem.

DAVID: Great. I couldn't have put it better. I see you've been absorbing what I've been telling you about should statements! I hope *all* my readers are as smart and attentive as you are! Are there any other reasons you think you *should have* behaved differently?

YOU: Uh, well, let me see. How about: I *should have* been more assertive because I wouldn't have had to pay more than I owed?

DAVID: Okay. Now what's the rational response to that? What is illogical about that argument?

YOU: Well, since I'm human I won't always do the right thing.

DAVID: Exactly. In fact, the following syllogism may help you. First premise: All human beings make mistakes, like sometimes paying too much. Do you agree with me so far?

YOU: Yes.

DAVID: And what are you?

YOU: A human being.

DAVID: And what follows?

YOU: I should make mistakes.

DAVID: Right.

That should be enough should removal techniques for you. Oops! I just did it myself! Let me say—it would be nice if you found those methods helpful. I think you'll find that by reducing this mental tyranny, you'll feel better because you won't be berating yourself. Instead of feeling guilty, you can use your energy to make necessary changes and enhance your self-control and productivity.

3. Learn to Stick to Your Guns. One of the big disadvantages of being guilt-prone is that others can and will use this guilt to manipulate you. If you feel obligated to please everyone, your family and friends will be able to coerce you effectively into doing many things that may not be in your best self-interest. To cite a trivial example, how many social invitations have you halfheartedly accepted so as not to hurt someone's feelings? In this case the price you pay for saying yes when you really would have preferred to say no is not great. You only end up wasting one evening. And there is a payoff. You will avoid feeling guilty, and you can fantasize that you are an especially nice person. Furthermore, if you try to decline the invitation, the disappointed host may say, "But we are *expecting* you. Do you mean you are going to let the old gang down? Aw, come on." And *then* what would you say? How would you feel?

Your obsession with pleasing others becomes more tragic when your decisions become so dominated by guilt that you end up trapped and miserable. The irony is that, more often than not, the consequences of letting someone manipulate you with guilt end up being destructive not only to you but to the other person. Although your guilt-motivated actions are often based on your idealism, the inevitable effects of giving in turn out to be quite the opposite.

For example, Margaret was a happily married twenty-seven-year-old woman whose obese brother, a gambler, tended to take advantage of her in a variety of ways. He borrowed money when he ran short and often forgot to repay it. When he was in town (often for several months at a time) he assumed it was his right to eat dinner with her family every night, to drink up the liquor, and to use her new car whenever he wanted. She rationalized giving in to his demands by saying: "If I asked him for a favor or needed his help, he'd do the same for me. After all, a loving brother and sister *should* help each other out. And besides, if I tried to say no to him he'd explode and I might lose him. Then I'd feel like *I* did something wrong."

At the same time, she was able to see the negative consequences of continually giving in: (1) She was supporting his dependent, self-defeating life-style and gambling addiction; (2) She felt trapped and taken advantage of; (3) The basis of the relationship was not love but blackmail—she was constantly having to say yes to his demands to avoid the tyranny of his temper and her own sense of guilt.

Margaret and I did some role-playing so she could learn to say no and stick to her guns in a tactful but firm manner. I played Margaret's role, and she pretended to be her brother:

BROTHER (played by Margaret): Are you using the car tonight?

MARGARET (played by me): I'm not planning to now.

BROTHER: Do you mind if I borrow it later?

MARGARET: I'd prefer that you don't.

BROTHER: Why not? You're not going to use it. It'll just be sitting there.

MARGARET: Do you feel I'm obliged to loan it to you?

BROTHER: Well, I'd do the same for you if I had a car and you needed it.

MARGARET: I'm glad you feel that way. Although I'm not planning to use the car, I'd like to have it available in case I decide to go somewhere later on.

BROTHER: But you're not planning to use it! Haven't we been brought up to *help* each other?

MARGARET: Yes we have. Do you think that means I always have to say yes to you? We both do a great deal for each other. You have made a lot of use of my car and from now on I'd feel more comfortable if you'd begin to arrange your own transportation.

BROTHER: I'm just planning to use it for an hour, so I'll get it back in case you need it. It's very important and it's only a half mile away, so I won't wear your car out, don't worry.

MARGARET: It sounds like it is something important to you. Perhaps you can arrange some other transportation. Could you walk that distance?

BROTHER: Oh, that's fine! If that's how you feel, don't come to *me* for any favors!

MARGARET: It sounds like you're pretty mad because I'm not doing what you want. Do you feel I'm always obliged to say yes?

BROTHER: You and your philosophy! Shove it! I refuse to listen to any more of this hogwash! (Begins to storm off).

MARGARET: Let's not talk about it any further then. Maybe in a couple of days you'll feel more like talking about it. I think we do need to talk things over.

After this dialogue we reversed roles so that Margaret could practice being more assertive. When I played her brother's role, I gave her as tough a time as I could, and she learned how to handle me. This practice boosted her courage. She felt it was helpful to keep certain principles in mind when standing up to her brother's manipulations. These were: (1) She could remind him it was her right not to say yes to all his demands. (2) She could find a grain of truth in his arguments (the disarming technique) so as to take the wind out of his sails, but she could then come back to her position that love did not mean always giving in. (3) She was to adopt a strong, decisive and uncompromising position as tactfully as possible. (4) She was not to buy into his role as a weak, inadequate little boy who couldn't stand on his own feet. (5) She was not to respond to his anger by getting angry herself, because this would reinforce his belief he was a victim who was being unjustly deprived by a cruel, selfish witch. (6) She had to risk the possibility he would temporarily withdraw and thwart her by refusing to talk to her or to consider her point of view. When he did this, she was to let him storm off but she could let him know there were some things she wanted to talk over with him later on when he was more in the mood to communicate.

When Margaret did confront him she found he was not nearly as tough a customer as she imagined. He actually seemed relieved and began to act more adult when she put some limits on the relationship.

If you choose to apply this technique, you will have to be determined to stick to your guns because the other guy (or gal) may try to bluff you into believing that you're mortally wounding them by not giving in to their requests. Remember that the hurt you inflict in the long run by not following your best self-interest is usually far greater.

Practicing ahead of time is the key to success. A friend will usually be happy to role-play with you and provide some useful feedback. If such a person is not available to you, or you feel too shy to ask, write out an imaginary dialogue of the type illustrated. This will go a long way to firing up the appropriate circuits in your brain so you'll have the necessary courage and skill to say no diplomatically but forcefully and make it stick when the time actually comes!

4. Antiwhiner Technique. This is one of the most surprising, delightfully effective methods in this book. It works like a charm in situations where someone—usually a loved one—makes you feel frustrated, guilty, and helpless through whining, complaining, and nagging. The typical pattern works like this: The whiner complains to you about something or someone. You feel the sincere desire to be helpful, so you make a suggestion. The person immediately squashes your suggestion and complains again. You feel tense and inadequate, so you try harder and make another suggestion. You get the same response. Anytime you try to break loose from the conversation, the other person implies he or she is being abandoned, and you are flooded with guilt.

Shiba lived with her mother while she completed graduate school. Shiba loved her mother, but found her constant harangues about her divorce, the lack of money, etc., so intolerable she sought treatment. I taught her the antiwhiner method the first session, as follows: Regardless of what her mother said, Shiba was to find some way to *agree* (the disarming technique), and then instead of offering advice, she was to say something genuinely complimentary. Shiba initially found this approach astonishing and rather bizarre because it differed radically from her usual approach. In the following dialogue, I asked Shiba to play the role of Mother while I played her role so I could demonstrate this technique:

SHIBA (as her mother): Do you know that during the divorce proceedings it came out that your dad sold his share in the business, and I was the last person to know about it?

DAVID (as Shiba): That's absolutely correct. You didn't hear about it until the divorce proceedings. You really deserve better.

SHIBA: I don't know what we're going to do for money. How am I going to put your brothers through college?

DAVID: That *is* a problem. We are short on money.

SHIBA: It was just like your father to pull something like this. His head isn't screwed on straight.

DAVID: He never was too good at budgeting. You've always been much better at that.

SHIBA: He's a louse! Here we are on the verge of poverty. What if I get sick? We'll end up in the poorhouse!

DAVID: You're right! It's no fun *at all* to live in the poorhouse. I agree with you completely.

Shiba reported that in her role as Mother she found it was "no fun" to complain because I kept agreeing with her. We did a role-reversal so she could master the technique.

In fact, it is your urge to *help* complainers that maintains the monotonous interaction. Paradoxically, when you agree with their pessimistic whining, they quickly run out of steam. Perhaps an explanation will make this seem less puzzling. When people whine and complain, they are usually feeling irritated, overwhelmed, and insecure. When you try to *help* them, this sounds to them like criticism because it implies they aren't handling things properly. In contrast, when you agree with them and add a compliment, they feel *endorsed*, and they then usually relax and quiet down.

5. Moorey Moaner Method. A useful modification of this technique was proposed by Stirling Moorey, a brilliant

British medical student who studied with our group in Philadelphia and sat in with me during therapy sessions during the summer of 1979. He worked with a chronically severely depressed fifty-two-year-old sculptor named Harriet with a heart of gold. Harriet's problem was her friends would often bend her ear with gossip and personal problems. She found these problems upsetting because of her excessive capacity for empathy. Because she wouldn't know how to help her friends, she felt trapped and resentful until she learned the "Moorey Moaner Method." Stirling simply instructed her to find a way to agree with what the person was saying, and then to distract the moaner by finding something positive in the complaint and commenting on it. Here are several examples:

1. MOANER: Oh, what in the world can I ever do about my daughter? I'm afraid she's been smoking pot again.

   RESPONSE: There sure is a lot of pot going around these days. Is your daughter still doing that outstanding art work? I heard she recently got an important award.

2. MOANER: My boss didn't give me my raise, and my last raise was nearly a year ago. I've been here for twenty years, and I think I deserve better.

   RESPONSE: You certainly do have seniority here and you've made tremendous contributions. Tell me, what was it like when you first started working twenty years ago? I'll bet things were a lot different then.

3. MOANER: My husband never seems to have enough time at home. Every night he's out with that darned bowling league.

   RESPONSE: Weren't you also doing some bowling recently? I heard you got some pretty high scores yourself!

Harriet mastered the Moorey Moaner Method quickly and reported a dramatic change in her mood and outlook because it gave her a simple, effective way to handle a problem that had been very real and overwhelming. When she returned for the next session, her depression—which had crippled her for over a decade—had lifted and was entirely gone. She was bubbling and joyous, and heaped well-deserved praise on Stirling's head. If you have a similar problem with your mother, mother-in-law, or friends, try Stirling's method. Like Harriet, you'll soon be smiling!

6. Developing Perspective. One of the commonest distortions that leads to a sense of guilt is personalization—the misguided notion that you are ultimately responsible for other people's feelings and actions or for naturally occurring events. An obvious example would be your sense of guilt when it rained unexpectedly on the day of a large picnic you had organized to honor the retiring president of your club. In this case you could probably shake your absurd reaction off without a great deal of effort because you clearly cannot control the weather.

Guilt becomes much more difficult to overcome when someone suffers substantial pain and discomfort and insists it results from their personal interaction with you. In such cases it can be helpful to clarify the extent to which you can realistically assume responsibility. Where does your responsibility end and the other person's begin? The technical name for this is "disattribution," but you might call it putting things into perspective.

Here's how it works. Jed was a mildly depressed college student whose twin brother, Ted, was so seriously depressed he dropped out of school and began to live like a recluse with his parents. Jed felt guilty about his brother's depression. Why? Jed told me he had always been more outgoing and hardworking than his brother. Consequently, from early childhood he always made better grades and had more friends than Ted. Jed reasoned that the social and academic success he enjoyed caused his brother to feel inferior and left out. Consequently, Jed concluded that he was the cause of Ted's depression.

He then carried this line of reasoning to its illogical extreme and hypothesized that by feeling depressed himself, he might help Ted stop feeling depressed and inferior through some type of reverse (or perverse) psychology. When he went home for the holidays, Jed avoided the usual social activities, minimized his academic success, and emphasized how blue he was feeling. Jed made sure he gave his brother the loud and clear message that he too was down and out.

Jed took his plan so seriously that he was quite hesitant to apply the mood-control techniques I was trying to teach him. In fact, he was downright *resistant* at first because he felt guilty about getting better and feared his recovery might have a devastating impact on Ted.

Like most personalization errors, Jed's painful illusion that he was at fault for his brother's depression contained enough half-truths to sound persuasive. After all, his brother probably had felt inferior and inadequate since early childhood and undoubtedly did harbor some jealous resentment of Jed's success and happiness. But the crucial questions were: Did it follow that Jed *caused* his brother's depression, and could Jed effectively reverse the situation by making himself miserable?

In order to help him assess his role in a more objective way, I suggested Jed use the triple-column technique (Figure 8–4). As a result of the exercise, he was able to see that his guilty thoughts were self-defeating and illogical. He reasoned that Ted's depression and sense of inferiority were ultimately caused by Ted's distorted thinking and not by his own happiness or success. For Jed to try to correct this by making himself miserable was as illogical as trying to put out a fire with gasoline. As Jed grasped this, his guilt and depression rapidly lifted, and he was soon back to normal functioning.

Figure 8–4.

Automatic Thoughts	*Cognitive Distortion*	*Rational Responses*
1. I am part of the cause for Ted's depression due to our relationship since early childhood. I have always worked harder and been more successful.	1. Jumping to conclusions (mind reading); personalization	1. I myself am not the cause of Ted's depression. It is Ted's illogical thoughts and attitudes that are causing his depression. The only responsibility that I can take is that of being part of the environment that Ted is interpreting in a negative, distorted manner.
2. I feel it would upset Ted if I told him I was having a good time at school while he is home alone doing nothing.	2. Jumping to conclusions (fortune teller error)	2. It might cheer Ted up and give him some hope if he knows I'm feeling better and having a good time. It probably only depresses Ted more if I act as miserable as he does because this takes away his hope.
3. If Ted is sitting around doing nothing, it is my responsibility to correct the situation.	3. Personalization	3. I can encourage him to do things, but I cannot force him. Ultimately this is his responsibility.

Figure 8–4. cont.

Automatic Thoughts	*Cognitive Distortion*	*Rational Responses*
4. I will be doing something for him by not doing anything for myself. In fact, it will help him if I am depressed.	4. Jumping to conclusions (mind reading)	4. My actions are totally independent of his actions. There is no reason to think that my depression will be helpful to him. He has even told me he doesn't want me to be dragged down. If he sees that I am improving, this might actually encourage him. I can possibly be a good role model for him by showing him that I can be happy. I can't eliminate his sense of inadequacy by botching up my life.

Part III

"Realistic" Depressions

Chapter 9

Sadness Is Not Depression

"Dr. Burns, you seem to be claiming that distorted thinking is the only cause of depression. But what if my problems are real?" This is one of the most frequent questions I encounter during lectures and workshops on cognitive therapy. Many patients raise it at the start of treatment, and list a number of "realistic" problems which they are convinced cause "realistic depressions." The most common are:

bankruptcy or poverty;
old age (some people also view infancy, childhood, adolescence, young adulthood and mid-life as periods of inevitable crisis);
permanent physical disability;
terminal illness;
the tragic loss of a loved one.

I'm sure you could add to the list. However, none of the above can lead to a "realistic depression." There is, in fact, no such thing! The real question here is how to draw the line between desirable and undesirable negative feelings. What is the difference between "healthy sadness" and depression?

The distinction is simple. Sadness is a normal emotion created by realistic perceptions that describe a negative event involving loss or disappointment in an undistorted way

Depression is an illness that *always* results from thoughts that are distorted in some way. For example, when a loved one dies, you validly think, "I lost him (or her), and I will miss the companionship and love we shared." The feelings such a thought creates are tender, realistic, and desirable. Your emotions will enhance your humanity and add depth to the meaning of life. In this way you *gain* from your loss.

In contrast, you might tell yourself, "I'll never again be happy because he (or she) died. It's unfair!" These thoughts will trigger in you feelings of self-pity and hopelessness. Because these emotions are based entirely on distortion, they will defeat you.

Either depression or sadness can develop after a loss or a failure in your efforts to reach a goal of great personal importance. Sadness comes, however, without distortion. It involves a flow of feeling and therefore has a time limit. It never involves a lessening of your self-esteem. Depression is frozen—it tends to persist or recur indefinitely, and always involves loss of self-esteem.

When a depression clearly appears after an obvious stress, such as ill health, the death of a loved one, or a business reversal, it is sometimes called a "reactive depression." At times it can be more difficult to identify the stressful event that triggered the episode. Those depressions are often called "endogenous" because the symptoms seem to be generated entirely out of thin air. In both cases, however, the cause of the depression is identical—your distorted, negative thoughts. It has no adaptive or positive function whatsoever, and represents one of the worst forms of suffering. Its only redeeming value is the growth you experience when you recover from it.

My point is this: When a genuinely negative event occurs, your emotions will be created exclusively by your thoughts and perceptions. Your feelings will result from the meaning you attach to what happens. A substantial portion of your suffering will be due to the *distortions* in your thoughts. When you eliminate these distortions, you will find that coping with the "real problem" will become less painful.

Let's see how this works. One clearly realistic problem involves serious illness, such as a malignancy. It is unfortunate that the family and friends of the afflicted person are often so convinced that it is normal for the patient to feel depressed, they fail to inquire about the cause of the depression, which more often than not turns out to be completely reversible. In fact, some of the *easiest* depressions to resolve are those found in people facing probable death. Do you know why? These courageous individuals are often "supercopers" who haven't made misery their life-style. They are usually willing to help themselves in any way they can. This attitude rarely fails to transform apparently irreversible and "real" difficulties into opportunities for personal growth. This is why I find the concept of "realistic depressions" so personally abhorrent. The attitude that depression is necessary strikes me as destructive, inhuman, and victimizing. Let's get down to some specifics, and you can judge for yourself.

Loss of Life. Naomi was in her mid-forties when she received a report from her doctor that a "spot" had appeared on her chest X ray. She was a firm believer that going to doctors was a way of asking for trouble, so she procrastinated many months in checking this report out. When she did, her worst suspicions were validated. A painful needle biopsy confirmed the presence of malignant cells, and subsequent lung removal indicated that a spread of the cancer had already occurred.

This news hit Naomi and her family like a hand grenade. As the months wore on, she became increasingly despondent over her weakened state. Why? It was not so much the physical discomfort from the disease process or the chemotherapy, although these were genuinely uncomfortable, but the fact that she was sufficiently weak that she had to give up the daily activities that had meant a great deal to her sense of identity and pride. She could no longer work around the house (now her husband had to do most of the chores), and she had to give up her two part-time jobs, one of which was volunteer reading for the blind.

You might insist, "Naomi's problems are *real.* Her misery is not caused by distortion. It's caused by the situation."

But was her depression so inevitable? I asked Naomi why her lack of activity was so upsetting. I explained the concept of "automatic thoughts," and she wrote down the following negative cognitions: (1) I'm not contributing to society; (2) I'm not accomplishing in my own personal realm; (3) I'm not able to participate in *active* fun; and (4) I am a drain and drag on my husband. The emotions associated with these thoughts were: anger, sadness, frustration, and guilt.

When I saw what she had written down, my heart leaped for joy! These thoughts were no different from the thoughts of physically healthy depressed patients I see every day in my practice. Naomi's depression was *not* caused by her malignancy, but the malignant *attitude* that caused her to measure her sense of worth by the amount she produced! Because she had always equated her personal worth with her achievements, the cancer meant—"You're over the hill! You're ready for the refuse heap!" This gave me a way to intervene!

I suggested that she make a graph of her personal "worth" from the moment of birth to the moment of death (see Figure 9-1, page 235). She saw her worth as a constant, estimating it at 85 percent on an imaginary scale from 0 to 100 percent. I also asked her to estimate her *productivity* over the same period on a similar scale. She drew a curve with low productivity in infancy, increasing to a maximum plateau in adulthood, and finally decreasing again later in life (see Figure 9-1). So far, so good. Then two things suddenly dawned on her. First, while her illness had reduced her productivity, she still contributed to herself and her family in numerous small but nevertheless important and precious ways. Only all-or-nothing thinking could make her think her contributions were a zero. Second, and much more important, she realized her personal worth was constant and steady; it was a *given* that was unrelated to her achieve-

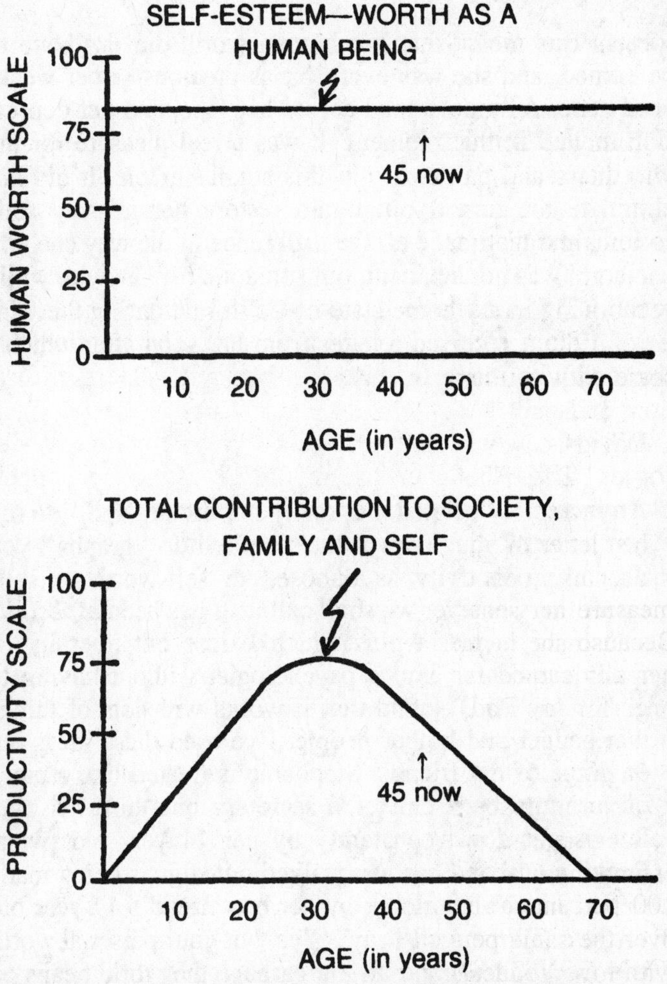

Figure 9–1. Naomi's worth and work graphs. In the upper figure Naomi plotted her human "worth" from the time of her birth to the time of her death. She estimated this at 85 percent. In the lower figure she plotted her estimated productivity and achievement over the course of her life. Her productivity began low in childhood, reached a plateau in adulthood, and would ultimately fall to zero at the time of death. This graph helped her comprehend that her "worth" and "achievement" were unrelated and had no correlation with each other.

ments. This meant that her human worth did *not* have to be earned, and she was every bit as precious in her weakened state. A smile spread across her face, and her depression melted in that moment. It was a real pleasure for me to witness and participate in this small miracle. It did *not* eliminate the tumor, but it did restore her missing self-esteem, and that made all the difference in the way she *felt*.

Naomi was not a patient, but someone I spoke with while vacationing in my home state of California during the winter of 1976. I received a letter from her soon after which I share with you here:

David—

An incredibly belated, but really important "P.S." to my last letter to you. To wit: the simple little "graphs" you did of productivity as opposed to self-worth or self-esteem or whatever we shall call it: It has been *especially* sustaining to me, a plus which I dose out liberally! It really turned me into a psychologist without having to go for my Ph.D. I find that it works with lots of things that badger and bother people. I've tried these ideas out on some of my friends. Stephanie is treated like a piece of furniture by a chit of a secretary one-third her age; Sue is put down constantly by her 14 year old twins; Becky's husband has just walked out; Ilga is being made to feel like an interloper by her boy friend's 17 year old son, etc. To them all I say "Yes, but your personal worth is a *CONSTANT*, and all the garbage the world heaps on you doesn't touch it!" Of course in many cases I realize it's an over-simplification and cannot be an anodyne for all things, but boy is it helpful and useful!

Again, thank you, sir!

As ever,
Naomi

She died in pain but with dignity six months later.

Loss of Limb. Physical handicaps represent a second category of problems felt to be "realistic." The afflicted individual—or the family members—automatically assume that the limitations imposed by old age or by a physical disability, such as an amputation or blindness, necessarily imply a decreased capacity for happiness. Friends tend to offer understanding and sympathy, thinking this represents a humane and "realistic" response. The case can be quite the opposite, however. The emotional suffering may be caused by twisted thinking rather than by a twisted body. In such a situation, a sympathetic response can have the undesirable effect of reinforcing self-pity as well as feeding into the attitude that the handicapped individual is doomed to less joy and satisfaction than others. In contrast, when the afflicted individual or family members learn to correct the distortions in their thinking, a full and gratifying emotional life can frequently result.

For example, Fran is a thirty-five-year-old married mother of two, who began to experience symptoms of depression around the time her husband's right leg became irreversibly paralyzed because of a spinal injury. For six years she sought relief from her intensifying sense of despair, and received a variety of treatments in and out of hospitals, including antidepressant drugs as well as electroshock therapy. Nothing helped. She was in a severe depression when she came to me, and she felt her problems were insoluble.

In tears she described the frustration she experienced in trying to cope with her husband's decreased mobility:

> Every time I see other couples doing things we can't do tears come to my eyes. I look at couples taking walks, jumping in the swimming pool or the ocean, riding bikes together, and it just hurts. Things like that would be pretty tough for me and John to do. They take it for granted just like we used to. Now it would be so good and wonderful if we could do it. But you know, and I know, and John knows—we can't.

At first, I too had the feeling Fran's problem was realistic. After all, they *couldn't* do many things that most of us can do. And the same could be said of old people, as well as those who are blind or deaf or who have had a limb amputated.

In fact, when you think of it, we *all* have limitations. So perhaps we should all be miserable . . . ? As I puzzled over this, Fran's distortion suddenly came to my mind. Do you know what it is? Look at the list on page 42 right now and see if you can pick it out . . . that's right, the distortion that led to Fran's needless misery was the mental filter. Fran was picking out and dwelling on each and every activity that was unavailable to her. At the same time the many things she and John *could* or might do together did not enter her conscious mind. No wonder she felt life was empty and dreary.

The solution turned out to be surprisingly simple. I proposed the following to Fran: "Suppose at home between sessions you were to make a list of all the things that you and John *can* do together. Rather than focus on things that you and John *can't* do, learn to focus on the ones you *can* do. I, for example, would love to go to the moon, but I don't happen to be an astronaut, so it's not likely I'll ever get the opportunity. Now, if I focused on the fact that in my profession and at my age it is extremely unlikely I could ever get to the moon, I could make myself very upset. On the other hand, there are many things I *can* do, and if I focus on these, then I won't feel disappointed. Now, what would be some things you and John *can* do as a couple?"

FRAN: Well, we enjoy each other's company still. We go out to dinner, and we're buddies.

DAVID: Okay. What else?

FRAN: We go for rides together, we play cards. Movies, Bingo. He's teaching me how to drive . . .

DAVID: You see, in less than thirty seconds you've already listed six things you can do together. Suppose I gave you between now and next session to continue the list. How many items do you think you could come up with?

FRAN: Quite a lot of them. I could come up with things we've never thought of, maybe something unusual like skydiving.

DAVID: Right. You might even come up with some more adventurous ideas. Keep in mind that you and John might in fact be able to do many things you are assuming you can't do. For example, you told me you can't go to the beach. You mentioned how much you'd like to go swimming. Could you go to a beach that's a little more secluded so you wouldn't have to feel quite so self-conscious? If I were on a beach and you and John were there, his physical disability wouldn't make one darn bit of difference to me. In fact, I recently visited a fine beach on the North Shore of Lake Tahoe in California with my wife and her family. As we were swimming, we suddenly happened upon a cove that had a nude beach, and here were all these young people with no clothes on. Of course, I didn't actually *look* at any of them, I want you to understand! But in spite of this I did happen to notice that one young man had his right leg missing from the knee down, and he was there having fun with the rest of them. So I'm not absolutely convinced that just because someone is crippled or missing a limb they can't go to the beach and have fun. What do you think?

Some people might scoff at the idea that such a "difficult and real" problem could be so easily resolved, or that an intractable depression like Fran's could turn around in response to such a simple intervention. She did in fact report a complete disappearance of her uncomfortable feelings and said she felt the best she had in years at the end of the session. In order to maintain such improvement, she will obviously need to make a consistent effort to change her thinking patterns over a period of time so she can overcome her bad habit of spinning an intricate mental web and getting trapped in it.

Loss of Job. Most people find the threat of a career reversal or the loss of livelihood a potentially incapacitating emotional blow because of the widespread assumption in Western culture that individual worth and one's capacity for happiness are directly linked with professional success. Given this value system, it seems obvious and realistic to anticipate that emotional depression would be inevitably linked with financial loss, career failure, or bankruptcy.

If this is how you feel, I think you would be interested in knowing Hal. Hal is a personable forty-five-year-old father of three, who worked for seventeen years with his wife's father in a successful merchandising firm. Three years before he was referred to me for treatment, Hal and his father-in-law had a series of disputes about the management of the firm. Hal resigned in a moment of anger, thus giving up his interests in the company. For the next three years, he bounced around from job to job, but had difficulty finding satisfactory employment. He didn't seem to be able to succeed at anything and began to view himself as a failure. His wife started working full time to make ends meet, and this added to Hal's sense of humiliation because he had always prided himself on being the breadwinner. As the months and years rolled on, his financial situation worsened, and he experienced increasing depression as his self-esteem bottomed out.

When I first met Hal, he had been attempting to work for three months as a trainee in commercial real-estate sales. He had rented several buildings, but had not yet finalized a sale. Because he was working on a strict commission basis, his income during this break-in period was quite low. He was plagued by depression and procrastination. He would at times stay at home in bed all day, thinking to himself, "What's the use? I'm just a loser. There's no point in going to work. It's less painful to stay in bed."

Hal volunteered to permit the psychiatric residents in our training program at the University of Pennsylvania to observe one of our psychotherapy sessions through a one-way mirror. During this session, Hal described a conversation in the locker room of his club. A well-to-do friend had informed Hal of his interest in the purchase of a particular building. You might think he would have jumped for joy on learning this, since the commission from such a sale would have given his career, confidence, and bank account a much needed boost. Instead of pursuing the contact, Hal procrastinated several weeks. Why? Because of his thought, "It's too complicated to sell a commercial property. I've never done this before. Anyway, he'll probably back out at the last minute. That would mean I couldn't make it in this business. It would mean I was a failure."

Afterward, I reviewed the session with the residents. I wanted to know what they thought about Hal's pessimistic, self-defeating attitudes. They felt that Hal did in fact have a good aptitude for sales work, and that he was being unrealistically hard on himself. I used this as ammunition during the next session. Hal admitted that he was more critical of himself than he would ever be toward anyone else. For example, if one of his associates lost a big sale, he'd simply say, "It's not the end of the world; keep plugging." But if it happened to him he'd say, "I'm a loser." Essentially, Hal admitted he was operating on a "double standard"—tolerant and supportive toward other people but harsh, critical, and punitive toward himself. You may have the same tendency. Hal initially defended his double standard by arguing it would be helpful to him:

HAL: Well, first of all, the responsibility and interest that I have in the other person is not the same as the responsibility that I have for myself.

DAVID: Okay. Tell me more.

HAL: If they don't succeed, it's not going to be bread off my table, or create any negative feelings within my family unit. So the only reason I'm interested in them is because it's nice to have everybody succeed, but there . . .

DAVID: Wait—wait—wait! You're interested in them because it's nice to have them succeed?

HAL: Yeah. I said . . .

DAVID: The standard you apply to them is one that you think would help them succeed?

HAL: Right.

DAVID: And is the standard you apply to yourself the one that will help you succeed? How do you feel when you say, "One missed sale means I'm a failure"?

HAL: Discouraged.

DAVID: Is this helpful?

HAL: Well, it hasn't produced positive results, so apparently it's not helpful.

DAVID: And is it *realistic* to say "One missed sale and I'm a failure"?

HAL: Not really.

DAVID: So why are you using this all-or-nothing standard on yourself? Why would you apply helpful and realistic standards to these other people who you don't care so much about and self-defeating, hurtful standards to yourself who you do care something about?

Hal was beginning to grasp that it wasn't helping him to live by a double standard. He judged himself by harsh rules that he would never apply to anyone else. He initially defended this tendency—as many demanding perfectionists will—by claiming it would *help* him in some way to be so much harder on himself than on others. However, he then quickly owned up to the fact that his personal standards were actually unrealistic and self-defeating because if he did try to sell the building and didn't succeed, he would view it as a catastrophe. His bad habit of all-or-nothing thinking was the key to the fear that paralyzed him and kept him from trying. Consequently, he spent most of his time in bed, moping.

Hal asked for some specific guidelines concerning things he might do to rid himself of his perfectionistic double standards so that he could judge all individuals, including himself, by *one* objective set of standards. I proposed that as a first step, Hal might use the automatic-thought, rational-response technique. For example, if he were sitting at home procrastinating about work, he might be thinking, "If I don't go to work early and stay all day and get caught up on all my work, there's no point in even trying. I might as well lie in bed." After writing this down, he would substitute a rational response, "This is just all-or-nothing thinking, and it's baloney. Even going to work for a half day could be an important step and might make me feel better."

Hal agreed to write down a number of upsetting thoughts before the next therapy session at those times he felt worthless and down on himself. (See Figure 9–2, page 244.) Two days later he received a layoff notice from his employer, and he came to the next session highly convinced his self-critical thoughts were absolutely valid and realistic. He'd been unable to come up with a single rational response. The notice implied that his failure to show up at work necessitated his release from his job. During the session, we discussed how he could learn to talk back to his critical voice.

DAVID: Okay, now let's see if we can write down some answers to your negative thoughts in the Rational Response column. Can you think of any answer based on what we talked about last session? Con-

Figure 9–2. Hal's homework for recording and challenging his self-critical thoughts. He wrote down the Rational Responses during the therapy session (see text).

Negative Thoughts (SELF-CRITICISM)	*Rational Responses* (SELF-DEFENSE)
1. I am lazy.	1. I have worked hard much of my life.
2. I enjoy being ill.	2. It's not fun.
3. I am inadequate. I am a failure.	3. I've had some degree of success. We've had a good home. We've reared three outstanding children. People admire and respect me. I have involved myself in community activity.
4. This lying around doing nothing represents the real me.	4. I am experiencing symptoms of an illness. It's not the "real me."
5. I could have done more.	5. At least I did more than most people. It's meaningless and pointless to say, "I could have done more" because anyone could say this.

sider your statement "I am inadequate." Would this in any way result from your all-or-nothing thinking and perfectionistic standards?

The answer might be clearer to you if we do a role-reversal. It's sometimes easier to speak objectively about someone else. Suppose I came to you with your story and told you that I was employed by my wife's father. Three years ago we had a fight. I felt I was being taken advantage of. I walked out. I've kinda been feeling blue ever since that time, and I've been tossing around from job to job. Now I've been fired from a job that was purely on a commission basis, and that's really a double defeat for me. In the first place, they didn't pay me anything, and then in the second place, they didn't even figure I was worth that much, so they fired me. I've concluded that I'm inadequate—an inadequate human being. What would you say to me?

HAL: Well, I . . . assuming that you'd gotten up to that point, say the first forty years or more of your life, you obviously *were* doing something.

DAVID: Okay, write that down in the Rational Response column. Make a list of all the good, adequate things you did for the first forty years of your life. You've earned money, you've raised children who were successful, etc., etc.

HAL: Okay. I can write down that I've had some success. We've had a good home. We've reared three outstanding children. People admire and respect me, and I have involved myself in community activities.

DAVID: Okay, now those are all the things you've done. How do you reconcile this with your belief that you are inadequate?

HAL: Well, I could have done more.

DAVID: Great! I was certain you'd figure out a clever way to disqualify your good points. Now write that down as another negative thought: "I could have done more." Beautiful!

HAL: Okay, I've written it down as number five.

DAVID: Okay, now what's the answer to that one? (long silence)

DAVID: What is it? What's the distortion in that thought?

HAL: You're a tricky bugger!

DAVID: What is the answer?

HAL: At least I did more than most people.

DAVID: Right, and what percent do you believe that?

HAL: That I believe one hundred percent.

DAVID: Great! Put it down in the Rational Response column. Now, let's go back to this "I could have done more." Suppose you were Howard Hughes sitting up in his tower, with all those millions and billions. What could you say to yourself to make yourself unhappy?

HAL: Well, I'm trying to think.

DAVID: Just read what you wrote down on the paper.

HAL: Oh. "I could have done more."

DAVID: You can always say that, can't you?

HAL: Yeah.

DAVID: And that's why a lot of people who have won fame and fortune are unhappy. It's just an example of perfectionistic standards. You can go on and on and on, and no matter how much achievement you experience, you can always say, "I could have done more." This is an arbitrary way of punishing yourself. Do you agree or not?

HAL: Well, yeah. I can see that. It takes more than one element really to be happy. Because if it was money, then every millionaire and billionaire would be euphoric. But there are more circumstances that involve being happy or satisfied with yourself than making money. That's not the drive that paralyzes me. I've never had a drive to go after money.

DAVID: What were your drives? Did you have a drive to raise a family?

HAL: That was very important to me. Very important. And I participated in the rearing of the children.

DAVID: And what would you do in raising your children?

HAL: Well, I would work with them, teach them, play with them.

DAVID: And how did they come out?

HAL: I think they're great!

DAVID: Now, you were writing down, "I'm inadequate. I'm a failure." How can you reconcile this with the fact that your aim was to raise three children and you did it?

HAL: Again, I guess I wasn't taking that into account.

DAVID: So how can you call yourself a failure?

HAL: I have not functioned as a wage earner . . . as an effective money-maker for several years.

DAVID: Is it realistic to call yourself a "failure" based on that? Here's a man who has had a depression for three years, and he finds it difficult to go to work, and now it's realistic to call him a failure? People with depressions are failures?

HAL: Well, if I knew more of what caused depression, I would be better able to make a value judgment.

DAVID: Well, we're not going to know the ultimate cause of depression for some time yet. But our understanding is that the immediate cause of depression is punitive, hurtful statements that you hit yourself with. Why this happens more to some people than others we don't know. The biochemical and genetic influences have not yet been worked out. Your upbringing undoubtedly contributed, and we can deal with that in another session if you like.

HAL: Since there is no final proof yet of the ultimate cause of depression, can't we think of that in terms of a failure in itself? I mean, we don't know where it's coming from . . . It must be something wrong with me that caused it . . . some way that I have failed myself that causes the depression.

DAVID: What evidence do you have for that?

HAL: I don't. It's just a possibility.

DAVID: Okay. But to make an assumption as punishing as that . . . *anything* is a possibility. But there is no *evidence* for that. When patients get over depressions, then they become just as productive as they ever were. Seems to me that if their problem was that they were failures, when they got over the depression they would still be failures. I've had college professors and corporate presidents who have come to me. They were just sitting and staring at the wall, but it was because of their depression. When they got over the depression, they started giving conferences and managing their businesses like before. So how can you possibly say that depression is due to the fact that they are failures? Seems to me that it's more the other way around—that the failure is due to the depression.

HAL: I can't answer that.

DAVID: It's *arbitrary* to say that you're a failure. You have had a depression, and people with depression don't do as much as when they are undepressed.

HAL: Then I'm a successful depressive.

DAVID: Right! Right! And part of being a successful depressive means to get better. So I hope that's what we're doing now. Imagine that you had pneumonia for the past six months. You wouldn't have earned any dough. You could also say, "This makes me a failure." Would that be realistic?

HAL: I don't see how I could claim that. Because I certainly wouldn't have willfully created the pneumonia.

DAVID: Okay, can you apply the same logic to your depression?

HAL: Yeah, I can see it. I don't honestly feel that my depression was willfully induced either.

DAVID: Of course it wasn't. Did you *want* to bring this on?

HAL: Oh boy, no!

DAVID: Did you consciously *do* anything to bring it on?

HAL: Not that I know of.

DAVID: And if we knew what was causing depression, then we could put the finger someplace. Since we don't know, isn't it silly to blame Hal for his own depression? What we do know is that depressed people get this negative view of themselves. And they feel and behave in accord with this negative vision of everything. You didn't bring that on purposely or *choose* to be incapacitated. And when you get over that vision and when you have switched back to a nondepressed way of looking at things, you are going to be just as productive or more so than you've ever been, if you're typical of other patients that I've worked with. You see what I mean?

HAL: Yeah, I *can* see.

It was a relief for Hal to realize that although he had been financially unsuccessful for several years, it was nonsensical to label himself as "a failure." This negative self-image and his sense of paralysis resulted from his all-or-nothing thinking. His sense of worthlessness was based on his tendency to focus only on the negatives in his life (the mental filter) and to overlook the many areas where he had experienced success (discounting the positive). He was able to see that he was aggravating himself unnecessarily by saying, "I could have done more," and he realized that financial value is not the same as human worth. Finally, Hal was able to admit that the *symptoms* he was experiencing—lethargy and procrastination—were simply manifestations of a temporary disease process and not indications of his "true self." It was absurd for him to think his depression was just punishment for some personal inadequacy, any more than pneumonia would be.

At the end of the session, the Beck Depression Inventory test indicated that Hal had experienced a 50 percent improvement. In the weeks that followed, he continued to help himself, using the double-column technique. As he trained himself to talk back to his upsetting thoughts, he was able to reduce the distortions in his harsh way of eva'uating himself, and his mood continued to improve.

Hal left the real-estate business and opened a paperback bookstore. He was able to break even; but in spite of considerable personal effort, he was unable to show enough profit to justify continuing beyond the first year's trial period. Thus, the marks of external success had not changed appreciably during this time. In spite of this, Hal managed to avoid significant depression and maintained his self-esteem. The day he decided to "throw in the towel" on the bookstore, he was still below the zero point financially, but his self-respect did *not* suffer. He wrote the following brief essay which he decided to read each morning while he was looking for a new job:

Why Am I Not Worthless?

As long as I have something to contribute to the well-being of myself and others, I am not worthless.

As long as what I do can have a positive effect, I am not worthless.

As long as my being alive makes a difference to even one person, I am not worthless (and this one person can be me if necessary).

If giving love, understanding, companionship, encouragement, sociability, counsel, solace means anything, I am not worthless.

If I can respect my opinions, my intelligence, I am not worthless. If others also respect me, that is a bonus.

If I have self-respect and dignity, I am not worthless.

If helping to contribute to the livelihood of my employees' families is a plus, I am not worthless.

If I do my best to help my customers and vendors through my productivity and creativity, I am not worthless.

If my presence in this milieu does makes a difference to others, I am not worthless.

I am not worthless. I am eminently worthwhile!

Loss of a Loved One. One of the most severely depressed patients I treated early in my career was Kay, a thirty-one-year-old pediatrician whose younger brother had committed suicide in a grisly way outside her apartment six weeks earlier. What was particularly painful for Kay was that she held herself responsible for his suicide, and the arguments she proposed in support of this point of view were quite convincing. Kay felt she was confronted by an excruciating problem that was entirely realistic and insoluble. She felt that she too deserved to die and was actively suicidal at the time of referral.

A frequent problem that plagues the family and friends of an individual who successfully commits suicide is the sense of guilt. There is a tendency to torture yourself with such thoughts as, "Why didn't I prevent this? Why was I so stupid?" Even psychotherapists and counselors are not immune to such reactions and may castigate themselves: "It's really my fault. If *only* I had talked to him differently in that last session. Why didn't I pin him down on whether or not he was suicidal? I should have intervened more forcefully. I murdered him!" What adds to the tragedy and irony is that in the vast majority of instances, the suicide occurs because of the victim's distorted belief that he has some insoluble problem which, viewed from a more objective perspective, would seem much less overwhelming and certainly not worth suicide.

Kay's self-criticism was all the more intense because she felt that she had gotten a better break in life than her brother, and so she had gone out of her way to try to compensate for this by providing emotional and financial support for him during his long bout with depression. She arranged for his psychotherapy, helped pay for it, and even got him an apartment near hers so that he could call her whenever he was very down.

Her brother was a physiology student in Philadephia. On the day of his suicide, he called Kay to ask about the effects of carbon monoxide on the blood for a talk he was to give in class. Because Kay is a blood specialist, she thought the question was innocent and gave him the information without thinking. She didn't talk to him very long because she was preparing a major lecture to deliver the following morning at the hospital where she worked. He used her information to make his fourth and final attempt outside her apartment window while she was preparing her lecture. Kay held herself responsible for his death.

She was understandably miserable, given the tragic situation she confronted. During the first few therapy sessions she outlined why she blamed herself and why she was convinced that she would be better off dead: "I had assumed the responsibility for my brother's life. I failed, so I feel I am responsible for his death. It proves that I did not adequately support him as I should have. I should have known that he was in an acute situation, and I failed to intervene. In retrospect, it's obvious that he was getting suicidal again. He'd had three prior serious suicide attempts. If I had just asked him when he called me, I could have saved his life.

I was angry with him on many occasions during the month before he died, and in all honesty he could be a burden and a frustration at times. At one time I remember feeling annoyed and saying to myself that perhaps he *would* be better off dead. I feel terrible guilt for this. Maybe I *wanted* him to die! I *know* that I let him down, and so I feel that I deserve to die."

Kay was convinced that her guilt and agony were appropriate and valid. Being a highly moral person with a strict Catholic upbringing, she felt that punishment and suffering were expected of her. I knew there was something fishy about her line of reasoning, but I couldn't quite penetrate her illogic for several sessions because she was bright and persuasive and made a convincing case against herself. I almost began to buy her belief that her emotional pain was "realistic." Then, the key that I hoped might free her from her mental prison suddenly dawned on me. The error she was making was number ten discussed in Chapter 3—personalization.

At the fifth therapy session, I used this insight to challenge the misconceptions in Kay's point of view. First of all, I emphasized that if she were responsible for her brother's death, she would have had to be the cause of it. Since the cause of suicide is not known, even by experts, there was no reason to conclude that she was the cause.

I told her that if we had to guess the cause of his suicide, it would be his erroneous conviction that he was hopeless and worthless and that his life was not worth living. Since she did not control his thinking, she could not be responsible for the illogical assumptions that caused him to end his life. They were his errors, not hers. Thus, in assuming responsibility for his mood and actions, she was doing so for something that was not within her domain of control. The most that anyone could or would expect of her was to try to be a helping agent, as she had been within the limits of her ability.

I emphasized that it was unfortunate she did not have the

knowledge necessary to prevent his death. If it had dawned on her that he was about to make a suicide attempt, she *would* have intervened in whatever manner possible. However, since she did not have this knowledge, it was not possible for her to intervene. Therefore, in blaming herself for his death she was illogically assuming that she could predict the future with absolute certainty, and that she had all the knowledge in the universe at her disposal. Since both these expectations were highly unrealistic, there was no reason for her to despise herself. I pointed out that even professional therapists are not infallible in their knowledge of human nature, and are frequently fooled by suicidal patients in spite of their presumed expertise.

For all these reasons, it was a major error to hold herself responsible for his behavior because she was not ultimately in control of him. I emphasized that she *was* responsible for her own life and well-being. At this point it dawned on her that she was acting irresponsibly, *not* because she "let him down" but because she was allowing herself to become depressed and was contemplating her own suicide. The responsible thing to do was to *refuse* to feel any guilt and to end the depression, and then to pursue a life of happiness and satisfaction. This would be acting in a responsible manner.

This discussion was followed by a rapid improvement in her mood. Kay attributed this to a profound change in her attitude. She realized we had exposed the misconceptions that made her want to kill herself. She then elected to remain in therapy for a period of time in order to work on enhancing the quality of her own life, and to dispel the chronic sense of oppression that had plagued her for many years prior to her brother's suicide.

Sadness Without Suffering. The question then arises, What is the nature of "healthy sadness" when it is not at all contaminated by distortion? Or to put it another way—does sadness really need to involve suffering?

While I cannot claim to know the definitive answer to this question, I would like to share an experience which occurred when I was an insecure medical student, and I was on my clinical rounds on the urology service in the hospital at Stanford University Medical Center in California. I was assigned to an elderly man who recently had had a tumor successfully removed from his kidney. The staff anticipated his rapid discharge from the hospital, but his liver function suddenly began to deteriorate, and it was discovered that the tumor had metastasized to his liver. This sad complication was untreatable, and his health began to fail rapidly over several days. As his liver function worsened, he slowly began to get groggier, slipping toward an unconscious state. His wife, aware of the seriousness of the situation, came and sat by his side night and day for over forty-eight hours. When she was tired, her head would fall on his bed, but she never left his side. At times she would stroke his head and tell him, "You're my man and I love you." Because he was placed on the critical list, the members of his large family, including children, grandchildren, and great-grandchildren, began to arrive at the hospital from various parts of California.

In the evening the resident in charge asked me to stay with the patient and attend the case. As I entered the room, I realized that he was slipping into a coma. There were eight or ten relatives there, some of them very old and others very young. Although they were vaguely aware of the seriousness of his condition, they had not been informed of just how grave the imminent situation was. One of his sons, sensing the old gentleman was nearing the end, asked me if I would be willing to remove the catheter which was draining his bladder. I realized the removal of the catheter would indicate to the family that he was dying, so I went to ask the nursing staff if this would be appropriate to do. The nursing staff told me that it would because he was indeed dying. After they showed me how to remove a catheter, I went back to the patient and did this while the family waited. Once I was done, they realized that a certain support had been removed, and the son said, "Thank you. I know it was uncomfortable for him, and he would have appreciated this." Then the son turned to me as if to confirm the meaning of the sign and asked, "Doctor, what is his condition? What can we expect?"

I felt a sudden surge of grief. I had felt close to this gentle, courteous man because he reminded me of my own grandfather, and I realized that tears were running down my cheeks. I had to make a decision either to stand there and let the family see my tears as I spoke with them or to leave and try to hide my feelings. I chose to stay and said with considerable emotion, "He is a beautiful man. He can still hear you, although he is nearly in a coma, and it is time to be close to him and say good-bye to him tonight." I then left the room and wept. The family members also cried and sat on the bed, while they talked to him and said good-bye. Within the next hour his coma deepened until he lost consciousness and died.

Although his death was profoundly sad for the family and for me, there was a tenderness and a beauty to the experience that I will never forget. The sense of loss and the weeping reminded me—"You can love. You can care." This made the grief an elevating experience that was entirely devoid of pain or suffering for me. Since then, I have had a number of experiences that brought me to tears in this same way. For me the grief represents an elevation, an experience of the highest magnitude.

Because I was a medical student, I was concerned that my behavior might be seen as inappropriate by the staff. The chairman of the department later took me aside and informed me that the patient's family had asked him to extend their appreciation to me for being available to them and for helping make the occasion of his passing intimate and beautiful. He told me that he too had always felt strongly toward this particular individual, and showed me a painting of a horse the elderly man had done which was hanging on his wall.

The episode involved a letting go, a feeling of closure, and a sense of good-bye. This was in no way frightening or terrible; but in fact, it was peaceful and warm, and added a sense of richness to my experience of life.

Part IV

Prevention and Personal Growth

Chapter 10

The Cause of It All

When your depression has vanished, it's a temptation to *enjoy* yourself and relax. Certainly you're entitled. Toward the end of therapy, many patients tell me they feel the best they've ever felt in their lives. It sometimes seems that the more hopeless and severe and intractable the depression seemed, the more extraordinary and delicious the taste of happiness and self-esteem once it is over. As you begin to feel better, your pessimistic thinking pattern will recede as dramatically and predictably as the melting of winter's snow when spring arrives. You may even wonder how in the world you came to believe such unrealistic thoughts in the first place. This profound transformation of the human spirit never ceases to amaze me. Over and over I have the opportunity to observe this magical metamorphosis in my daily practice.

Because your change in outlook can be so dramatic, you may feel convinced that your blues have vanished forever. But there is an invisible residue of the mood disorder that remains. If this is not corrected and eliminated, you will be vulnerable to attacks of depression in the future.

There are several differences between *feeling* better and *getting* better. Feeling better simply indicates that the painful symptoms have temporarily disappeared. Getting better implies:

1. Understanding *why* you got depressed.
2. Knowing *why* and *how* you got better. This involves a mastery of the particular self-help techniques that worked specifically for you so that you can reapply them and make them work again whenever you choose.
3. Acquiring self-confidence and self-esteem. Self-confidence is based on the knowledge that you have a good chance of being reasonably successful in personal relationships and in your career. Self-esteem is the capacity to experience maximal self-love and joy whether or not you are successful at any point in your life.
4. Locating the deeper causes of your depression.

Parts I, II, and III of this book were designed to help you achieve the first two goals. The next several chapters will help you with the third and fourth goals.

Although your distorted negative thoughts will be substantially reduced or entirely eliminated after you have recovered from a bout of depression, there are certain "silent assumptions" that probably still lurk in your mind. These silent assumptions explain in large part *why* you became depressed in the first place and can help you predict *when* you might again be vulnerable. And they contain therefore the key to relapse prevention.

Just what is a silent assumption? A silent assumption is an equation with which you define your personal worth. It represents your value system, your personal philosophy, the stuff on which you base your self-esteem. Examples: (1) "If someone criticizes me, I feel miserable because this automatically means there is something wrong with me." (2) "To be a truly fulfilled human being, I must be loved. If I am alone, I am bound to be lonely and miserable." (3) "My worth as a human being is proportional to what I've achieved." (4) "If I don't perform (or feel or act) perfectly, I have failed." As you will learn, these illogical assumptions can be utterly self-defeating. They create a vulnerability that predisposes you to uncomfortable mood swings. They represent your psychological Achilles' heel.

In the next several chapters you will learn to identify and evaluate your own silent assumptions. You might find that an addiction to approval, love, achievement, or perfection forms the basis of your mood swings. As you learn to expose and challenge your own self-defeating belief system, you will lay the foundation for a personal philosophy that is valid and self-enhancing. You will be on the road to joy and emotional enlightenment.

In order to unearth the origins of your mood swings, most psychiatrists, as well as the general public, assume that a long and painfully slow (several years) therapeutic process is necessary, after which most patients would find it difficult to explain the cause of their depression. One of the greatest contributions of cognitive therapy has been to circumvent this.

In this chapter you will learn two different ways to identify silent assumptions. The first is a startlingly effective method called the "vertical-arrow technique," which allows you to probe your inner psyche.

The vertical-arrow technique is actually a spin-off of the double-column method introduced in Chapter 4, in which you learned how to write down your upsetting automatic thoughts in the left-hand column and substitute more objective rational responses. This method helps you feel better because you deprogram the distortions in your thinking patterns. A brief example is shown in Figure 10–1. It was written by Art, the psychiatric resident described in Chapter 7, who became upset after his supervisor tried to offer a constructive criticism.

Putting the lie to his upsetting thoughts reduced Art's feelings of guilt and anxiety, but he wanted to know how and why he made such an illogical interpretation in the first place. Perhaps you've also begun to ask yourself—is there a *pattern* inherent in my negative thoughts? Is there some psychic kink that exists on a deeper level of my mind?

Figure 10–1.

Automatic Thoughts		*Rational Responses*
1. Dr. B said the patient found my comment abrasive. He probably thinks I'm a lousy therapist.	→	1. Mind reading; mental filter; labeling. Just because Dr. B pointed out my error it doesn't follow he thinks I'm a "lousy therapist." I'd have to ask him to see what he really thinks, but on many occasions he has praised me and said I had outstanding talent.

Art used the vertical-arrow technique to answer these questions. First, he drew a short downward arrow directly *beneath* his automatic thought (see Figure 10–2, page 265). This downward arrow is a form of shorthand which tells Art to ask himself, "If this automatic thought were actually true, what would it mean to me? Why would it be upsetting to me?" Then Art wrote down the next automatic thought that immediately came to mind. As you can see, he wrote, "If Dr. B. thinks I'm a lousy therapist, it would mean I *was* a lousy therapist because Dr. B. is an expert." Next Art drew a second downward arrow beneath this thought and repeated the same process so as to generate yet another automatic thought, as shown in Figure 10–2. Every time he came up with a new automatic thought, he immediately drew a vertical arrow beneath it and asked himself, "If that were true, why would it upset me?" As he did this over and over, he was able to generate a chain of automatic thoughts, which led to the silent assumptions that gave rise to his problems. The downward-arrow method is analogous to peeling successive layers of skin off an onion to expose

Figure 10–2. Exposing the silent assumption(s) that give rise to your automatic thoughts with the use of the vertical-arrow method. The downward arrow is a form of shorthand for the following questions: "If that thought were true, why would it upset me? What would it mean to me?" The question represented by each downward arrow in the example appears in quotation marks next to the arrow. This is what you might ask yourself if you had written down the automatic thought. This process leads to a chain of automatic thoughts that will reveal the root cause of the problem.

Automatic Thoughts	*Rational Responses*
1. Dr. B. probably thinks I'm a lousy therapist. →	
↓ "If he *did* think this, why would it be upsetting to me?"	
2. That would mean I *was* a lousy therapist because he's an expert. →	
↓ "Suppose I *was* a lousy therapist, what would this mean to me?"	
3. That would mean I was a total failure. It would mean I was no good. →	
↓ "Suppose I *was* no good. Why would this be a problem? What would it mean to me?"	
4. Then the word would spread and everyone would find out what a bad person I was. Then no one would respect me. I'd get drummed out of the medical society, →	

Figure 10–2. cont.

Automatic Thoughts		*Rational Responses*
and I'd have to move to another state. ↓ "And what would that mean?"		
5. It would mean I was worthless. I'd feel so miserable I'd want to die.	→	

the ones beneath. It is actually quite simple and straightforward, as you will see in Figure 10–2.

You will notice that the vertical-arrow technique is the *opposite* of the usual strategy you use when recording your automatic thoughts. Ordinarily you substitute a rational response that shows why your automatic thought is *distorted* and *invalid* (Figure 10–1). This helps you change your thinking patterns in the here and now so that you can think about life more objectively and feel better. In the vertical-arrow method you imagine instead that your distorted automatic thought is absolutely valid, and you look for the *grain of truth* in it. This enables you to penetrate the core of your problems.

Now review Art's chain of automatic thoughts in Figure 10–2 and ask yourself—what are the silent assumptions that predispose him to anxiety, guilt, and depression? There are several:

1. If someone criticizes me, they're bound to be correct.
2. My worth is determined by my achievement.
3. One mistake and the whole is ruined. If I'm not successful at *all* times, I'm a total zero.
4. Others won't tolerate my imperfection. I have to be

perfect to get people to respect and like me. When I goof up, I'll encounter fierce disapproval and be punished.

5. This disapproval will mean I am a bad, worthless person.

Once you have generated your own chain of automatic thoughts and clarified your silent assumptions, it is crucial to pinpoint the distortions and substitute rational responses as you usually do (see Figure 10–3, page 268).

The beauty of the downward-arrow method is that it is inductive and Socratic: Through a process of thoughtful questioning, you discover on your own the beliefs that defeat you. You unearth the origin of your problems by repeating the following questions over and over: "If that negative thought were true, what would it mean to me? Why would it upset me?" *Without introducing some therapist's subjective bias* or personal beliefs or theoretical leanings, you can *objectively* and systematically go right to the root of your problems. This circumvents a difficulty that has plagued the history of psychiatry. Therapists from all schools of thought have been notorious for interpreting patients' experiences in terms of preconceived notions that may have little or no experimental validation. If you don't "buy" your therapist's explanation of the origin of your problems, this is likely to be interpreted as "resistance" to the "truth." In this subtle way, your troubles get forced into your therapist's mold regardless of what you say. Imagine the bewildering array of explanations for suffering that you would hear if you went to a religious counselor (spiritual factors), a psychiatrist in a Communist country (the social-political-economic environment), a Freudian analyst (internalized anger), a behavior therapist (a low rate of positive reinforcement), a drug-oriented psychiatrist (genetic factors and brain-chemistry imbalance), a family therapist (disturbed interpersonal relationships), etc.!

A word of caution when you apply the vertical-arrow method. You will short-circuit the process if you write down

Figure 10–3. After eliciting his chain of automatic thoughts, using the downward-arrow method, Art identified the cognitive distortions and substituted more objective responses.

Automatic Thoughts		*Rational Responses*
1. Dr. B. probably thinks I'm a lousy therapist. ↓ "If he *did* think this, why would it be upsetting to me?"	→	1. Just because Dr. B. pointed out my error it doesn't follow he thinks I'm a "lousy therapist." I'd have to ask him to see what he really thinks, but on many occasions he has praised me and said I had outstanding talent.
2. That would mean I *was* a lousy therapist because he's an expert. ↓ "Suppose I *was* a lousy therapist, what would this mean to me?"	→	2. An expert can only point out my specific strengths and weaknesses as a therapist. Any time anyone labels me as "lousy" they are simply making a global, destructive, useless statement. I have had a lot of success with most of my patients, so it *can't* be true I'm "lousy" no matter who says it.
3. That would mean I was a total failure. It would mean I was no good. ↓ "Suppose I *was* no good. Why would this be a problem? What would it mean to me?"	→	3. Overgeneralization. Even if I was relatively unskilled and ineffective as a therapist, it wouldn't mean I was "a total failure" or "no good." I have many other interests, strengths, and desirable qualities that aren't related to my career.
4. Then the word would spread and everyone		4. This is absurd. If I made a mistake, I can correct

Figure 10–3. cont.

Automatic Thoughts		*Rational Responses*
would find out what a bad person I was. Then no one would respect me. I'd get drummed out of the medical society, and I'd have to move to another state. ↓ "And what would that mean?"	→	it. "The word" isn't going to spread around the state like wildfire just because I made an error! What are they going to do, publish a headline in the newspaper: "NOTED PSYCHIATRIST MAKES MISTAKE"?
5. It would mean I was worthless. I'd feel so miserable I'd want to die.	→	5. Even if everyone in the world disapproves of me or criticizes me, it can't make me worthless because *I'm not worthless.* If I'm not worthless, I must be quite worthwhile. So, what is there to feel miserable about?

thoughts that contain descriptions of your emotional reactions. Instead, write down the negative thoughts that *cause* your emotional reactions. Here's an example of the *wrong* way to do it:

First Automatic Thought: My boyfriend didn't call me this weekend as he promised he would.

↓ "Why is that upsetting to me? What does it mean to me?"

Second Automatic Thought: Oh, it's awful and terrible because I can't stand it.

This is useless. We already *know* you feel awful and terrible. The question is—what *thoughts* automatically crossed your mind that *caused* you to feel so upset? What would it mean to you if he *had* neglected you?

Here's the correct way to do it:

1. My boyfriend didn't call me this weekend as he promised he would.

↓ "Why would that be upsetting to me? What does it mean to me?"

2. That means he's neglecting me. That means he really doesn't love me.

↓ "And suppose that were true. What would that mean to me?"

3. That would mean there's something wrong with me. Otherwise he'd be more attentive.

↓ "And suppose that were true. What would that mean to me?"

4. That would mean I was going to be rejected.

↓ "And if I were in fact rejected, what then? What would that mean to me?"

5. That would mean I was unlovable and I would always be rejected.

↓ "And if that happened, why would it upset me?"

6. That would mean I'd end up alone and miserable.

Thus, by pursuing the *meaning* rather than your *feelings*, your silent assumptions became obvious: (1) If I'm not loved I'm not worthwhile; and (2) I'm bound to be miserable if I'm alone.

This is *not* to say your feelings aren't important. The whole point is to deliver the real McCoy—valid emotional transformation.

The Dysfunctional Attitude Scale (DAS). Because of the crucial importance of eliciting the silent assumptions that give rise to your mood swings, a second, simpler method for eliciting them called the "Dysfunctional Attitude Scale"

(DAS) has been developed by a member of our group, Dr. Arlene Weissman. She has compiled a list of one hundred self-defeating attitudes that commonly occur in individuals predisposed to emotional disorders. Her research has indicated that while negative automatic thoughts are reduced dramatically between episodes of depression, a self-defeating belief system remains more or less constant during episodes of depression and remission. Dr. Weissman's studies confirm the concept that your silent assumptions represent a predisposition to emotional turbulence that you carry with you at all times.

Although a complete presentation of the lengthy Dysfunctional Attitude Scale would be beyond the scope of this book, I have selected a number of the more common attitudes and have added several others which will be useful. As you fill out the questionnaire, indicate how much you agree or disagree with each attitude. When you are finished, an answer key will let you score your answers and generate a profile of your personal value systems. This will show your areas of psychological strength and vulnerability.

Answering the test is quite simple. After each of the thirty-five attitudes, put a check in the column that represents your estimate of how you think *most* of the time. Be sure to choose only one answer for each attitude. Because we are all different, there is no "right" or "wrong" answer to any statement. To decide whether a given attitude is typical of your own philosophy, recall how you look at things *most of the time*.

EXAMPLE:

	Agree Strongly	*Agree Slightly*	*Neutral*	*Disagree Slightly*	*Disagree Very Much*
35. People who have the marks of success (good looks, social status, wealth, or fame) are bound to be happier than those who do not.		✓			

In this example the checkmark in the Agree Slightly column indicates that the statement is somewhat typical of the attitudes of the person completing the inventory. Now go ahead.

The Dysfunctional Attitude Scale*

	Agree Strongly	*Agree Slightly*	*Neutral*	*Disagree Slightly*	*Disagree Very Much*
1. Criticism will obviously upset the person who receives the criticism.					

*Copyright 1978, Arlene Weissman.

	Agree Strongly	*Agree Slightly*	*Neutral*	*Disagree Slightly*	*Disagree Very Much*
2. It is best to give up my own interests in order to please other people.					
3. I need other people's approval in order to be happy.					
4. If someone important to me expects me to do something, then I really should do it.					
5. My value as a person depends greatly on what others think of me.					
6. I cannot find happiness without being loved by another person.					

	Agree Strongly	*Agree Slightly*	*Neutral*	*Disagree Slightly*	*Disagree Very Much*
7. If others dislike you, you are bound to be less happy.					
8. If people whom I care about reject me, it means there is something wrong with me.					
9. If a person I love does not love me, it means I am unlovable.					
10. Being isolated from others is bound to lead to unhappiness.					
11. If I am to be a worthwhile person, I must be truly outstanding in at least one major respect.					

	Agree Strongly	*Agree Slightly*	*Neutral*	*Disagree Slightly*	*Disagree Very Much*
12. I must be a useful, productive, creative person or life has no purpose.					
13. People who have good ideas are more worthy than those who do not.					
14. If I do not do as well as other people, it means I am inferior.					
15. If I fail at my work, then I am a failure as a person.					
16. If you cannot do something well, there is little point in doing it at all.					

	Agree Strongly	*Agree Slightly*	*Neutral*	*Disagree Slightly*	*Disagree Very Much*
17. It is shameful for a person to display his weaknesses.					
18. A person should try to be the best at everything he undertakes.					
19. I should be upset if I make a mistake.					
20. If I don't set the highest standards for myself, I am likely to end up a second-rate person.					
21. If I strongly believe I deserve something, I have reason to expect that I should get it.					

	Agree Strongly	*Agree Slightly*	*Neutral*	*Disagree Slightly*	*Disagree Very Much*
22. It is necessary to become frustrated if you find obstacles to getting what you want.					
23. If I put other people's needs before my own, they should help me when I need something from them.					
24. If I am a good husband (or wife), then my spouse is bound to love me.					
25. If I do nice things for someone, I can anticipate that they will respect me and treat me just as well as I treat them.					

	Agree Strongly	*Agree Slightly*	*Neutral*	*Disagree Slightly*	*Disagree Very Much*
26. I should assume responsibility for how people feel and behave if they are close to me.					
27. If I criticize the way someone does something and they become angry or depressed, this means I have upset them.					
28. To be a good, worthwhile, moral person, I must try to help everyone who needs it.					

	Agree Strongly	*Agree Slightly*	*Neutral*	*Disagree Slightly*	*Disagree Very Much*
29. If a child is having emotional or behavioral difficulties, this shows that the child's parents have failed in some important respect.					
30. I should be able to please everybody.					
31. I cannot expect to control how I feel when something bad happens.					
32. There is no point in trying to change upsetting emotions because they are a valid and inevitable part of daily living.					

	Agree Strongly	*Agree Slightly*	*Neutral*	*Disagree Slightly*	*Disagree Very Much*
33. My moods are primarily created by factors that are largely beyond my control, such as the past, or body chemistry, or hormone cycles, or biorhythms, or chance, or fate.					
34. My happiness is largely dependent on what happens to me.					
35. People who have the marks of success (good looks, social status, wealth, or fame) are bound to be happier than those who do not.					

Now that you have completed the DAS, you can score it in the following way. Score your answer to each of the thirty-five attitudes according to this key:

Agree Strongly	*Agree Slightly*	*Neutral*	*Disagree Slightly*	*Disagree Very Much*
−2	−1	0	+1	+2

Now add up your score on the first five attitudes. These measure your tendency to measure your worth in terms of the opinions of others and the amount of approval or criticism you receive. Suppose your scores on these five items were +2; +1; −1; +2; 0. Then your total score for these five questions would be +4.

Proceed in this way to add up your score for items 1 through 5, 6 through 10, 11 through 15, 16 through 20, 21 through 25, 26 through 30, and 31 through 35, and record these as illustrated in the following example:

SCORING EXAMPLE:

Value System	*Attitudes*	*Individual Scores*	*Total Scores*
I. Approval	1 through 5	+2, +1, −1, +2, 0	+4
II. Love	6 through 10	−2, −1, −2, −2, 0	−7
III. Achievement	11 through 15	+1, +1, 0, 0, −2	0
IV. Perfectionism	16 through 20	+2, +2, +1, +1, +1	+7

SCORING EXAMPLE continued:

Value System	*Attitudes*	*Individual Scores*	*Total Scores*
V. Entitlement	21 through 25	+1, +1, −1, +1, 0	+2
VI. Omnipotence	26 through 30	−2, −1, 0, −1, +1	−3
VII. Autonomy	31 through 35	−2, −2, −1, −2, −2	−9

RECORD YOUR ACTUAL SCORES HERE:

Value System	*Attitudes*	*Individual Scores*	*Total Scores*
I. Approval	1 through 5		
II. Love	6 through 10		
III. Achievement	11 through 15		
IV. Perfectionism	16 through 20		
V. Entitlement	21 through 25		
VI. Omnipotence	26 through 30		
VII. Autonomy	31 through 35		

Each cluster of five items from the scale measures one of seven value systems. Your total score for each cluster of five items can range from +10 to −10. Now plot your total scores on each of the seven variables so as to develop your "personal-philosophy profile" as follows:

SCORING EXAMPLE:

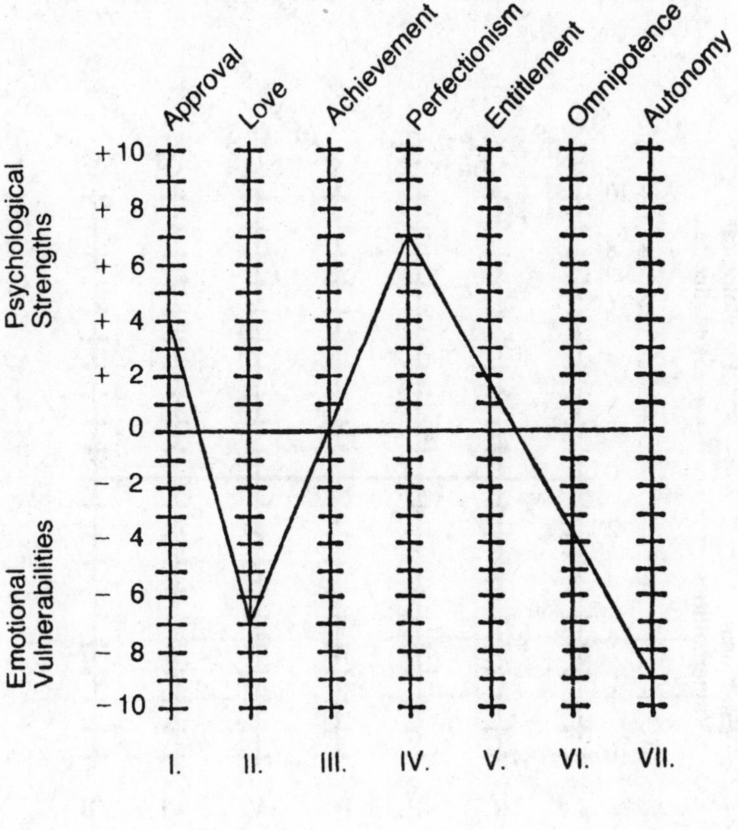

As you can see, a positive score represents an area where you are psychologically *strong*. A negative score represents an area where you're emotionally *vulnerable*.

This individual has strengths in the areas of approval, perfectionism, and entitlement. His vulnerabilities lie in the areas of love, omnipotence, and autonomy. The meanings of these concepts will be described. First, plot your own personal-philosophy profile here.

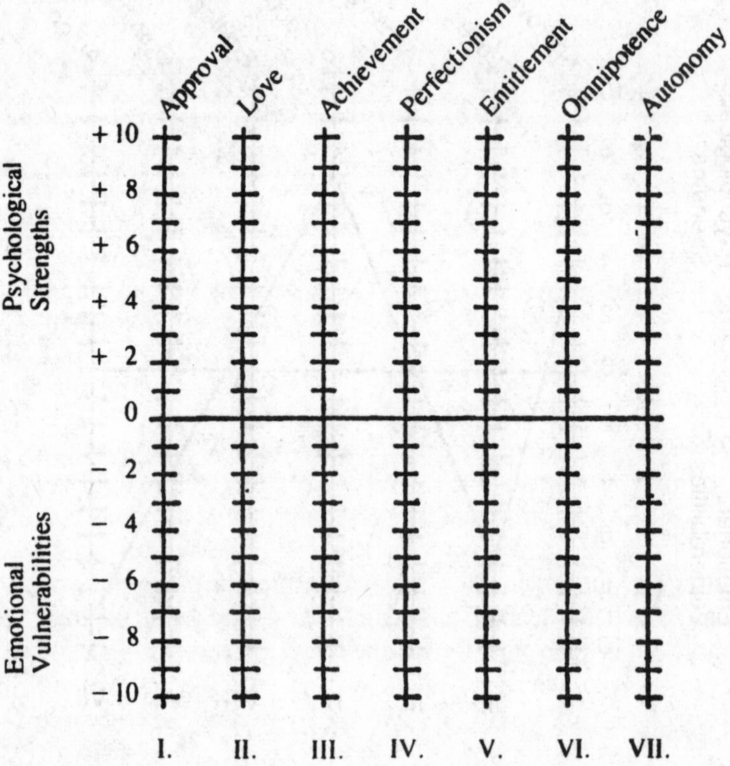

Interpreting Your DAS Scores

I. Approval. The first five attitudes on the DAS test probe your tendency to measure your self-esteem based on how people react to you and what they think of you. A positive score between zero and ten indicates you are independent, with a healthy sense of your own worth even when confronted with criticism and disapproval. A negative score between zero and minus ten indicates you are excessively dependent because you evaluate yourself through other people's eyes. If someone insults you or puts you down, you automatically tend to look down on yourself. Since your emotional well-being is exquisitely sensitive to what you imagine people think of you, you can be easily manipulated, and you are vulnerable to anxiety and depression when others criticize you or are angry with you.

II. Love. The second five attitudes on the test assess your tendency to base your worth on whether or not you are loved. A positive score indicates you see love as desirable, but you have a wide range of other interests you also find gratifying and fulfulling. Hence, love is not a requirement for your happiness or self-esteem. People are likely to find you attractive because you radiate a healthy sense of self-love and are interested in many aspects of living.

A negative score indicates you are a "love junkie." You see love as a "need" without which you cannot survive, much less be happy. The closer your score is to minus ten, the more dependent on love you are. You tend to adopt inferior, put-down roles in relationships with people you care about for fear of alienating them. The result of this, more often than not, is that they lose respect for you and consider you a burden because of your attitude that without their love you would collapse. As you sense that people drift away from you, you become gripped by a painful, terrifying withdrawal syndrome. You realize you may not be able to "shoot up" with your daily dose of affection and attention. You then become consumed by the driving compulsion to "get love." Like most junkies, you may even resort to coercive, manipulative behavior to get your "stuff." Ironically, your needy, greedy love addiction drives many people away, thus intensifying your loneliness.

III. Achievement. Your score on attitudes 11 through 15 will help you measure a different type of addiction. A negative score indicates you are a workaholic. You have a constricted sense of your own humanity, and you see yourself as a commodity in the marketplace. The more negative your score, the more your sense of self-worth and your capacity for joy are dependent on your productivity. If you go on vacation, if your business slumps, if you retire or become ill and inactive, you will be in danger of an emotional crash. Economic and emotional depressions will seem identical to you. A positive score, in contrast, indicates that you enjoy creativity and productivity, but do not see them as an exclusive or necessary road to self-esteem and satisfaction.

IV. Perfectionism. Items 16 through 20 measure your tendency to perfectionism. A negative score indicates you are hooked on searching for the Holy Grail. You demand perfection in yourself—mistakes are taboo, failure is worse than death, and even negative emotions are a disaster. You're supposed to look, feel, think, and behave superbly at all times. You sense that being less than spectacular means burning in the flames of hell. Although you drive yourself at an intense pace, your satisfactions are meager. Once you do achieve a goal, another more distant goal instantly replaces it, so you never experience the reward of getting to the top of the mountain. Eventually you begin to wonder why the promised payoff from all your effort never seems to materialize. Your life becomes a joyless, tedious treadmill. You are living with unrealistic, impossible personal standards, and you need to reevaluate them. Your problem does *not* lie in your performance, but in the yardstick you use to measure it. If you bring your expectations in line with reality, you will be regularly *pleased* and *rewarded* instead of *frustrated*.

A positive score suggests you have the capacity to set meaningful, flexible, appropriate standards. You get great satisfaction from processes and experiences, and you are not exclusively fixated on outcomes. You don't have to be outstanding at everything, and you don't always have to "try your best." You don't fear mistakes, but you see them as golden opportunities to learn and to endorse your humanity. Paradoxically, you are likely to be much more productive than your perfectionistic associates because you do not become compulsively preoccupied with detail and correctness. Your life is like a flowing river or a geyser compared with your rigid perfectionistic friends who appear more like icy glaciers.

V. Entitlement. Attitudes 21 through 25 measure your sense of "entitlement." A negative score indicates that you feel "entitled" to things—success, love, happiness, etc. You expect and demand that your wants be met by other people and by the universe at large because of your inherent goodness or hard work. When this does *not* happen—as is often the case—you are locked into one of two reactions. Either you feel depressed and inadequate or you become irate. Thus, you consume enormous amounts of energy being frustrated, sad, and mad. Much of the time you see life as a sour, rotten experience. You complain loudly and often, but you do little to solve problems. After all, you're *entitled* to have them solved, so why should you have to put out any effort? As a result of your bitter, demanding attitudes, you invariably get far *less* of what you want from life.

A positive score suggests you don't feel automatically entitled to things, so you *negotiate* for what you want and often get it. Because of your awareness that other people are unique and different, you realize there is no inherent reason why things should always go your way. You experience a negative outcome as a disappointment but not a tragedy because you are a percentage player, and you don't expect perfect reciprocity or "justice" at all times. You are patient and persistent, and you have a high frustration tolerance. As a result, you often end up ahead of the pack.

VI. Omnipotence. Attitudes 26 through 30 measure your tendency to see yourself as the center of your personal universe and to hold yourself responsible for much of what goes on around you. A negative score indicates you often make the personalization error discussed in Chapters 3 and 6. You blame yourself inappropriately for the negative actions and attitudes of others who are not really under your control. Consequently, you are plagued by guilt and self-condemnation. Paradoxically, the attitude that you should be omnipotent and all-powerful cripples you and leaves you anxious and ineffectual.

A positive score, in contrast, indicates you know the joy that comes from accepting that you are *not* the center of the universe. Since you are *not* in control of other adults, you are not ultimately responsible for them but only for yourself. This attitude does not isolate you from others. Quite the opposite is true. You relate to people effectively as a friendly collaborator, and you are not threatened when they disagree with your ideas or fail to follow your advice. Because your attitude gives people a sense of freedom and dignity, you paradoxically become a human magnet. Others often want to be close to you because you have relinquished any attempt to control them. People frequently listen to and respect your ideas because you do not polarize them with an angry insistence they *must* agree with you. As you give up your drive for *power*, people repay you by making you a person of *influence*. Your relationships with your children and friends and associates are characterized by mutuality instead of dependency. Because you don't try to dominate people, they admire, love, and respect you.

VII. Autonomy. Items 31 through 35 measure your autonomy. This refers to your ability to find happiness within yourself. A positive score indicates that all your moods are ultimately the children of your thoughts and attitudes. You assume responsibility for your feelings because you recognize they are ultimately created by you. This *sounds* as if you might be lonely and isolated because you realize that all meaning and feelings are created only in your head. Paradoxically, however, this vision of autonomy frees you from the petty confines of your mind and delivers the world to you with a full measure of all the satisfaction, mystery, and excitement that it can offer.

A negative score suggests you are still trapped in the belief that your potential for joy and self-esteem comes from the outside. This puts you at a great disadvantage because everything outside is ultimately beyond your control. Your moods end up the victim of external factors. Do you want this? If not, you can eventually free yourself from this attitude as surely as a snake sheds its skin, but you will have to work at it with the various methods outlined in this book. When it's finally your turn to experience the transformation to autonomy and personal responsibility, you will be amazed—or awestruck—or pleased—or delightfully overwhelmed. It's well worth a major personal commitment.

In the following chapters a number of these attitudes and value systems will be examined in detail. As you study each one, ask yourself: (1) Is it to my advantage to maintain this particular belief? (2) Is this belief really true and valid? (3) What specific steps can I take that will allow me to rid myself of attitudes that are self-defeating and unrealistic, and substitute others that are more objective and more self-enhancing?

Chapter 11

The Approval Addiction

Let's consider your belief that it would be *terrible* if someone disapproved of you. Why does disapproval pose such a threat? Perhaps your reasoning goes like this: "If one person disapproves of me, it means that everyone would disapprove of me. It would mean there was something wrong with me."

If these thoughts apply to you, your moods will shoot up every time you are being stroked. You reason, "I got some positive feedback so I can feel good about myself."

Why is this illogical? Because you are overlooking the fact that it is only your thoughts and beliefs which have the power to elevate your spirits. Another person's approval has no ability to affect your mood unless you believe what he or she says is valid. But if you believe the compliment is earned, it is *your belief* which makes you feel good. You must validate external approval before you experience mood elevation. This validation represents your personal self-approval.

Suppose you were visiting the psychiatric ward of a hospital. A confused, hallucinating patient approaches you and says, "You are wonderful. I had a vision from God. He told me the thirteenth person to walk through the door would be the Special Messenger. You are the thirteenth, so I know you are God's Chosen One, the Prince of Peace, the Holy of Holies. Let me kiss your shoe." Would this extreme approval elevate your mood? You'd probably feel nervous and uncomfortable. That's because you don't believe what the patient is saying is valid. You discredit the comments. It is only *your* beliefs about yourself that can affect the way you feel. Others can say or think whatever they want about you, good or bad, but only your thoughts will influence your emotions.

The price you pay for your addiction to praise will be an extreme vulnerability to the opinions of others. Like any addict, you will find you must continue to feed your habit with approval in order to avoid withdrawal pangs. The moment someone who is important to you expresses disapproval, you will crash painfully, just like the junkie who can no longer get his "stuff." Others will be able to use this vulnerability to manipulate you. You will have to give in to their demands more often than you want to because you fear they might reject or look down on you. You set yourself up for emotional blackmail.

You may come to see that your addiction to approval is not to your advantage, but still believe that other people *really do* have the right to judge not only the merit of what you do and say but also your worth as a human being. Imagine that you made a second visit to the psychiatric hospital ward. This time a different hallucinating patient approaches you and says, "You're wearing a red shirt. This shows you are the Devil! You are evil!" Would you feel bad because of this criticism and disapproval? Of course not. Why would these disapproving words not upset you? It's simple—because you don't believe the statements are true. You must "buy into" the other person's criticism—and believe that you are in fact no good—in order to feel bad about yourself.

Did it ever occur to you that if someone disapproves of you, it might be *his* or *her* problem? Disapproval often reflects other people's irrational beliefs. To take an extreme example, Hitler's hateful doctrine that Jews were inferior did not reflect anything about the inner worth of the people he intended to destroy.

There will, of course, be many occasions when disapproval will result from an actual error on your part. Does it follow that you are a worthless, no-good person? Obviously not. The other person's negative reaction can only be directed toward a *specific* thing you did, not at your worth. A human being *cannot* do wrong things *all* the time!

Let's look at the other side of the coin. Many well-known criminals have had bands of fervent admirers regardless of how repulsive and abhorrent their crimes. Consider Charles Manson. He promoted sadism and murder, yet was regarded as a messiah by his numerous followers, who seemed to do whatever he suggested. I want to make it abundantly clear that I am not advocating atrocious behavior, nor am I an admirer of Charles Manson. But ask yourself these questions: If Charles Manson did not end up totally rejected for what he did or said, what have *you* ever done that was so terrible that you will be rejected by everyone? And do you still believe in the equation: approval = worth? After all, Charles Manson enjoyed the intense adulation of his "family." Did the approval he received make him an especially worthy person? This is obvious nonsense.

It's a fact that approval *feels good*. There's nothing wrong with that; it's natural and healthy. It is also a fact that disapproval and rejection usually taste bitter and unpleasant. This is human and understandable. But you are swimming in deep, turbulent waters if you continue to believe that approval and disapproval are the proper and ultimate yardsticks with which to measure your worth.

Did you ever criticize someone? Did you ever disagree with a friend's opinion? Did you ever scold a child because of his or her behavior? Did you ever snap at a loved one when you were feeling irritable? Did you ever choose not to associate with someone whose behavior was distasteful to you? Then ask yourself—when you disagreed, or criticized, or disapproved—were you making the ultimate moral judgment that the other person was a totally worthless, no-good human being? Do you have the power to make such sweeping judgments about other people? Or were you simply expressing the fact that you held a different point of view and were upset with what the other person did or said?

For example, in the heat of anger you may have blurted out to your spouse, "You're no damn good!" But when the flame cools down a day or two later, didn't you admit to yourself that you were exaggerating the extent of his or her "badness"? Sure, your loved one may have many faults, but isn't it absurd to think your outburst of disapproval or criticism makes him or her totally and forever worthless? If you admit your disapproval does not contain enough moral atomic power to devastate the meaning and value of another person's life, why give *their* disapproval the power to wipe out *your* sense of self-worth? What makes *them* so special? When you tremble in terror because someone dislikes you, you magnify the wisdom and knowledge that person possesses, and you have simultaneously sold yourself short as being unable to make sound judgments about yourself. Of course, someone might point out a flaw in your behavior or an error in your thinking. I hope they will because you can learn this way. After all, we're all imperfect, and others have the *right* to tell us about it from time to time. But are you obliged to make yourself miserable and hate yourself every time someone flies off the handle or puts you down?

The Origin of the Problem. Where did you get this approval addiction in the first place? We can only speculate that the answer may lie in your interactions with people who were important to you when you were a child. You may have had a parent who was unduly critical when you misbehaved, or who was irritable even at times when you weren't doing anything particularly wrong. Your mother may have snapped, "You're *bad* for doing that!" or your father may have blurted out, "You're *always* goofing up. You'll never learn."

As a small child you probably saw your parents as gods.

They taught you how to speak and tie your shoes, and *most* of what they told you was valid. If Daddy said, "You will be killed if you walk out into traffic," this was *literally true*. Like most children, you might have assumed that nearly everything your parents said was true. So when you heard "You're *no good*" and "You'll *never learn*," you literally *believed* it and this hurt badly. You were too young to be able to reason, "Daddy is *exaggerating* and *overgeneralizing*." And you didn't have the emotional maturity to see that Daddy was irritable and tired that day, or perhaps had been drinking and wanted to be left alone. You couldn't determine whether his outburst was *his* problem or yours. And if you were old enough to suggest he was being unreasonable, your attempts to put things into a sane perspective may have been rapidly deprogrammed and discouraged with a swift smack on the behind.

No wonder you developed the bad habit of automatically looking down on yourself every time someone disapproved of you. It wasn't your fault that you picked up this tendency as a child, and you can't be blamed for growing up with this blind spot. But it *is* your responsibility as an adult to think the issue through realistically, and to take specific steps to outgrow this particular vulnerability.

Just how does this fear of disapproval predispose you to anxiety and depression? John is an unmarried, soft-spoken fifty-two-year-old architect who lives in fear of criticism. He was referred for treatment because of a severe recurring depression, which had not diminished in spite of several years of therapy. One day when he was feeling particularly good about himself, he approached his boss enthusiastically with some new ideas about an important project. The boss snapped, "Later, John. *Can't you see I'm busy!*" John's self-esteem collapsed instantly. He dragged himself back to his office, drowning in despair and self-hatred, telling himself he was no good. "How could I have been so thoughtless?" he asked himself.

As John shared this episode with me, I asked him the simple and obvious questions, "Who was the one who was acting goofy—you or your boss? Were you actually behaving in an inappropriate manner, or was your boss acting irritable and unpleasant?" After a moment's reflection, he was able to identify the true culprit. The possibility that the boss was acting obnoxiously had not occurred to him because of his automatic habit of blaming himself. He felt relief when he suddenly realized he had absolutely nothing to be ashamed of in how he had acted. His boss, who was aloof, was probably under pressure himself and off the mark that day.

John then raised the question, "Why am I always struggling so hard for approval? Why do I fall apart like this?" He then remembered an event that occurred when he was twelve. His only sibling, a younger brother, had tragically died after a long bout with leukemia. After the funeral he overheard his mother and grandmother talking in the bedroom. His mother was weeping bitterly and said, "Now I've got *nothing* to live for." His grandmother responded, "Shush. Johnny is just down the hall! He might hear you!"

As John shared this with me, he began to weep. He *had* heard these comments, and they meant to him, "This proves I'm not worth much. My brother was the important one. My mother doesn't really love me." He never let on that he had been listening, and through the years he tried to push the memory out of his mind by telling himself, "It really isn't important whether or not she loves me anyway." But he struggled intensely to please his mother with his achievements and his career in a desperate bid to win her approval. In his heart he didn't believe he had any true worth, and perceived himself as inferior and unlovable. He tried to compensate for his missing self-esteem by earning other people's admiration and approval. His life was like a constant effort to inflate a balloon with a hole in it.

After recalling this incident, John was able to see the irrationality of his reaction to the comments he had overheard in the hall. His mother's bitterness, and the emptiness she felt, were a natural part of the grieving process that any parent goes through when a child dies. Her comments had *nothing to do with John*, but only with her temporary depression and despair.

Putting this memory into a new perspective helped John see how illogical and self-defeating it was to link his worth to the opinions of others. Perhaps you too are beginning to see that your belief in the importance of external approval is highly unrealistic. Ultimately you, and only you, can make yourself consistently happy. No one else can. Now, let's review some simple steps that you can take to put these principles into practice so you can transform your desire for self-esteem and self-respect into an emotional reality.

The Path to Independence and Self-Respect

Cost-Benefit Analysis. The first step in overcoming your belief in any of the self-defeating assumptions from the DAS test is to perform a cost-benefit analysis. Ask yourself, what are the advantages and disadvantages of telling myself that disapproval makes me less worthwhile? After listing all the ways this attitude hurts you and helps you, you will be in a position to make an enlightened decision to develop a healthier value system.

For example, a thirty-three-year-old married woman named Susan found she was overly involved with church and community activities because she was a responsible and capable worker and was frequently selected for various committees. She felt enormously pleased every time she was chosen for a new job and she feared saying no to any request because that would mean risking someone's disapproval. Because she was terrified about letting people down, she became more and more addicted to the cycle of giving up her own interests and desires in order to please others.

The DAS test and the "Vertical Arrow Technique" described in the previous chapter revealed one of her silent assumptions to be: "I must always do what people expect me to do." She seemed reluctant to give up this belief, so she performed a cost-benefit analysis (Figure 11–1). Be-

Figure 11–1. The Cost-Benefit Method for Evaluating "Silent Assumptions." ASSUMPTION: "I must always do what people expect me to do."

Advantages of Believing This	*Disadvantages of Believing This*
1. If I'm able to meet people's expectations, I can feel I'm in control. This feels good.	1. I sometimes compromise and end up doing things that are not in my best interest that I don't really want to do.
2. When I please people I will feel secure and safe.	2. This assumption keeps me from testing relationships—I never know if I could be accepted just for me. Thus, I always have to earn love and the right to be close to people by doing what people want me to do. I become like a slave.
3. I can avoid a lot of guilt and confusion. I don't have to think things out, since all I have to do is what others want me to.	3. It gives people too much power over me—they can coerce me with the threat of disapproval.
4. I don't have to worry about people being upset with me or looking down on me.	4. It makes it hard for me to know what I really want. I'm not used to setting priorities for myself and making independent decisions.
5. I can avoid conflict and I don't have to be assertive and speak up for myself.	5. When people do disapprove of me, as is inevitable at times, then I conclude I've done something to displease them, and I experience severe guilt and depression. This puts my moods under the control of other people instead of myself.
	6. What other people want me to do may not always be what's best for me, since they often have their own interests at heart. Their expectations for me may not always be realistic and valid.
	7. I end up seeing other people as so weak and fragile that they are dependent on me and would be hurt and miserable if I let them down.

Figure 11–1. cont.

Advantages of Believing This	*Disadvantages of Believing This*
	8. Because I fear taking risks and having someone upset with me, my life becomes static. I don't feel motivated to change, to grow or to do things differently so as to enhance my range of experiences.

cause the disadvantages of her approval addiction greatly outweighed the advantages, she became much more open to changing her personal philosophy. Try this simple technique with regard to one of your self-defeating assumptions about disapproval. It can be an important first step to personal growth.

Rewrite the Assumption. If, based on your cost-benefit analysis, you see that your fear of disapproval hurts you more than it helps, the second step is to rewrite your silent assumption so that it becomes more realistic and more self-enhancing (you can do this with any of the 35 attitudes on the DAS test that represent areas of psychological vulnerability for you). In the above example, Susan decided to revise her belief as follows: "It can be enjoyable to have someone approve of me, but I don't need approval in order to be a worthwhile person or to respect myself. Disapproval can be uncomfortable, but it doesn't mean I'm less of a person."

The Self-Respect Blueprint. As a third step it might help you to write a brief essay entitled "Why It Is Irrational and Unnecessary to Live in Fear of Disapproval or Criticism." This can be your personal blueprint for achieving greater self-reliance and autonomy. Prepare a list of all the reasons why disapproval is unpleasant but not fatal. A few have already been mentioned in this chapter, and you might review them before you begin to write. In your essay include only what seems convincing and helpful to you. Make sure you believe each argument you write down so your new sense of independence will be realistic. *Don't* rationalize! For example, the statement, "If someone disapproves of me, I don't need to get upset because they're really not the kind of person I'd care to have as a friend," won't work because it's a distortion. You are trying to preserve your self-esteem by writing the other person off as no good. Stick with what you know to be the truth.

As new ideas come to you, add them to your list. Read it over every morning for several weeks. This might be a first step in helping you trim other people's negative opinions and comments about you down to life-size.

Here are a few ideas that have worked well for a lot of people. You might use some of them in your own essay.

1. Remember that when someone reacts negatively to you, it may be his or her irrational thinking that is at the heart of the disapproval.
2. If the criticism is valid, this need not destroy you. You can pinpoint your error and take steps to correct it. You can *learn* from your mistakes, and you don't have to be ashamed of them. If you are human, then you *should* and *must* make mistakes at times.
3. If you have goofed up, it does not follow that you are a BORN LOSER. It is impossible to be wrong *all* the time or even *most* of the time. Think about the thousands of things you have done *right* in your life! Furthermore, you can change and grow.
4. Other people cannot judge your worth as a human being, only the validity or merit of specific things you do or say.
5. Everyone will judge you differently no matter how well you do or how badly you might behave. Disapproval cannot spread like wildfire, and one rejection cannot lead to a never-ending series of rejections. So even if worse comes to worst and you do get rejected by someone, you can't end up totally alone.

6. Disapproval and criticism are usually uncomfortable, but the discomfort will pass. Stop moping. Get involved in an activity you've enjoyed in the past even though you feel certain it's absolutely pointless to start.
7. Criticism and disapproval can upset you *only* to the extent that you "buy into" the accusations being brought against you.
8. Disapproval is rarely permanent. It doesn't follow that your relationship with the person who disapproves of you will necessarily end just because you are being criticized. Arguments are a part of living, and in the majority of cases you can come to a common understanding later on.
9. If you are criticizing someone else, it doesn't make that person totally bad. Why give another individual the power and right to judge you? We're all just human beings, not Supreme Court justices. Don't magnify other people until they are larger than life.

Can you come up with some additional ideas? Think about this topic over the next few days. Jot your ideas down on a piece of paper. Develop your own philosophy about disapproval. You'll be surprised to find how much this can help you change your perspective and enhance your sense of independence.

Verbal Techniques. In addition to learning to think differently about disapproval, it can be a lot of help to learn to behave differently toward individuals who express disapproval. As a first step, review the assertive methods such as the disarming technique presented in Chapter 6. Now we will discuss some additional approaches to help you build your skills in coping with disapproval.

First of all, if you fear someone's disapproval, have you ever thought of asking the person if he or she, in fact, *does* look down on you? You might be pleasantly surprised to learn that the disapproval existed only in your head. Although it requires some courage, the payoff can be tremendous.

Remember Art, the psychiatrist described in Chapter 6, who was receiving training at the University of Pennsylvania? Art had no suspicion that a particular patient of his might be suicidal. The patient had no history or symptoms of depression, but felt hopelessly trapped in an intolerable marriage. Art received a call one morning that his patient had been found dead with a bullet hole through his head. Although the suspicion of homicide was raised, the probable cause of death was suicide. Art had never lost a patient in this way. His reaction included sadness, because of his fondness for this particular patient, and anxiety, for fear that his supervisor and peers would disapprove of him and look down on him for his "mistake" and lack of foresight. After discussing the death with his supervisor, he asked frankly, "Do you feel I have let you down?" His supervisor's response conveyed a sense of warmth and empathy, not rejection. Art was relieved when his supervisor told him that he too had experienced a similar disappointment in the past. He emphasized that this was an opportunity for Art to learn to cope with one of the professional hazards of being a psychiatrist. By discussing the case and refusing to give in to his fear of disapproval, Art learned that he *had* made an "error"—he had overlooked the fact that a feeling of "hopelessness" can lead to suicide in individuals who are not clinically depressed. But he also learned that others did not demand perfection of him, and that he wasn't expected to guarantee a successful outcome for any patient.

Suppose it had not turned out so well and his supervisor or peers had condemned him for being thoughtless or incompetent. What then? The worst possible outcome would have been rejection. Let's talk about some strategies for coping with the worst conceivable eventuality.

Rejection Is Never Your Fault! Aside from bodily injury or a destruction of your assets, the greatest pain a person can try to inflict on you is through rejection. This threat is the source of your fear when you are being "put down."

There are several types of rejection. The most common and obvious is called "adolescent rejection," although it is not limited to the adolescent age-group. Suppose you have a romantic interest in someone you are dating or have met, and it turns out you're not his or her cup of tea. Perhaps it's your looks, race, religion, or personality style that are the problem. Or maybe you are too tall, short, fat, thin, old, young, smart, dull, aggressive, passive, etc. Since you don't fit that person's mental image of an ideal mate closely enough, he or she rebuffs your advances and gives you the cold shoulder.

Is this your fault? Obviously not! The individual is simply turning you down because of subjective preferences and tastes. One person may like apple pie better than cherry pie. Does this mean that cherry pie is inherently undesirable? Romantic interests are almost infinitely variable. If you are one of those toothpaste-commercial types who is blessed with what our culture defines as "good looks" and an appealing personality, it will be much easier for you to attract potential dates and mates. But you will learn this mutual attraction is a far cry from developing a loving permanent relationship, and even the beautiful and handsome types will have to cope with rejection sometimes. No one can turn on each and every person they meet.

If you are only average or below average in appearance and personality, you will have to work harder initially to attract people, and you may have to cope with more frequent turndowns. You will have to develop your social skills and master some powerful secrets of making people feel attracted to you. These are: (1) Don't sell yourself short by looking down on yourself. Refuse to persecute yourself. Boost your self-esteem to the hilt with the methods outlined in Chapter 4. If you love yourself, people will respond to this sense of joy you radiate and want to be close to you. (2) Express genuine compliments to people. Instead of waiting around nervously to find out if they will like you or reject you, like them first and let them know about it. (3) Show an interest in other people by learning about what turns them on. Get them to talk about what excites them most, and respond to their comments in an upbeat manner.

If you persevere along these lines, you will eventually discover there *are* people who find you attractive, and you in turn will discover you have a great capacity for happiness. Adolescent rejection is an uncomfortable nuisance, but it's not the end of the world and it's not your fault.

"Ah ha!" you retort. "But how about the situation where a lot of people reject you because you turn them off with your abrasive mannerisms? Suppose you're conceited and self-centered. Certainly that's your fault, isn't it?" This is a second type of rejection, which I call "angry rejection." Again, I think you will see that it's not your fault if you are angrily rejected because of a personal fault.

In the first place, other people aren't obliged to reject you just because they find things about you they don't like—they have other options. They can be assertive and point out what they don't like about your behavior, or they can learn not to let it bother them so much. Of course, they have the right to avoid and reject you if they want, and they are free to choose any friends they prefer. But this doesn't mean that you are an inherently "bad" human being, and it is definitely not the case that everyone will react to you in the same negative way. You will experience a spontaneous chemistry with some people, whereas you will tend to clash with others. This is no one's fault, it's just a fact of life.

If you have a personality quirk that alienates more people than you would like—such as being excessively critical or losing your temper frequently—it would definitely be to your advantage to modify your style. But it's ridiculous to blame yourself if someone rejects you based on this imperfection. We're all imperfect, and your tendency to fault yourself—or to "buy into" the hostility that someone else directs at you—is self-defeating and pointless.

The third type of rejection is "manipulative rejection."

In this case the other person uses the threat of withdrawal or rejection to manipulate you in some way. Unhappy spouses, and even frustrated psychotherapists, sometimes resort to this ploy to coerce you into changing. The formula goes like this: "Either you do such and such or we're all through!" This is a highly irrational and usually self-defeating way of trying to influence people. Such manipulative rejection is simply a culturally taught coping pattern, and it is usually ineffective. It rarely leads to an enhanced relationship because it generates tension and resentment. What it really indicates is a low frustration tolerance and poor interpersonal skills on the part of the individual making the threat. It certainly isn't *your* fault that they do this, and it usually isn't to your advantage to let yourself be manipulated this way.

So much for the theoretical aspects. Now, what can you say and do when you are actually getting rejected? One effective way to learn is to use role-playing. To make the dialogue more entertaining and challenging, I will play the role of the rejector and confront you with the worst things about you I can think of. Since I'm acting caustic and insulting, begin by asking if I am in fact rejecting you because of the way I've been treating you lately:

YOU: Dr. Burns, I notice you've been acting somewhat cool and distant. You seem to be avoiding me. When I try to talk to you, you either ignore me or snap at me. I wonder if you're upset with me or if you've had thoughts of rejecting me.

Comment: You don't accuse me initially of rejecting you. That would put me on the defensive. Furthermore, I might *not* be rejecting you—I might be upset about the fact that nobody's buying my book, so I'm just generally irritable. Just for practice, let's assume the worst—that I am trying to dump you.

DAVID: I'm glad we got it out in the open. I have in fact decided to reject you.

YOU: Why? Apparently I've been turning you off a lot.

DAVID: You're a no-good piece of rot.

YOU: I can see you're upset with me. Just what have I been doing wrong?

Comment: You avoid defending yourself. Since *you* know you are not a "piece of rot," there's no point in insisting to me that you're not. It will just fire me up more, and our dialogue will quickly deteriorate into a shouting match. (This "empathy method" was presented in detail in Chapter 6.)

DAVID: Everything about you stinks.

YOU: Can you be specific? Did I forget to use deodorant? Are you upset by the way I talk, something I've said lately, my clothes, or what?

Comment: Again, you resist getting sucked into an argument. By urging me to pinpoint what I dislike about you, you are forcing me to fire my best shot and say something meaningful or end up looking like an ass.

DAVID: Well, you hurt my feelings when you put me down the other day. You don't give a damn about me. I'm just a "thing" to you, not a human being.

Comment: This is a common criticism. It tips you off that the rejector basically cares for you, but feels deprived and fears losing you. The rejector decides to lash out at you to protect his shaky self-esteem. The rejector might also say you're too stupid, too fat, too selfish, etc. *Whatever* the nature of the criticism, your strategy is now twofold: (a) Find some grain of truth in the criticism and let the rejector know you agree in part (see the "disarming technique,'' Chapter 6); (b) apologize or offer to try to correct any actual error you actually did make (see ''feedback and negotiation,''Chapter 6).

YOU: I'm really sorry I said something that rubbed you the wrong way. What was it?

DAVID: You told me I was a no-good jerk. So I've had it with you—this is the end.

YOU: I can see that was a thoughtless, hurtful comment I made. What other things have I said that hurt your feelings? Was that all? Or have I done this many times? Go ahead and say all the bad things you think about me.

DAVID: You're unpredictable. You can be sweet as sugar, and then all of a sudden you're cutting me to shreds with your sharp tongue. When you get mad, you turn into a foul-mouthed pig. I can't stand you, and I can't see how anyone else puts up with you. You're arrogant and cocky, and don't give a damn about anyone but yourself. You're a selfish snot, and it's time you woke up and learned the hard way. I'm sorry I've got to be the one to put you down, but it's the only way you're going to learn. You have no real feelings for anyone but yourself, and we're through for good!

YOU: Well, I can see there are numerous problems in our relationship we've never looked at, and it sounds like I've really been missing the boat. I can see that I have been acting irritable and thoughtless. I can see how unpleasant I've been and how uncomfortable it's been for you. Tell me more about this side of me.

Comment: You then continue to extract negative comments from the rejector. Avoid being defensive and continue to find some grain of truth in what the rejector says. After you have elicited all the criticisms and agreed with whatever was true about them, you are ready to fire the sharpest arrow straight into the rejector's balloon. Point out that you have acknowledged your imperfections and that you are willing to try to correct your errors. Then ask the rejector why he is rejecting you. This maneuver will help you see why rejection is never your fault! You are responsible for your errors, and you will assume responsibility for trying to correct them. But if someone rejects you for your imperfections, that's their goofiness, not yours! Here's how this works.

YOU: I can see I've done and said a number of things you don't like. I'm certainly willing to try to correct these problems to the greatest extent possible. I can't promise miracles, but if we work at it together, I see no reason why things can't improve. Just by talking this way, our communications are already better. So why are you going to reject me?

DAVID: Because you infuriate me.

YOU: Well, sometimes differences come up between people, but I don't see that this has to destroy our relationship. Are you rejecting me because you feel infuriated or what?

DAVID: You're a no-good bum, and I refuse to talk to you again.

YOU: I'm sorry you feel that way. I'd much prefer to continue our friendship in spite of these hurt feelings. Do we need to break off entirely? Maybe this discussion was just what we needed to understand each other better. I don't really know why you've decided to reject me. Can you tell me why?

DAVID: Oh, no! I'm not being tricked by you. You goofed up once too often, and that's it! No second chances! Good-bye!

Comment: Now whose goofy behavior is this? Yours or the one who is rejecting you? Whose fault is it that the rejection occurs? After all, you offer to try to correct your errors and to improve the relationship through frank communication and compromise. So how can you be blamed for the rejection? Obviously you can't.

Using the above approach may not prevent all actual rejections, but you will enhance the probability of a positive outcome sooner or later.

Recovering from Disapproval or Rejection. You actually have been disapproved of or rejected in spite of your efforts to improve the relationship with the other person. How can you most quickly overcome the emotional upset you understandably feel? First, you must realize that life goes on, so this particular disappointment need not impair the quality of your happiness forever. Following the rejection or disapproval it will be your *thoughts* which are doing the emotional damage, and if you fight these thoughts and stubbornly refuse to give in to distorted self-abuse, the upset will pass.

One method which might be quite helpful is one that has aided people who experience prolonged grief reactions following the loss of a loved one. If bereaved individuals schedule periods each day to allow themselves to be flooded by the painful memories and thoughts of the deceased loved one, this can accelerate and complete the grieving process. If you do this when you are alone, it will be most helpful. Sympathy from another person often backfires; some studies have reported that it prolongs the painful period of mourning.

You can use this "grieving" method to cope with rejection or disapproval. Schedule one or more periods of time each day—five to ten minutes are probably enough—to think all the sad, angry, and despairing thoughts you want. If you feel sad, cry. If you feel mad, pound a pillow. Keep flooding yourself with painful memories and thoughts for the full time period you have set aside. Bitch, moan and complain nonstop! When your scheduled sad period is over, STOP IT and carry on with life until your next scheduled cry session. In the meantime, if you have negative thoughts, write them down, pinpoint the distortions, and substitute rational responses as outlined in previous chapters. You may find this will help you gain partial control over your disappointment and hasten your return to full self-esteem more quickly than you anticipated.

Turning on the "Inner Light"

The key to emotional enlightenment is the knowledge that only your thoughts can affect your moods. If you are an approval addict, you are in the bad habit of flicking your inner switch *only* when someone else shines their light on you first. And you mistakenly confuse their approval with your own self-approval because the two occur almost simultaneously. You mistakenly conclude that the other person has made you feel good! The fact that you do at times enjoy praise and compliments proves that *you know how to approve of yourself*! But if you are an approval addict, you have developed the self-defeating habit of endorsing yourself *only* when someone you respect approves of you first.

Here's a simple way to break that habit. Obtain the wrist counter described in earlier chapters and wear it for at least two or three weeks. Every day try to notice positive things about yourself—things you do well whether or not you get an external reward. Each time you do something you approve of, click the counter. For example, if you smile warmly at an associate one morning, click whether he scowls or smiles back. If you make that phone call you were putting off—click the counter! You can "endorse" yourself for big or trivial things. You can even click it if you *remember* positive things you did in the past. For example, you might recall the day you got your driver's license or your first job. Click the counter whether or not you have a positive emotional arousal. Initially you may have to *force* yourself to notice good things about yourself, and it may seem mechanical. Persist anyway because after several days I think you will notice that the inner light is beginning to glow—dimly at first and then more brightly. Every night look at the digits on the counter and record the total number of personal endorsements on your daily log. After two or three weeks, I suspect you will begin to learn the art of self-respect, and you will feel much better about yourself. This simple procedure can be a big first step toward achieving independence and self-approval. It sounds easy—and it is. It's surprisingly powerful, and the rewards will be well worth the small amount of time and effort involved.

Chapter 12

The Love Addiction

The "silent assumption" which often goes hand in hand with the fear of disapproval is "I cannot be a truly happy and fulfilled human being unless I am loved by a member of the opposite sex. True love is necessary for ultimate happiness."

The *demand* or *need* for love before you can feel happy is called "dependency." Dependency means that you are unable to assume responsibility for your emotional life.

The Disadvantages of Being a Love Junkie. Is being loved an absolute necessity or a desirable option?

Roberta is a thirty-three-year-old single woman who moped around her apartment evenings and weekends because she told herself, "It's a couple's world. Without a man I am nothing." She came to my office attractively groomed, but her comments were bitter. She was brimming with resentment because she was sure that being loved was as crucial as the oxygen she breathed. However, she was so needy and greedy that this tended to drive people away.

I suggested that she start by preparing a list of the advantages and disadvantages of believing that "without a man (or woman) I am nothing." The disadvantages on Roberta's list were clear-cut: "(1) This belief makes me despondent since I have no lover. (2) Furthermore, it takes away any incentive I might have to do things and go places. (3) It makes me feel lazy. (4) It brings on a sense of self-pity. (5) It robs me of self-pride and confidence, and makes me envious of others and bitter. (6) Finally, it brings on self-destructive feelings and a terrible fear of being alone."

Then she listed what she thought were the advantages of believing that being loved was an absolute necessity for happiness: "(1) This belief will bring me a companion, love, and security. (2) It will give purpose to my life and a reason to live. (3) It will give me events to look forward to." These advantages reflected Roberta's belief that telling herself she couldn't live without a man would somehow bring a companion into her life.

Were these advantages real or imaginary? Although Roberta had believed for many years that she couldn't exist without a man, this attitude still hadn't brought a desirable mate. She admitted that making men so totally important in her life was not the magic charm that would bring one to her doorstep. She acknowledged that clinging and dependent individuals often demand so much attention from other people and appear so needy that they have great difficulty not only initially attracting people of the opposite sex but also maintaining an ongoing relationship. Roberta was able to grasp the idea that people who have found happiness within themselves are usually the most desirable to members of the opposite sex and become like magnets because they are at peace and generate a sense of joy. Ironically, it is usually the dependent woman, the "manaholic," who ends up alone.

This really isn't so surprising. If you take the position you "need" someone else for a sense of worth, you broadcast the following: "Take me! I have no inherent worth! I can't stand myself!" No wonder there are so few buyers! Of course, your unstated demand does not endear people to you either: "Since you're *obliged* to love me, you're rotten shit if you don't."

You may cling to your dependency because of the erroneous notion that if you do achieve independence, others will see you as a rejecting person and you will end up alone. If this is your fear, you are equating dependency with warmth. Nothing could be farther from the truth. If you are lonely and dependent, your anger and resentment stem from the fact that you feel deprived of the love you believe you are entitled to receive from others. This attitude drives you farther into isolation. If you are more independent, you are not *obliged* to be alone—you simply have the capacity to feel happy when you are alone. The more independent you are, the more secure you will be in your feelings. Furthermore, your moods will not go up and down at someone else's mercy. After all, the amount of love that someone can feel for you is often quite unpredictable. They may not appreciate everything about you, and they may not act in an affectionate way all the time. If you are willing to learn to love yourself, you will have a far more dependable and continuous source of self-esteem.

The first step is to find out if you *want* independence. All of us have a much greater chance of achieving our goals if we understand what they are. It helped Roberta to realize that her dependency was condemning her to an empty existence. If you are still clinging to the notion that it is desirable to be "dependent," list the advantages, using the double-column technique. Spell out how you benefit if you let love determine your personal worth. Then in order to assess the situation objectively, write down the counterarguments, or rational responses, in the right-hand column. You may learn that the advantages of your love addiction are partially or totally illusory. Figure 12–1 shows how a woman with a problem similar to Roberta's assessed these issues. This written exercise motivated her to look within herself for what she had been seeking in others, and enabled her to see that her dependency was the real enemy because it incapacitated her.

Perceiving the Difference Between Loneliness and Being Alone. As you read the previous section you may

Figure 12–1. An Analysis of the Presumed "Advantages" of Being a "Love Junkie."

Advantages of Being Dependent on Love to Be Happy	*Rational Responses*
1. Someone will take care of me when I'm hurt.	1. This is also true of independent people. If I am in an auto accident, they will take me to an emergency room. The doctors will care for me whether I am a dependent or independent person. It is nonsense that only dependent people get help when they are hurt.
2. But if I am dependent, I won't have to make decisions.	2. But as a dependent person, I will have much less control over my life. It is unreliable to depend on other people to make decisions for me. For example, do I want someone else to tell me what to wear today or what to eat for dinner? They might not choose the thing that is my first choice.
3. But as an independent person, I might make the wrong decision. Then I'd have to pay the consequences.	3. So pay the consequences—you can learn from your mistakes if you are independent. No one can be perfect, and there are no guarantees of absolute certainty in life. The uncertainty can be part of the spice of life. It's how I cope—not whether I am right all the time—that forms the basis of self-respect. And besides, I will be able to take the credit when things work out well.
4. But if I am a dependent person, I won't have to think. I can just react to things.	4. Independent people can also choose not to think if they want to. There is no rule that says that only dependent people have the right to stop thinking.

Figure 12–1. cont.

Advantages of Being Dependent on Love to Be Happy	*Rational Responses*
5. But if I am dependent, I will be gratified. It will be like eating candy. It feels good to have someone to care for me and to lean on.	5. Candy gets nauseating after a while. The person I choose to depend on may not be willing to love and stroke me, and take care of me forever. He may get tired of it after a while. And if he withdraws from me either through anger or resentment, I will then feel miserable because I'll have nothing else to rely on. They will be able to manipulate me if I am dependent, just like a slave or robot.
6. But if I am a dependent person, I will be loved. Without love I couldn't live.	6. As an independent person, I can learn to love myself and this may make me even more desirable to others, and if I can learn to love myself, I can *always* be loved. My dependency in the past has driven others away from me more frequently than it has attracted people to me. Babies can't survive without love and support, but I won't die without love.
7. But some men are looking for dependent women.	7. There's some truth to this, but relationships which are based on dependency frequently fall apart and culminate in divorce because you are asking the other person to give you something which they are not in the position to give: namely, self-esteem and self-respect. Only I can make myself happy, and if I rely on someone else to do this for me, I am likely to be bitterly disappointed in the end.

have concluded that it would be to your advantage if you could learn to regulate your moods and find happiness within yourself. This would give you the capacity to feel as alive when you are alone as when you are with someone you love. But you may be thinking, "That all sounds well and good, Dr. Burns, but it is not realistic. The truth is that it is undeniably emotionally inferior to be alone. All my life I have known that love and happiness are identical, and all my friends agree. You can philosophize until you're blue in the face. But when it comes down to the bottom line, love is where it's at and being alone is a curse!"

In fact, many people are convinced that love makes the world go around. You see this message in ads, you hear it in popular songs, you read it in poems.

You can however convincingly disprove your assumption that love is necessary before you can experience happiness. Let's take a hard look at the equation, alone = lonely.

Consider, first, that we get many of life's basic satisfactions by ourselves. For example, when you climb a mountain, pick a flower, read a book, or eat a hot fudge sundae, you do not require someone else's company for these experiences to be enjoyable. A physician can enjoy the satisfaction of treating a patient whether or not he and the patient are involved in a meaningful personal relationship. When writing a book, an author is generally by himself or herself. As most students know, you do most of your learning when you are alone. The list of pleasures and satisfactions that you can enjoy when alone is endless.

This indicates that many sources of gratification are accessible to you whether or not you are with someone else. Can you add to that list? What are some pleasures that you can have alone? Do you ever listen to good music on your stereo? Do you enjoy gardening? Jogging? Carpentry? Hiking? A lonely bank teller named Janet, who was recently separated from her husband, enrolled in a creative dancing class and found (to her surprise) that she could derive enormous pleasure from practicing by herself at home. As she became caught up in the rhythm of the movements, she felt at peace with herself in spite of the fact that she had no one to love.

Perhaps you are thinking now, "Oh, Dr. Burns, is that your point? Well, it's *trivial*! Of course, I can experience temporary moments of mediocre distraction by doing things when I'm alone. This might take the edge off the blues, but those things are just some crumbs from the table that might keep me from starving totally. I want the banquet, the real thing! Love! True and complete happiness!"

That was exactly what Janet told me before she enrolled in the dancing class. Because she assumed it was miserable to be alone, it hadn't occurred to her to do enjoyable things and care for herself during the separation from her husband. She had been living according to a double standard whereby if she was with her husband, she would go to great lengths to plan pleasurable activities, but when she was alone, she would simply mope and do very little. This pattern obviously functioned as a self-fulfilling prophecy, and she did in fact find it unpleasant to be alone. Why? Simply because she failed to treat herself in a caring way. It had never occurred to her to challenge her lifelong assumption that all her activities would be unsatisfactory unless she had someone to share them with. On another occasion, instead of heating a TV dinner after work, Janet decided to plan a special meal, just as if she were going to entertain a man she cared a lot about. She carefully prepared her dinner and set the table with candles. She began with a glass of fine wine. After dinner she read a good book and listened to her favorite music. To her amazement, she found the evening a total pleasure. The next day, which was Saturday, Janet decided to go to the art museum alone. She was surprised to discover that she got more enjoyment out of this excursion alone than she had in the past when dragging her reluctant and disinterested husband along.

As a result of adopting an active, compassionate attitude toward herself, Janet discovered for the first time in her life that she could not only make it on her own but could really enjoy herself.

As is so often the case, she began to generate an infectious joy of living that caused many individuals to feel attracted to her, and she began to date. In the meantime her husband began to get disillusioned with his girl friend and wanted his wife back. He noticed Janet was happy as a lark without him, and at this point the tables began to turn. After Janet told him she no longer wanted him back, he suffered a severe depression. She ultimately established a very satisfying relationship with another man and remarried. The key to her success was simple—as a first step, she proved that she could develop a relationship with herself. After this, the rest was easy.

The Pleasure-Predicting Method

I don't expect you to rely on my word on this topic, or even on the reports of others like Janet who have learned how to experience the joys of self-reliance. Instead, I propose you perform a series of experiments, just as Janet did, to test out your belief that "being alone is a curse." If you are willing to do this, you can arrive at the truth in an objective, scientific manner.

To help you, I have developed the "Pleasure-Predicting Sheet" shown in Figure 12–2. This form is divided into a series of columns in which you predict and record the actual amount of satisfaction you derive from various work and recreational activities you engage in when alone, as well as from those you share with other people. In the first column, record the date of each experiment. In the second column, write down several activities that you plan to do as a part of that day's experiments. I suggest that you carry out a series of forty or fifty experiments over a two- to three-week period. Choose activities that would ordinarily give you a sense of accomplishment or pleasure, or which have the potential for learning or personal growth. In the third column, record who you do the activity with. If you do it alone, write "self" in this column. (This word will remind

Figure 12–2. The Pleasure-Predicting Sheet.

Date	*Activity for Satisfaction. (Sense of Achievement or Pleasure)*	*Who Did You Do This With? (If Alone, Specify Self)*	*Predicted Satisfaction (0–100%). (Write This Before the Activity)*	*Actual Satisfaction (0–100%). (Record This After the Activity)*
8/18/99	Visit arts and crafts center	self	20%	65%
8/19/99	Go to rock concert	self	15%	75%
8/26/99	Movie	Sharon	85%	80%
8/30/99	Party	Many invited guests	60%	75%
9/2/99	Read novel	self	75%	85%
9/6/99	Jogging	self	60%	80%
9/9/99	Go shopping for blouse at boutique	self	50%	85%
9/10/99	Go to market	mother	40%	30% (argument)
9/10/99	Walk to the park	Sharon	60%	70%
9/14/99	Date	Bill	95%	80%
9/15/99	Study for exam	self	70%	65%
9/16/99	Go for driving test	mother	40%	95% (passed test!)
9/16/99	Ride bicycle to ice cream store	self	80%	95%

you that you are never really alone, since you are always with yourself!) In the fourth column, predict the satisfaction you think you will derive from this activity, estimating it on a scale of between 0 and 100 percent. The higher the number, the greater the anticipated satisfaction. Fill in the fourth column *before* you do each planned activity, not after!

Once you have filled in the columns, proceed with the activities. Once they are completed, record the actual satisfaction in the last column, using the same 0- to 100-percent rating system.

After you have performed a series of such experiments, you will be able to interpret the data you have collected. You can learn many things. First, by comparing the predicted satisfaction (column four) with the actual satisfaction (column five), you will be able to find out how accurate your predictions are. You may find that you typically underestimate the amount of satisfaction you anticipate experiencing, especially when doing things alone. You might also be surprised to learn that activities with others are not always as satisfying as anticipated. In fact, you may even find that there are many times when it was *more* enjoyable to be alone, and you might discover that the highest ratings you received when you were alone were equal to or higher than those for activities involving others. It can be helpful to compare the amount of satisfaction you derived from work activities versus pleasurable activities. This information can help you achieve an optimal balance between work and fun as you continue to plan your activities.

Questions are probably now crossing your mind, "Suppose I do something and it *isn't* as satisfying as I predicted? Or suppose I make a low prediction and it really comes out that way?" In this case try to pinpoint the automatic negative thoughts that dampen the experience for you. Then talk back to these thoughts. For example, a lonely sixty-five-year-old woman whose children were all grown and married decided to enroll in an evening course. All the other students were of college freshman age. She felt tense the first week of classes because of her thought, "They probably think I'm an old bag with no right to be here.'' When she reminded herself she had no idea what the other students thought of her, she felt some relief. After talking to another student, she found out that some of them admired her gumption. She then felt much better, and her satisfaction levels began to climb.

Now let's see how the Pleasure-Predicting Sheet can be used to overcome dependency. Joanie was a fifteen-year-old high-school student who had suffered from a chronic depression for several years after her parents moved to a new town. She had difficulty making friends in the new high school, and believed, as many teenage girls do, that she had to have a boyfriend and be a member of the ''in crowd'' before she could be happy. She spent nearly all her free time at home alone, studying and feeling sorry for herself. She resisted and resented the suggestion she start going out and doing things because she claimed there would simply be no point in doing them alone. Until a circle of friends magically dropped into her lap, she seemed determined to sit and brood.

I persuaded Joanie to use the Pleasure-Predicting Sheet. Figure 12–2 shows that Joanie scheduled a variety of activities, such as visiting an arts and crafts center on a Saturday, going to a rock concert, etc. Because she did them alone, she anticipated they would be unrewarding, as indicated by her low predictions in column four. She was surprised to find she actually did have a reasonably good time. As this pattern tended to repeat itself, she began to realize that she was predicting things in an unrealistic negative way. As she did more and more on her own, her mood began to improve. She still *wanted* friends, but no longer felt condemned to misery when she was alone. Because she proved she could make it on her own, her self-confidence went up. She then became more assertive with her peers, and invited several people to a party. This helped her develop a network of friends, and she found that boys as well as girls in her high-school class were interested in her. Joanie continued to use the Pleasure-Predicting Sheet to evaluate the levels of satisfaction she experienced in dates and activities with her new friends. She was surprised to find that they were comparable to the enjoyment levels she experienced in doing things alone.

There is a difference between wanting and needing something. Oxygen is a *need*, but love is a *want*. I repeat: LOVE IS NOT AN ADULT HUMAN NEED! It's okay to *want* a loving relationship with another human being. There is nothing wrong with that. It is a delicious pleasure to be involved in a good relationship with someone you love. But you do not *need* that external approval, love, or attention in order to survive *or* to experience maximal levels of happiness.

Attitude Modification. Just as love, companionship, and marriage are not necessary for happiness and self-esteem, they are not sufficient either. The proof of this is the millions of men and women who are married and miserable. If love were the antidote to depression, then I would soon be out of business because the vast majority of the suicidal individuals I treat are in fact loved very dearly by their spouses, children, parents, and friends. Love is not an effective antidepressant. Like tranquilizers, alcohol, and sleeping pills, it often makes the symptoms worse.

In addition to restructuring your activities more creatively, challenge the upsetting negative thoughts that flow through your mind when you are alone.

This was helpful to Maria, a lovely thirty-year-old single woman, who found that when she did activities on her own, she sometimes soured the experience unnecessarily by telling herself, "Being alone is a curse." In order to combat the feelings of self-pity and resentment this thought created, she wrote a list of counterarguments (see Figure 12–3, page 323). She reported this was very helpful in breaking the cycle of loneliness and depression.

Over a year after terminating my work with her I sent her an early draft of this chapter, and she wrote back: "Last night I read very thoroughly the chapter . . . It proves that

Figure 12–3. "Being alone is a curse." Counterarguments: The advantages of being alone.

1. Being alone gives a person the opportunity to explore what she or he really thinks, feels, and knows.
2. Being alone gives the person a chance to try all sorts of new things that might be harder to try if one had ties to a housemate, spouse, etc.
3. Being alone forces you to develop your personal strengths.
4. Being alone enables you to put aside excuses for taking responsibility for yourself.
5. Being a woman alone is better than being a woman with an unsuitable male mate. The same applies to a man.
6. Being a woman alone can be an opportunity to develop into a full human being and not be an appendage to a man.
7. Being a woman alone can be helpful in making you more understanding of the problems women in different situations face. This can help you learn to be more supportive of other women and can enable you to develop more meaningful relationships with them. The same could also apply to men and their understanding of various male problems.
8. Being a woman alone can show a woman that even if she later lived with a man, she need not be constantly afraid of his leaving her or dying. She knows that she can live alone and has the potential for happiness within herself; thus, the relationship can be one of mutual enhancement rather than one of mutual dependency and demandingness.

it is not being alone that is so bad or so good, but rather *how one thinks* regarding that or any other condition of being. *Thoughts* are so powerful! They can make or break you, right? . . . It is almost funny, but now I am almost afraid to 'have a man.' I do rather well, maybe better, without one . . . Dave, did you ever think you would hear this from me?''

The double-column technique can be especially useful in helping you overcome the negative thinking pattern that makes you fear standing on your own two feet. For example, a divorced woman with one child contemplated suicide because her lover—a married man—had broken off with her. She had an intensely negative self-image, and didn't believe that she would ever be capable of sustaining an ongoing relationship. She was sure she would always end up a reject and a loner. She wrote in her journal the following thoughts as she contemplated a suicide attempt:

> The empty place in the bed next to me silently mocks me. I am alone—alone—my greatest fear, my most dreaded fate, a reality. I am a woman alone and in my mind that means I am nothing. The logic I am operating on goes something like this:
>
> 1. If I were desirable and attractive there would be a man beside me now.
> 2. There is no man beside me.
> 3. Therefore I am undesirable and unattractive.
> 4. Therefore there is no point in living.

She went on to ask herself in her journal, "Why do I need a man? A man would solve all my problems. He would take care of me. He would give my life direction and most importantly he would provide me with a reason to get out of bed each morning when all I now want to do is put my head under the covers and sink into oblivion."

She then utilized the double-column technique as a way of challenging the upsetting thoughts in her mind. She labeled the left-hand column "Accusations of My Dependent Self," and labeled the right-hand column "Counterarguments of My Independent Self." She then carried out a dialogue with herself to determine what the truth of the matter really was (see Figure 12–4, page 325).

After doing the written exercise, she decided to read it over each morning in order to develop the motivation to get out of bed. She wrote the following outcome in her personal diary:

Figure 12–4.

Accusations of My Dependent Self	*Counterarguments of My Independent Self*
1. I need a man.	1. Why do you need a man?
2. Because I can't cope on my own.	2. Have you been coping so far in life?
3. Okay. But I'm lonely.	3. Yes, but you have a child and you do have friends, and you have enjoyed being with them very much.
4. Yes, but they don't count.	4. They don't count because you dismiss them.
5. But people will think no man wants me.	5. People will think what they want to think. What is important is what you think. Only your thoughts and beliefs can affect your moods.
6. I think I am nothing without a man.	6. What did you accomplish having a man that you couldn't accomplish on your own?
7. Actually nothing. Everything important I've done on my own.	7. Then why do you need a man?
8. I guess I don't need a man. I just want one.	8. It's fine to want things. They just can't become so important that life loses its meaning without them.

I learned to see that there is a big difference between wanting and needing. I want a man but I no longer feel that I must have a man to survive. By maintaining a more realistic inner dialogue with myself and by looking at my own strengths, by listing and reading and reading again the things that I have obtained on my own, I slowly am beginning to develop a sense of confidence in my ability to handle what might come. I find that I am taking better care of myself. I am treating myself as I would have treated a beloved friend in the past with kindness and compassion, with a tolerance for flaws and an appreciation of assets. Now I can view a difficult situation not as a pestilence especially contrived to plague me but as an opportunity to practice the skills I am learning, to challenge my negative thoughts, to reaffirm my strengths and to enhance my confidence in my ability to deal with life.

Chapter 13

Your Work Is Not Your Worth

A third silent assumption that leads to anxiety and depression is "My worth as a human being is proportional to what I have achieved in my life." This attitude is at the core of Western culture and the Protestant work ethic. It sounds innocent enough. In fact, it is self-defeating, grossly inaccurate, and malignant.

Ned, the physician described in earlier chapters, called me at home one recent Sunday evening. He had been feeling panicky all weekend. His upset was triggered by plans to attend the twentieth reunion of his college class (he graduated from an Ivy League college). He had been invited to give the keynote address to the alumni. Why was Ned in such a state of apprehension? He was concerned that he might meet up with some classmate at his reunion who had achieved more than he had. He explained why this was so threatening: "It would mean I was a failure."

Ned's exaggerated preoccupation with his achievements is particularly common among men. While women are not immune to career concerns, they are more likely to be depressed after the loss of love or approval. Men, in contrast, are especially vulnerable to concerns about career failure because they've been programmed from childhood to base their worth on their accomplishments.

The first step in changing any personal value is to determine if it works more to your advantage or disadvantage. Deciding that it will not really help you to measure your worth by what you produce is the crucial first step in changing your philosophy. Let's begin with a pragmatic approach, a cost-benefit analysis.

Clearly, there *can be* some advantage to equating your self-esteem with your accomplishments. In the first place, you can say "I'm okay" and feel good about yourself when you have achieved something. For example, if you win a golf game, you can pat yourself on the back and feel a little smug and superior to your partner because he missed his putt on the last hole. When you go jogging with a friend and he runs out of breath before you do, you can puff up with pride and tell yourself, "He's a good guy for sure, but *I'm just a little better!*" When you make a big sale at work, you can say, "I'm producing today. I'm doing a good job. My boss will be pleased and *I can respect myself.*" Essentially, your work ethic allows you to feel you've earned personal worth and the right to feel happy.

This belief system may make you especially motivated to produce. You might put extra effort into your career because you're convinced this will give you extra worthiness units, and you will therefore see yourself as a more desirable person. You can avoid the horrors of being "just average." In a nutshell, you may work harder to win, and when you win you may like yourself better.

Let's look at the other side of the coin. What are the disadvantages of your philosophy of "worth equals achievement"? First, if your business or career is going well, you may become so preoccupied with it that you may inadvertently cut yourself off from other potential sources of satisfaction and enjoyment as you slave away from early morning to late night. As you become more and more of a workaholic, you will feel excessively driven to produce because if you fail to keep up the pace, you will experience a severe withdrawal characterized by inner emptiness and despair. In the absence of achievement, you'll feel worthless and bored because you'll have no other basis for self-respect and fulfillment.

Suppose as a result of illness, business reversal, retirement, or some other factor beyond your control, you find you are unable to produce at the same high level for a period of time. Now you may pay the price of a severe depression, triggered by the conviction that because you are less productive it means you are no good. You'll feel like a tin can that's been used and is now ready for the trash. Your lack of self-esteem might even culminate in a suicide attempt, the ultimate payment for measuring your worth exclusively by the standards of the marketplace. Do you want this? Do you need this?

There may be other prices to pay. If your family suffers from your neglect, a certain resentment may build up. For a long time they may hold it in, but sooner or later you'll get the bill. Your wife has been having an affair and is talking about divorce. Your fourteen-year-old son has been arrested for burglary. When you try to talk with him, he snubs you: "Where've you been all these years, Dad?" Even if these unfortunate developments do not happen to you, you will still have one great disadvantage—the lack of true self-esteem.

I have recently begun treating a very successful businessman. He claims to be one of the top money earners in the world in his profession. Yet he is victimized by episodic states of fear and anxiety. What if he should fall off the pinnacle? What if he had to give up his Rolls-Royce Silver Cloud and drive a Chevrolet instead? That would be unbearable! Could he survive? Could he still love himself? He doesn't know if he could find happiness without the glamour or glory. His nerves are constantly on edge because he can't answer these questions. What would *your* answer be? Would you still respect and love yourself if you experienced a substantial failure?

As with any addiction, you find that greater and greater doses of your "upper" will be needed in order to become "high." This tolerance phenomenon occurs with heroin, "speed" (amphetamines), alcohol, and sleeping pills. It also happens with riches, fame, and success. Why? Perhaps because you automatically set your expectations higher and higher once you have achieved a particular level. The excitement quickly wears off. Why doesn't the aura last? Why do you keep needing more and more? The answer is obvious: Success does not guarantee happiness. The two are not identical and are not causally related. So you end up chasing a mirage. Since your *thoughts* are the true key to your moods and not success, the thrill of victory fades quickly. The old achievements soon become old hat—you begin to feel sadly bored and empty as you stare at your trophy case.

If you do not get the message that happiness does not reliably and necessarily follow from success, you may work even harder to try to recapture the feeling you once had from being on top. This is the basis for your addiction to work.

Many individuals seek guidance or therapy because of the disillusionment that begins to dawn on them in their middle or later years. Eventually these questions may confront you as well: What's my life all about? What's the meaning of it all? You may believe your success makes you worthwhile, but the promised payoff seems elusive, just beyond your grasp.

As you read the above paragraphs, you may suspect that the disadvantages of being a success junkie outweigh the advantages. But you may still believe it is basically *true* that people who are superachievers are more worthwhile—the big shots seem "special" in some way. You may be convinced that true happiness, as well as the respect of others, comes primarily from achievement. But is this really the case?

In the first place, consider the fact that most human beings are not great achievers, yet most people are happy and well respected. In fact, one could say that the majority of the people in the United States are loved and happy, yet by definition most of them are pretty much average. Thus, it *cannot* be the case that happiness and love come only through great achievement. Depression, like the plague, is no respecter of status and strikes those who live in fancy neighborhoods as often—if not more frequently—as it does those of average or below-average means. Clearly, happiness and great achievement have no necessary connection.

Does Work = Worth?

Okay, let's assume you've decided that it's not to your advantage to link your work and your worth, and you also admit that achievement will not reliably bring you love, respect, or happiness. You may still feel convinced that on *some level*, people who achieve a lot are somehow better than others. Let's take a hard look at this notion.

First, would you say that everybody who achieves is particularly worthwhile just because of their achievement? Adolf Hitler was clearly a great achiever at the height of his career. Would you say that made him particularly worthwhile? Obviously not. Of course, Hitler would have insisted he was a great human being because he was a successful leader and because he equated his worth and achievements. In fact, he was probably convinced that he and his fellow Nazis were supermen because they were achieving so much. Would you agree with them?

Perhaps you can think of a neighbor or someone you don't like very much who does achieve a lot and yet seems overly grasping and aggressive. Now, is that person especially worthwhile in your opinion just because he or she is an achiever? In contrast, perhaps you know someone you care for or respect who is not a particularly great achiever. Would you say that person is still worthwhile? If you answer yes, then ask yourself—if they can be worthwhile without great achievement, then why can't I be?

Here's a second method. If you insist your worth is determined by your achievement, you are creating a self-esteem equation: worth = achievement. What is the basis for making this equation? What objective proof do you have that it is valid? Could you experimentally measure people's worth as well as their achievement so as to find out if they were in fact equal? What units would you use to measure it? The whole idea is nonsense.

You can't prove the equation because it is just a stipulation, a *value system*. You're defining worth as achievement and achievement as worth. Why define them as each other? Why not say worth is worth and achievement is achievement? Worth and achievement are different words with different meanings.

In spite of the above arguments, you may still be convinced that people who achieve more are better in some way. If so, I'm going to hit you now with a most powerful method which, like dynamite, can shatter this attitude even when it appears to be etched in granite.

First, I would like you to play the role of Sonia (or Bob), an old friend from high-school days. You have a family and teach school. I have pursued a more ambitious career. In the dialogue you will assume that human worth is determined by achievement, and I will push the implications of this to their obvious, logical, and obnoxious conclusion. Are you ready? I hope so because you're about to be assaulted in a most unpleasant way by a belief you apparently still cherish.

DAVID: Sonia (or Bob), how are you doing?

YOU (playing the role of my old friend): Just fine, David. How are you?

DAVID: Oh, great. I haven't seen you since high school. What's been happening?

YOU: Oh, well, I got married, and I'm teaching at Parks High School and I have a little family at home. Things are great.

DAVID: Well, gee. I'm sorry to hear that. I turned out a lot better than you.

YOU: How's that? Come again?

DAVID: *I* went to graduate school and *I* got my Ph.D. and *I* have become quite successful in business. I'm earning a lot of money. In fact, I'm one of the wealthier people in town now. I've achieved a great deal. More than you by a long shot. I don't mean to insult you or anything, but I guess that means I'm a lot better person than you, huh?

YOU: Well, gee, Dave, I'm not sure what to say. I thought I was a rather happy person before I started to talk to you.

DAVID: I can understand that. You're at a loss for words, but you might as well face facts. I've got what it takes, and you don't. I'm *glad* you're happy, though. Mediocre, average people are entitled to a little happiness too. After all, I certainly don't begrudge you a few crumbs from the banquet table. But it's just too bad you couldn't have done more with your life.

YOU: Dave, you seem to have changed. You were such a nice person in high school. I get the feeling you don't like me anymore.

DAVID: Oh, no! We can still be friends as long as you admit you're an inferior, second-rate person. I just want to remind you to look up to me from now on, and I want you to realize that I'll look down on you because I'm more worthwhile. This follows from the assumption that we have—worth equals achievement. Remember that attitude you cherish? I've *achieved* more, so I'm worth more.

YOU: Well, I sure hope I don't run into you soon again, Dave. It's not been such a pleasure talking to you.

That dialogue cools most people off very quickly because it illustrates how the inferior-superior system follows logically from equating your worth with your achievement.

Actually, many people do feel inferior. The role-playing can help you see how ludicrous the assumption is. In the above dialogue, who was acting jerky? The happy housewife/schoolteacher or the arrogant businessman trying to make a case that he was better than other people? I hope this imaginary conversation will help you see clearly how screwball the whole system is.

If you like, we can do a role-reversal to put the icing on the cake. This time *you* play the role of the very successful person, and I want you to try to put me down as sadistically as you can. You can pretend to be the editor of *Cosmopolitan* magazine, Helen Gurley Brown.* I went to high school with you; I'm just an average high-school teacher now, and it's your job to argue that you're better than I am.

YOU (playing the role of Helen Gurley Brown): Dave, how have you been? It's been a long time.

DAVID: (playing the role of a high-school teacher): Well, fine. I have a little family, and I'm teaching high school here. I'm a physical education teacher and really enjoying life. I understand you've made it big.

YOU: Yeah. Well, I really have been kind of lucky. I'm editor of *Cosmopolitan* now. Perhaps you heard.

DAVID: Of course I have. I've seen you on TV on the talk shows plenty of times. I hear you make a huge income, and you even have your own agent.

YOU: Life's been good. Yeah. It's really been terrific.

DAVID: Now there's just one thing I heard about you that I really didn't understand. You were talking to a friend of ours, and you were saying how you're so much better than I am now that you've made

*This is a purely imaginary dialogue having no bearing on the real Helen Gurley Brown.

it big, whereas my career is just average. What did you mean by that?

YOU: Well, Dave, I mean, just think about all the things I've accomplished in my life. Here I am influencing millions, and whoever heard of Dave Burns in Philadelphia? I'm hobnobbing with the stars, and you're bouncing a basketball around in the court with a bunch of kids. Don't get me wrong. You're certainly a fine, sincere, average person. It's just that you never made it, so you might as well face facts!

DAVID: You've made a great impact, and you're a woman of influence and fame. I respect that a lot, and it sounds quite rewarding and exciting. But please forgive me if I'm dense. I just don't understand how that makes you a better person. How does that make me inferior to you or make you more worthwhile? With my little local mind, I must be missing something obvious.

YOU: Face it, you just sit around and interact with no particular purpose or destiny. I have charisma. I'm a mover and shaker. That gives me a bit of an edge, wouldn't you say?

DAVID: Well, I don't interact to *no* purpose, but my purposes may seem modest in comparison with yours. I teach phys ed, and I coach the local football games and that kind of thing. Your orbit is certainly big and fancy in comparison with mine. But I don't understand how that makes you a better person than I am, or how it follows that I'm inferior to you.

YOU: I'm just more highly developed and more elaborate. I think about more important things. I go on the lecture circuit, and people flock to hear me by the thousands. Famous authors work for me. Who do you lecture to? The local PTA?

DAVID: Certainly in achievement, money, and influence you're way ahead of me. You've done very well. You were very bright to begin with, and you've worked very hard. You're a big success now. But how does that make you more worthwhile than I am? You must forgive me, but I still don't grasp your logic.

YOU: I'm more *interesting*. It's like an amoeba versus a highly developed biological structure. Amoebas are kind of boring after a while. I mean your life must be like an amoeba's. You're just bumbling around aimlessly. I'm a more interesting, dynamic, desirable person; you're second-rate. You're the burnt toast; I'm the caviar. Your life is a bore. I don't see how I can say it more clearly.

DAVID: My life isn't as boring as you might think. Take a close look at it. I'd be surprised to hear what you have to say here because I can't find *anything* boring about my life. What I do is exciting and vital to me. The people I teach are every bit as important to me as the glamorous movie stars you interact with. But even if it *were* true that my life was more tedious and routine and less interesting than yours, how would that make you a better person or more worthwhile?

YOU: Well, I suppose it just really boils down to the fact that if you have an amoeba existence, then you can only judge it on the basis of your amoeba mentality. I can judge your situation, but you can't judge mine.

DAVID: What is the basis for your judgment? You can call me an amoeba, but I don't know what that means. You seem to be reduced to name-calling. All it means is that apparently my life is not especially interesting to you. Certainly I'm not nearly as successful or glamorous, but how does that make you a better or more worthwhile person?

YOU: I'm almost starting to give up.

DAVID: Don't give up here. Press on. Perhaps you *are* a better person!

YOU: Well, certainly society values me more. That's what makes me better.

DAVID: It makes you more highly valued by society. That's undoubtedly the case. I mean Johnny Carson hasn't contacted me for any appearances recently.

YOU: I've noticed that.

DAVID: But how does being more highly valued by society make you a more worthwhile person?

YOU: I'm earning a huge salary. I'm worth millions. Just how much *are* you worth, Mr. Schoolteacher?

DAVID: You clearly have more financial worth. But how does that make you a more *worthwhile human being*? How does commercial success make you a better person?

YOU: Dave, if you're not going to worship me, I'm not going to talk to you.

DAVID: Well, I don't see how that would make me less worthwhile either. Unless you have the idea that you're going to go around deciding who's worthwhile based on who worships you!

YOU: Of course I do!

DAVID: Does that go along with being editor of *Cosmopolitan*? If so, please tell me how you make these decisions. If I'm not worthwhile, I'd definitely like to know why so that I can give up feeling good and considering myself equal to other people.

YOU: Well, it must be that your orbit is rather small and dreary. While I'm on my Lear jet to Paris, you're in a crowded school bus going to Sheboygan.

DAVID: My orbit may be small, but it's very gratifying. I enjoy the teaching. I enjoy the kids. I like to see them develop. I like to see them learn. At times they make mistakes, and I have to let them know. There's a lot of real love and humanity that goes on there. A lot of drama. What about that seems dreary to you?

YOU: Well, there's not as much to learn. No real challenge. It seems to me that in a world as small as yours you learn just about everything there is to learn, and then you just repeat things over and over.

DAVID: Your work presents quite a challenge as it turns out. How could I know everything there is to know about even one student? They all seem complex and exciting to me. I don't think I have *anybody* figured out completely. Do you? Working with even one student is a complex challenge to *all* my abilities. Having so many young people to work with is a challenge beyond what I could ask for. I don't understand what you mean when you say my world is small and boring and everything is figured out.

YOU: Well, it just seems to me that you are unlikely to run into many people in your world who are going to develop as highly as I have.

DAVID: I don't know. Some of my students have high IQ's and may develop the same way you did, and some of them are mentally subnormal and will only develop to a modest level. Most are average and each one is fascinating to me. What did you mean when you said they were boring?

Why is it that only the great achievers are interesting to you?

YOU: I give in! Uncle!

I hope you did in fact "give in" when you played the role of the successful snob. The method I used to thwart your claim you were better than I was quite simple. Whenever you claimed you were a better or more worthy person because of some specific quality such as intelligence, influence, status, or whatever, I immediately *agreed* with you that you are better *in that particular quality* (or set of qualities) and then I asked you—"But how does that make you a better (or more worthwhile) *person*?" This question *cannot be answered*. It will take the wind out of the sails of *any* system of values that sets some people up as being superior to others.

The technical name for this method is "operationalization." In it you must *spell out* just what quality makes anyone more or less worthwhile than anyone else. You can't do it!

Of course, other people would rarely think or say such insulting things to you as were said in the dialogues. The real put-down goes on in your head. You are the one who's telling yourself your lack of status, or achievement, or popularity, or love, etc., makes you less worthwhile and desirable; so you're the one who's going to have to put an end to the persecution. You can do this in the following way: Carry on a similar dialogue with yourself. Your imaginary opponent, who we'll name the Persecutor, will try to argue that you are inherently inferior or less worthwhile because of some imperfection or lack. You simply assertively agree with the grain of truth in his criticism, but raise the question of how it follows that you are less worthwhile. Here are several examples:

1. Persecutor: You're not a very good lover. Sometimes you don't even get a firm erection. This means you're less of a man and an inferior person.

You: It certainly shows that I'm nervous about sex and not a particularly skilled or confident lover. But how does this make me less of a man or less of a person? Since only a man can feel nervous about an erection, this would seem to be an especially "manly" experience; doing it well makes you more of a man! Furthermore, there's a great deal more to being a man than just having sex.

2. Persecutor: You're not as hardworking or as successful as most of your friends. You're lazy and no good.
   You: This means I'm less ambitious and hardworking. I may even be less talented, but how does it follow that I'm "lazy and no good"?
3. Persecutor: You're not worth much because you're not outstanding in *anything*.
   You: I agree that I don't hold a single world championship. I'm not even second best at anything. In fact, at most things I'm pretty much average. How does it follow that I'm not worth much?
4. Persecutor: You're not popular, you don't even have many close friends, and no one cares about you much. You have no family and not even any casual lovers. So you're a loser. You're an inadequate person. There's obviously something wrong with you. You're worthless.
   You: It's true I have no lover at this time, and there are just a few friends I feel close to. How many do I need to be an "adequate person"? Four? Eleven? If I'm not popular, it may be that I'm relatively unskilled socially, and I may have to work harder at this. But how does it follow that I'm a "loser"? Why am I worthless?

I suggest you try out the method illustrated above. Write down the worst persecutory insults you can level at yourself and then answer them. It may be hard at first, but eventually the truth will dawn on you—you can be imperfect or unsuccessful or unloved by others, but *not* one iota less worthwhile.

Four Paths to Self-Esteem

You might ask, "How *can* I attain self-esteem if my worth doesn't come from my success or from love or approval? If you peel all these criteria away one by one and expose them as invalid bases for personal worth, it seems there will be nothing left. Just what is it that I have to do?" Here are four valid paths to self-esteem. Choose the one that seems most useful to you.

The first path is both pragmatic and philosophical. Essentially, you must acknowledge that human "worth" is just an abstraction; it doesn't exist. Hence, there is actually no such thing as human worth. Therefore, you cannot have it or fail to have it, and it cannot be measured. Worth is not a "thing," it is just a global concept. It is so generalized it has no concrete practical meaning. Nor is it a useful and enhancing concept. It is simply self-defeating. It doesn't do you any good. It only causes suffering and misery. So rid yourself *immediately* of *any* claim to being "worthy," and you'll *never have to measure up* again or fear being "worthless."

Realize that "worthy" and "worthless" are just empty concepts when applied to a human being. Like the concept of your "true self," your "personal worth" is just meaningless hot air. Dump your "worth" in the garbage can! (You can put your "true self" in there too, if you like.) You'll find you've got nothing to lose! Then you can focus on living in the here and now instead. What problems do you face in life? How will you deal with them? *That's* where the action is, not in the elusive mirage of "worth."

You may be afraid to give up your "self" or your "worth." What are you afraid of? What terrible thing will happen? Nothing! The following imaginary dialogue may make this clearer. Let's assume that I am worthless. I want you to rub it in and try to make me feel upset.

YOU: Burns, you're worthless!

DAVID: Of course I'm worthless. I fully agree. I realize that there is nothing about me that makes me "worthy." Love, approval, and achievement can't give me any "worth," so I'll accept the fact that *I have none*! Should this be a problem for me? Is something bad going to happen now?

YOU: Well, you must be miserable. You're just "no good."

DAVID: Assuming I am "no good," so what? What specifically do I have to be miserable about? Does being "worthless" put me at a disadvantage in some way?

YOU: Well, how can you respect yourself? How could anyone? You're just a scum!

DAVID: You may think I'm a scum, but I do respect myself, and so do lots of other people. I see no valid reason not to respect myself. *You* may not respect me, but I don't see that as a problem.

YOU: But worthless people *can't* be happy or have any fun. You're supposed to be depressed and despicable. My panel of experts met and determined that you're a total zero.

DAVID: So, call the papers and let them know. I can see the headline: "Philadelphia Physician Found to Be Worthless." If I'm really that bad off, it's reassuring because now I have nothing to lose. I can live my life fearlessly. Furthermore, I *am* happy and I *am* having fun, so being a "total zero" *can't* be bad. My motto is —"Worthless is Wonderful!" In fact, I'm thinking of having a T-shirt made up like that. Perhaps I'm missing out on something, though. Apparently you're *worthwhile*, whereas I'm not. What good does this "worth" do you? Does it make you better than people like me, or what?

The question may occur to you—"If I gave up my belief that success adds to my personal worth, then what would be the point in doing anything?" If you stay in bed all day, the probability that you will bump into something or someone that will make your day a little brighter is very small. Furthermore, there can be enormous satisfactions from daily living that are totally independent of any concept of personal worth. For example, as I am writing this I feel very turned on, but it isn't due to my belief that I am particularly "worthwhile" because I'm writing it. The exhilaration comes from the creative process, pulling ideas together, editing, watching clumsy sentences sharpen up, and wondering how you will react when you read this. This process is an exciting adventure. Involvement, commitment, and taking a risk can be quite stimulating. This is an adequate payoff, to my way of thinking.

You might also wonder—"What is the *purpose* and *meaning* of life without a concept of worth?" It's simple. Rather than grasp for "worth," aim for satisfaction, pleasure, learning, mastery, personal growth and communication with others every day of your life. Set realistic goals for yourself and work toward them. I think you will find this so abundantly gratifying you'll forget all about "worth," which in the last analysis has no more buying power than fool's gold.

"But I'm a humanistic or spiritual person," you might argue. "I've always been taught that *all* human beings have worth, and I just don't want to give up this concept." Very well, if you want to look at it that way, I'll agree with you, and this brings us to the second path to self-esteem. Acknowledge that everyone has one "unit of worth" from the time they are born until the time they die. As an infant you may achieve very little, and yet you are still precious and worthwhile. And when you are old or ill, relaxed or asleep, or just doing "nothing," you still have "worth." Your "unit of worth" can't be measured and can never change, and it is the same for everyone. During your lifetime, you can enhance your happiness and satisfaction through productive living, or you can act in a destructive manner and make yourself miserable. But your "unit of worth" is always there, along with your potential for self-esteem and joy. Since you can't measure it or change it, there is no point in dealing with it or being concerned about it. Leave that up to God.

Paradoxically, this solution comes down to the same bottom line as the previous solution. It becomes pointless and irresponsible to deal with your "worth," so you might as well focus on living life productively instead! What problems do you confront today? How will you go about solving them? Questions such as these are meaningful and useful, whereas rumination about your personal "worth" just causes you to spin your wheels.

Here is the third path to self-esteem: Recognize that there is only one way you can *lose* a sense of self-worth—by persecuting yourself with unreasonable, illogical negative thoughts. Self-esteem can be defined as the state that exists when you are not arbitrarily haranguing and abusing yourself but choose to fight back against those automatic thoughts with meaningful rational responses. When you do this effectively, you will experience a natural sense of jubilation and self-endorsement. Essentially, you don't have to get the river flowing, you just have to avoid damming it.

Since only distortion can rob you of self-esteem, this means that nothing in "reality" can take away your sense of worth. As evidence for this, many individuals under conditions of extreme and realistic deprivation do not experience a loss of self-esteem. Indeed, some individuals who were imprisoned by the Nazis during World War II refused to belittle themselves or buy into the persecutions of their captors. They reported an actual enhancement of self-esteem in spite of the miseries they were subjected to, and in some cases described experiences of spiritual awakening.

Here is the fourth solution: Self-esteem can be viewed as your decision to treat yourself like a beloved friend. Imagine that some VIP you respect came unexpectedly to visit you one day. How might you treat that person? You would wear your best clothes and offer your finest wine and food, and you would do everything you could to make him feel comfortable and pleased with his visit. You would be sure to let him know how highly you valued him, and how honored you were that he chose to spend some time with you. Now—why not treat *yourself* like that? Do it *all* the time if you can! After all, in the final analysis, no matter how impressed you are with your favorite VIP, you are more important to you than he is. So why not treat yourself at *least* as well? Would you insult and harangue such a guest with vicious, distorted put-downs? Would you peck away at his weaknesses and imperfections? Then why do this to yourself? Your self-torment becomes pretty silly when you look at it this way.

Do you have to *earn* the right to treat yourself in this loving, caring way? No, this attitude of self-esteem will be an *assertion* that you make, based on a full awareness and acceptance of your strengths and imperfections. You will fully acknowledge your positive attributes without false humility or a sense of superiority, and will freely admit to all your errors and inadequacies without any sense of inferiority or self-depreciation whatever. This attitude embodies the essence of self-love and self-respect. It does not have to be earned, and it *cannot* be earned in any way.

Escape from the Achievement Trap

You might be thinking, "All that philosophizing about achievement and self-worth is well and good. After all, Dr. Burns has a good career and a book on the market, so it's easy for him to tell *me* to forget about achievement. It sounds about as genuine as a rich man trying to explain to a beggar that money isn't important. The raw fact is, I *still feel bad* about myself when I do poorly, and I believe that life would be a whole lot more exciting and meaningful if I had more success. The truly happy people are the big shots, the executives. I'm only average. I've never done anything really outstanding, so I'm *bound* to be less happy and satisfied. If this isn't right, then prove it to me! Show me what I can do to change the way I feel, and only then will I be a true believer."

Let's review several steps you might take to liberate yourself from the trap of feeling you must perform in an outstanding manner in order to earn your right to feel worthwhile and happy.

Remember to Talk Back. The first useful method is to keep practicing the habit of talking back to those negative, distorted thoughts which cause you to feel inadequate. This will help you realize that the problem is not your actual performance, but the critical way in which you put yourself down. As you learn to evaluate what you do realistically, you will experience increased satisfaction and self-acceptance.

Here's how it worked for Len, a young man pursuing a career playing the guitar in rock bands. He sought treatment because he felt like a "second-rate" musician. From the time he was young, he was convinced he had to be a "genius" in order to be appreciated. He was easily hurt by criticism, and often made himself miserable by comparing himself with better-known musicians. He would feel deflated when he told himself, "I'm a nobody in comparison with X." He was certain that his friends and fans also viewed him as a mediocre person, and he concluded that he could never receive his fair share of the good things in life: praise, admiration, love, etc.

Len utilized the double-column technique to expose the nonsense and illogic in what he was saying to himself (Figure 13–1). This helped him to see that it was *not* a lack of musical talent that was the cause of his problems, but his unrealistic thinking patterns. As he began to correct this distorted thinking, his self-confidence improved. He described the effect of this: "Writing down my thoughts and answering them helped me to see how hard I was being on

Figure 13–1. Len's homework form for recording and answering his upsetting thoughts about being "the greatest."

Automatic Thoughts	*Rational Responses*
1. If I'm not "the greatest," it means I won't get any attention from people.	1. (All-or-nothing thinking). Whether or not I'm "the greatest," people *will* listen to me, they *will* see me perform, and many *will* respond positively to my music.
2. But *everybody* doesn't like the kind of music I play.	2. This is true of all musicians, even Beethoven or Bob Dylan. No musician can please everybody. Quite a few people do respond to my music. If I enjoy my music, then that should be enough.
3. But how can *I* enjoy my music if I know I'm not "the greatest"?	3. By playing music that turns me on, just as I always have! Besides, there's no such thing as "the world's greatest musician." So stop trying to be it!
4. But if I were *more* famous and talented, then I'd have *more* fans. How can I be happy on the sidelines when the big-name performers with charisma are in the spotlight?	4. How many fans and how many girl friends do I need before I'll be happy?
5. But I feel that no girl could really love me until I become a big-name talent.	5. Other people are loved who are just "average" in their work. Do I really have to be a big shot before someone will love me? Many of the guys I know get plenty of dates and they're not so unusual.

myself, and it gave me a sense there was something I could do to change. Instead of sitting there getting bombed by what I was telling myself, I suddenly had some antiaircraft artillery to fight back with."

Tune In to What Turns You On. One assumption which might be driving you to constant preoccupation with achievement is the idea that true happiness comes only through success in your career. This is unrealistic because the majority of life's satisfactions do not require great achievement at all. It takes no special talent to enjoy an average walk through the woods on an autumn day. You don't have to be "outstanding" to relish the affectionate hug of your young son. You can enjoy a good game of volley ball tremendously even though you're just an average player. What are some of life's pleasures that have turned you on? Music? Hiking? Swimming? Food? Travel? Conversation? Reading? Learning? Sports? Sex? You don't have to be famous or a top performer to enjoy these to the hilt. Here's how you can turn up the volume so that this kind of music comes in loud and clear.

Josh is a fifty-eight-year-old man with a history of destructive, manic mood swings as well as incapacitating depressions. When he was a child, Josh's parents emphasized over and over that his career was destined to be extraordinary, so he always felt he had to be number one. He eventually did make an exceptional contribution in his chosen field, electrical engineering. He won numerous awards, was appointed to presidential commissions, and was credited with many patents. However, as his cyclic mood disorder became increasingly severe, Josh began to have "high" episodes. During these periods, his judgment became grossly impaired and his behavior was so bizarre and disruptive that he had to be hospitalized on several occasions. Sadly, he came down off one high to learn he had lost his family as well as his prestigious career. His wife had filed for divorce, and he had been forced into an early retirement by the company he worked for. Twenty years of achievement went down the drain.

In the years that followed, Josh was treated with lithium and developed a modest consulting business. Eventually he was referred to me for treatment because he still experienced uncomfortable mood swings, especially depression, in spite of the lithium.

The crux of his depression was clear-cut. He was discouraged about his life because his career no longer measured up in terms of the money and prestige he had experienced in the past. While he had enjoyed the role of charismatic "charger" as a young man, he was now approaching sixty and felt alone and "over the hill." Because he still believed the only way to true happiness and personal worth was through superlative, creative achievements, he felt certain that his constricted career and modest life-style made him second-rate.

Since he was still a good scientist at heart, Josh decided to test his hypothesis that his life was destined to be mediocre by using the Pleasure-Predicting Sheet (described in previous chapters). Each day he agreed to schedule various activities that might give him a sense of pleasure, satisfaction, or personal growth. These activities could be related to his consulting business as well as hobbies and recreational pursuits. Before each activity he was to write down his prediction of how enjoyable it would be and mark it between 0 percent (no satisfaction at all) and 99 percent (the maximum enjoyment a human being can experience).

After filling out these forms for several days, Josh was surprised to find that life had just as much potential for joy and satisfaction as it ever had (see Figure 13–2). His discovery that work was at times quite rewarding and that numerous other activities could be just as enjoyable, if not more so, was a revelation to him. He was amazed one Saturday night when he went roller-skating with his girl friend. As they moved to the music, Josh found he began to tune into the beat and the melody, and as he became absorbed in the rhythm, he experienced a great sense of exhilaration. The data he collected on the Pleasure-Predicting Sheet indicated he didn't need a trip to Stockholm to receive the Nobel Prize to experience the ultimate in satisfaction—he didn't have to go any farther than the skat-

Figure 13–2. The Pleasure-Predicting Sheet.

Date	*Activity for Pleasure or Satisfaction*	*Who Did You Do This With? (If Alone, Specify Self)*	*Predicted Satisfaction (0–100%). (Record This* Before *the Activity)*	*Actual Satisfaction (0–100%). (Record This* After *the Activity)*
4/18/99	Work on consulting project	self	70%	75%
4/19/99	Take long walk before breakfast	self	40%	85%
4/19/99	Prepare written report	self	50%	50%
4/19/99	Make a ''missionary call'' on a potential customer	self	60%	40% (no new business)
4/20/99	Roller-skating	girl friend	50%	99%!

ing rink! His experiment proved that life was still filled with abundant opportunities for pleasure and fulfillment if he would enlarge his mental focus from a microscopic fixation on work and open himself up to the broad range of rich experiences that living can offer.

I am not arguing that success and achievement are undesirable. That would be unrealistic. Being productive and doing well can be enormously satisfying and enjoyable. However, it is neither *necessary* nor *sufficient* to be a great achiever in order to be maximally happy. You don't have to earn love or respect on the treadmill, and you don't have to be number one before you can feel fulfilled and know the meaning of inner peace and self-esteem. Now doesn't that make good sense?

Chapter 14

Dare to Be Average!—Ways to Overcome Perfectionism

I dare you to try to be "average." Does the prospect seem blah and boring? Very well—I dare you to try it for just one day. Will you accept the challenge? If you agree, I predict two things will happen. First, you won't be particularly successful at being "average." Second, in spite of this you will receive substantial satisfaction from what you do. More than usual. And if you try to keep this "averageness" up, I suspect your satisfaction will magnify and turn to joy. That's what this chapter is all about—learning to defeat perfectionism and enjoy the spoils of pure joy.

Think of it this way—there are two doors to enlightenment. One is marked "Perfection," and the other is marked "Average." The "Perfection" door is ornate, fancy, and seductive. It tempts you. You want very much to go through. The "Average" door seems drab and plain. Ugh! Who wants it?

So you try to go through the "Perfection" door and always discover a brick wall on the other side. As you insist on trying to break through, you only end up with a sore nose and a headache. On the other side of the "Average" door, in contrast, there's a magic garden. But it may never have occurred to you to open this door to take a look!

You don't believe me? I didn't think so, and you don't have to. I want you to maintain your skepticism! It's healthy—but at the same time I dare you to check me out. Prove me wrong! Put my claim to the test. Walk through that "Average" door just *one day* in your life. You may end up amazed!

Let me explain why. "Perfection" is man's ultimate illusion. It simply doesn't exist in the universe. There is no perfection. It's really the world's greatest con game; it promises riches and delivers misery. The harder you strive for perfection, the worse your disappointment will become because it's only an abstraction, a concept that doesn't fit reality. Everything can be improved if you look at it closely and critically enough—every person, every idea, every work of art, every experience, everything. So if you are a perfectionist, you are guaranteed to be a loser in whatever you do.

"Averageness" is another kind of illusion, but it's a benign deception, a useful construct. It's like a slot machine that pays a dollar fifty for every dollar you put in. It makes you rich—on all levels.

If you're willing to explore this bizarre-sounding hypothesis, let's begin. But beware—don't let yourself become *too* average because you may not be used to so much euphoria. After all, a lion can eat only so much meat after the kill!

Do you remember Jennifer, the perfectionistic writer-student mentioned in Chapter 4? She complained that friends and psychotherapists kept telling her to stop being such a perfectionist, but no one ever bothered to tell her how to go about doing this. This chapter is dedicated to Jennifer. She's not the only one who feels in a quandry about this. At my lectures and workshops, psychotherapists have often asked me to prepare a how-to-do-it manual that illustrates the fifteen techniques I have developed for overcoming perfectionism. Well—here's the manual. These methods work. You have nothing to fear or lose because the effects are not irreversible.

1. The best place to begin your fight against perfectionism is with your motivation for maintaining this approach. Make a list of the advantages and disadvantages of being perfectionistic. You may be surprised to learn that it is not actually to your advantage. Once you understand that it does *not* in fact help you in any way, you'll be much more likely to give it up.

Jennifer's list is shown in Figure 14–1. She concluded that her perfectionism was clearly not to her advantage. Now make *your* list. After you have completed it, read on.

2. Using your list of the advantages and disadvantages of perfectionism, you might want to do some experiments to test some of your assumptions about the advantages. Like many people, you may believe "Without my perfectionism I'd be nothing. I couldn't perform effectively." I'll bet you never put this hypothesis to the test because your belief in your inadequacy is such an automatic habit it has never even occurred to you to question it. Did you ever think that maybe you've been as successful as you are *in spite of* your perfectionism and not because of it! Here's an experiment that will allow you to come to the truth of the matter. Try altering your standards in various activities so you can see how your performance responds to high standards, middle standards, and low standards. The results may surprise you. I've done this with my writing, my psychotherapy with patients, and my jogging. And in all cases I have been pleasantly shocked to discover that by *lowering* my standards not only do I feel better about what I do but I tend to do it more effectively.

For example, I began jogging in January 1979 for the first time in my life. I live in a very hilly region, and initially I couldn't run more than two or three hundred yards without having to stop and walk because there are hills in all directions from my driveway. Each day I made it my aim to run a little less far than the day before. The effect of this was that I could always accomplish my goal easily. Then I would feel so good it would spur me on farther—and every step was gravy, more than I had aimed for. Over a period of months I built up to the point at which I could run seven

Figure 14–1. Jennifer's list of advantages and disadvantages of perfectionism. She concluded, "Clearly the disadvantages outweigh the one possible advantage."

Advantages of Perfectionism	*Disadvantages*
1. It can produce fine work. I'll try hard to come up with an exceptional result.	1. It makes me so "tight" and nervous I can't produce fine work.
	2. I become afraid and unwilling to risk the mistakes necessary to come up with a fine product.
	3. It makes me very critical of myself. I can't enjoy life because I can't admit my successes or allow myself to revel in them.
	4. I can't ever relax because I'll always be able to find something somewhere that *isn't perfect*, and then I'll get self-critical.
	5. Since I can never be perfect, I'll always be depressed.
	6. It makes me intolerant of others. I end up without many friends because people don't appreciate being criticized. I find so many faults in people I lose my capacity to feel warm and to like them.
	7. Another disadvantage is that my perfectionism keeps me from trying new things and making discoveries. I'm so afraid of making mistakes that I don't do much at all besides the same familiar things I'm good at. The result is that it narrows my world and makes me bored and restless because I have no new challenges.

miles over a steep terrain at a fairly rapid pace. I have never abandoned my basic principles—to try to accomplish less than the day before. Because of this rule I never feel frustrated or disappointed in my running. There have been many days when due to sickness or fatigue, I actually *didn't* run far or fast. Today, for example, I could only run a quarter mile because I had a cold and my lungs said NO FARTHER! So I told myself, "This is as far as I was *supposed* to go." I felt good because I achieved my goal.

Try this. Choose any activity, and instead of aiming for 100 percent, try for 80 percent, 60 percent, or 40 percent. Then see how much you enjoy the activity and how productive you become. Dare to aim at being average! It takes courage, but you may amaze yourself!

3. If you are a compulsive perfectionist you may believe that without aiming for perfection you couldn't enjoy life to the maximum or find true happiness. You can put this notion to the test by using the Antiperfectionism Sheet (Figure 14–2). Record the actual amount of satisfaction you get from a wide range of activities, such as brushing your teeth, eating an apple, walking in the woods, mowing the lawn, sunbathing, writing a report for work, etc. Now estimate how *perfectly* you did each activity between 0 and 100 percent, as well as marking how *satisfying* each was between 0 and 100 percent. This will help you break the illusory connection between perfection and satisfaction.

Here's how it works. In Chapter 4 I referred to a physician who was convinced he had to be perfect at all times. No matter how much he accomplished he would always raise his standards slightly higher, and then he'd feel miserable. I told him he was the Philadelphia all-or-nothing thinking champion! He agreed but protested he didn't know how to change. I persuaded him to do some research on his moods and accomplishments, using the Antiperfectionism Sheet. One weekend he did some plumbing at home because a pipe broke and flooded the kitchen. He was a novice plumber, but did manage to fix the leak and clean up the mess. On the sheet he recorded this as 99 percent satisfaction (see

Figure 14–2. The Antiperfectionism Sheet.

Activity	*Record How Effectively You Did This Between 0% and 100%*	*Record How Satisfying This Was Between 0% and 100%*
Fix broken pipe in kitchen	20% (I took a long time and made a lot of mistakes.)	99% (I actually did it!)
Give lecture to medical school class	98% (I got a standing ovation.)	50% (I usually get a standing ovation. I wasn't particularly thrilled with my performance.)
Play tennis after work	60% (I lost the match but played okay.)	95% (Really felt good. Enjoyed the game and the exercise.)
Edit draft of my latest paper for one hour	75% (I stuck with it and corrected many errors, and smoothed out the sentences.)	15% (I kept telling myself it wasn't *the definitive paper* and felt quite frustrated.)
Talk to student about his career options	50% (I didn't do anything special. I just listened to him and offered a few obvious suggestions.)	90% (He really seemed to appreciate our talk, so I felt turned on.)

Figure 14–2). Since it was the first time he'd ever tried to fix a pipe, he recorded his expertise as only 20 percent. He got the job done, but it was time-consuming and required considerable guidance from a neighbor. In contrast, he received low degrees of satisfaction from some activities he did an outstanding job on.

This experience with the Antiperfectionism Sheet persuaded him that he did not have to be perfect at something to enjoy it, and, furthermore, that striving for perfection and performing exceptionally did not guarantee happiness, but indeed tended to be associated more frequently with less satisfaction. He concluded he could either give up his compulsive drive for perfection and settle for joyous living and high productivity, or make his happiness of secondary importance and constantly push for greatness, and settle for emotional anguish and modest productivity. Which would you choose? Try out the Antiperfectionism Sheet and put yourself to the test.

4. Let's assume that you've decided to give up your perfectionism at least on a trial basis just to see what happens. However, you have the lingering notion that you *really could* be perfect in at least some areas if you tried hard enough, and that when you achieve this, something magical will happen. Let's take a hard look at whether this goal is realistic. Does a model of perfection *ever* really fit reality? Is there *anything* you have personally encountered that is so perfect it could not be improved?

To test this, look around you *right now* and see how things could be improved. For example, take someone's clothing, a flower arrangement, the color and clarity of a television picture, the quality of a singer's voice, the effectiveness of this chapter, *anything* at all. I believe you can *always* find some way in which something could be improved. When I first did this exercise, I was riding on a train. Most things, such as the dirty, rusty old tracks, were so obviously imperfect I could easily find many ways to improve them. Then I came to a problem area. A young black man had his hair in one of those fuzzy naturals. It looked perfectly smooth and sculptured, and I couldn't think of any way it could possibly be improved. I began to panic and saw my whole antiperfectionist philosophy going down the drain! Then I suddenly noticed some spots of gray on his head. I felt instant relief! His hair was imperfect after all! As I looked more closely, I noticed a few hairs that were too long and out of place. The closer I examined the young man, the more uneven hairs I could see—hundreds in fact! This helped convince me that any standard of perfection just doesn't fit reality. So why not give it up? You are guaranteed to be a sure loser if you maintain a standard for evaluating your performance that you can't *ever* meet. Why persecute yourself any longer?

5. Another method for overcoming perfectionism involves a confrontation with fear. You may not be aware that fear always lurks behind perfectionism. Fear is the fuel that drives your compulsion to polish things to the ultimate. If you choose to give up your perfectionism, you may initially have to confront this fear. Are you willing? There is, after all, a payoff in perfectionism—it protects you. It may protect you from risking criticism, failure, or disapproval. If you decide to start doing things less perfectly, at first you may feel as shaky as if a big California earthquake were about to hit.

If you don't appreciate the powerful role that fear plays in maintaining perfectionistic habits, the exacting behavior patterns of perfectionistic people can seem incomprehensible or infuriating. There is, for example, a bizarre illness known as "compulsive slowness," in which the victim becomes so totally bound up with getting things "just right" that simple everyday tasks can become totally consuming. An attorney with this brutal disorder became preoccupied with how his hair looked. For hours each day he would stand before a mirror with a comb and scissors trying to make adjustments. He became so involved in this, he had to cut back on his legal practice so he could have more and more time to work on his hair. Each day his hair got shorter and shorter because of all his furious clipping. Eventually it was only an eighth of an inch long all over his head. Then he became preoccupied with balancing the hairline along his forehead, and started shaving it to get it "just right." Each day the hairline receded farther and farther until eventually he had shaved his head totally bald! Then he felt a sense of relief and let it all grow back again, hoping it would come in "even." After the hair grew back, he would start clipping it again, and the whole cycle would be repeated. This ludicrous routine went on for years and left him a substantially disabled person.

His case may seem extreme but cannot be considered severe. Far worse forms of the disorder exist. Although the victims' strange habits may seem absurd, the effects are tragic. Like alcoholics, these individuals may sacrifice career and family to their miserable compulsions. You too may be paying heavily for your perfectionism.

What motivates these exacting, overcontrolled individuals? Are they insane? Usually not. What traps them in the senseless drive for perfection is fear. The moment they try to *stop* what they are doing, they are gripped by a powerful uneasiness that rapidly escalates to raw terror. This drives them back to their compulsive ritual in a pathetic attempt to find relief. Getting them to give up their perfectionistic malignancy is like trying to persuade a man hanging by his fingers from the edge of a cliff to let go.

You may have noticed compulsive tendencies in yourself to a much less severe degree. Have you ever pushed relentlessly to look for an important item like a pencil or a key you misplaced when you knew it was best to forget about it and wait for it to show up? You do this because it's *tough to stop*. The moment you try, you become uneasy and nervous. You feel somehow "not right" without the lost item, as if the whole meaning of your life were in the balance!

One method of confronting and conquering this fear is called "response prevention." The basic principle is simple and obvious. You *refuse* to give in to the perfectionistic habit, and you allow yourself to become flooded with fear and discomfort. Stubbornly stick it out and do not give in no matter how upset you become. Hang in there and allow your upset to reach its maximum. After a period of time the compulsion will begin to diminish until it disappears completely. At this point—which might require as much as several hours or as little as ten to fifteen minutes—you have won! You've defeated your compulsive habit.

Figure 14–3. The Response-Prevention Form. Record the degree of anxiety and any automatic thoughts every one or two minutes until you feel completely relaxed. The following experiment was performed by someone who wanted to end a bad habit of compulsively checking door locks.

Time	*Percent of Anxiety or Uneasiness*	*Automatic Thoughts*
4:00	80%	What if someone steals the car?
4:02	95%	This is ridiculous. Why not just go and make sure the car is okay?
4:04	95%	Someone may be in it right now. I can't stand this!
4:06	80%	
4:08	70%	
4:10	50%	
4:12	20%	This is boring. The car will probably be okay.
4:14	5%	
4:16	0%	Hey—I did it!

Let's take a simple example. Suppose you are in the habit of double-checking the house or car locks several times. Certainly it's okay to check things *once*, but more often than that is redundant and pointless. Drive your car to a parking lot, lock the doors, and walk away. Now—refuse to check them! You will feel uneasy. You'll try to persuade yourself to go back and "just make sure." DON'T. Instead, record your degree of anxiety every minute on the "Response-Prevention Form" (see Figure 14–3) until the anxiety has vanished. At this point, you win. Often, one such exposure is sufficient to break a habit permanently, or you may need numerous exposures as well as a booster shot from time to time. Many bad habits lend themselves to this format, including various "checking rituals" (checking to see if the stove is turned off or if the mail has fallen into the mailbox, etc.), cleaning rituals (compulsive hand-washing or excessive housecleaning), and others. If you are ready and willing to break free of these tendencies, I think you'll find the response-prevention technique quite helpful.

6. You may be asking yourself about the origin of the crazy fear that drives you to compulsive perfectionizing. You can use the vertical-arrow method described in Chapter 10 to expose the silent assumption that causes your rigid, tense approach to living. Fred is a college student who was so preoccupied with getting a term paper "just right" that he dropped out of college to work on it for an entire year to avoid the horrors of turning in a product he wasn't entirely satisfied with. Fred finally enrolled in college again when he felt ready to turn the paper in, but sought treatment for his perfectionism because he realized it might take too long to complete college this way!

He had his confrontation with fear when he was required to turn in another term paper at the end of his first semester back in school. This time the professor gave him the ultimatum of either turning it in by six P.M. on the due date, or getting docked one full grade for every day it was late. Since Fred had an adequate draft of the paper, he realized it wouldn't be wise for him to try to polish it and revise it, so he reluctantly turned it in at 4:55, knowing that there were a number of uncorrected typographical errors as well as some sections he wasn't entirely satisfied with. The moment he turned it in, his anxiety began to mount. Minute by minute it increased, and soon Fred was gripped by such a severe panic attack that he called me at home late in the evening. He was convinced that something terrible was about to happen to him because he had turned in an imperfect paper.

I suggested he use the vertical-arrow method to pinpoint just what he was so afraid of. His first automatic thought was, "I didn't do an excellent job on the paper." He wrote this down (see Figure 14–4, page 363), and then asked himself, "If that were true, why would it be a problem for

Figure 14–4. Fred used the vertical-arrow method to uncover the origin of his fears about turning in an "imperfect" paper for a class. This helped relieve some of the terror he was experiencing. The question next to each vertical arrow represents what Fred asked himself in order to uncover the next automatic thought at a deeper level. By unpeeling the onion in this way, he was able to expose the silent assumptions which represented the origin and root of his perfectionism (see text).

Automatic Thoughts	*Rational Responses*
1. I didn't do an excellent job on the paper. ↓ "If that were true, why would it be a problem for me?"	1. All-or-nothing thinking. The paper is pretty good even though it's not perfect.
2. The professor will notice all the typos and the weak sections. ↓ "And why would that be a problem?"	2. Mental filter. He probably will notice typos, but he'll read the whole paper. There are some fairly good sections.
3. He'll feel that I didn't care about it. ↓ "Suppose he does. What then?"	3. Mind reading. I don't know that he will think this. If he *did*, it wouldn't be the end of the world. A lot of students don't care about their papers. Besides I *do* care about it, so if he thought this he'd be wrong.
4. I'll be letting him down. ↓ "If that were true and he did feel that way, why would it be upsetting to me?"	4. All-or-nothing thinking; fortune teller error. I can't please everyone all the time. He's liked most of my work. If he does feel disappointed in this paper he can survive.
5. I'll get a D or an F on the paper. ↓ "Suppose I did—what then?"	5. Emotional reasoning; fortune teller error. I *feel* this way because I'm upset. But I can't predict the future. I might get a B or a C, but a D or an F isn't very likely.

Figure 14–4. cont.

Automatic Thoughts	*Rational Responses*
6. That would ruin my academic record. ↓ "And then what would happen?"	6. All-or-nothing thinking; fortune teller error. Other people goof up at times, and it doesn't seem to ruin their lives. Why can't I goof up at times?
7. That would mean I wasn't the kind of student I was supposed to be. ↓ "Why would that be upsetting to me?"	7. Should statement. Who ever laid down the rule I was "supposed" to be a certain way at all times? Who said I was predestined and morally obliged to live up to some particular standard?
8. People will be angry with me. I'll be a failure. ↓ "And suppose they *were* angry and I *was* a failure? Why would that be so terrible?"	8. The fortune teller error. If someone is angry with me, it's their problem. I can't be pleasing people all the time—it's too exhausting. It makes my life a tense, constricted, rigid mess. Maybe I'd do better to set my own standards and risk someone's anger. If I fail at the paper, it certainly doesn't make me "A FAILURE."
9. Then I would be ostracized and alone. ↓ "And then what?"	9. The fortune teller error. *Everyone* won't ostracize me!
10. If I'm alone, I'm bound to be miserable.	10. Disqualifying positive data. Some of my happiest times have been when I'm alone. My "misery" has nothing to do with being alone, but comes from the fear of disapproval and from persecuting myself for not living up to perfectionistic standards.

me?" This question generated the upsetting thought lurking behind it, as demonstrated in Figure 14–4. Fred wrote down the next thought that came to mind, and continued to use the downward-arrow technique to reveal his fears at a deeper and deeper level. He continued peeling the layers off the onion in this way until the deepest origin of his panic and perfectionism was uncovered. This required only a few minutes. His silent assumption then became obvious: (1) One mistake and my career will be ruined. (2) Others demand perfection and success from me, and will ostracize me if I fall short.

Once he wrote down his upsetting automatic thoughts, he was in a position to pinpoint his thinking errors. Three distortions appeared most often—all-or-nothing thinking, mind reading, and the fortune teller error. These distortions had trapped him in a rigid, coercive, perfectionistic, approval-seeking approach to life. Substituting rational responses helped him recognize how unrealistic his fears were and took the edge off his panic.

Fred was skeptical, however, because he wasn't entirely convinced a catastrophe was not about to strike. He needed some actual evidence to be convinced. Since he'd been keeping the elephants away by blowing the trumpet all his life, he couldn't be *absolutely* sure a stampede wouldn't occur once he decided to set the trumpet down.

Two days later Fred got the needed evidence: He picked up his paper, and there was an A – at the top. The typographical errors had been corrected by the professor, who wrote a thoughtful note at the end that contained substantial praise along with some helpful suggestions.

If you are going to let go of your perfectionism, then you may also have to expose yourself to a certain amount of initial unpleasantness just as Fred did. This can be your golden opportunity to learn about the origin of your fears, using the vertical-arrow technique. Rather than run from your fear, sit still and *confront* the bogeyman! Ask yourself, "What am I afraid of?" "What's the worst that could happen?" Then write down your automatic thoughts as Fred did, and call their bluff. It *will* be frightening, but if you tough it out and endure the discomfort, you will conquer your fears because they are ultimately based on illusions. The exhilaration you experience when you make this transformation from worrier to warrior can be the start of a more confident assertive approach to living.

The thought may have occurred to you—but suppose Fred *did* end up with a B, C, D, or an F? What then? In reality, this *usually* doesn't happen because in your perfectionism, you are in the habit of leaving yourself such an excessively wide margin of safety that you can usually relax your efforts considerably without a measurable reduction in the quality of the actual performance. However, failures *can* and *do* occur in life, and none of us is totally immune. It can be useful to prepare ahead of time for this possibility so that you can benefit from the experience. You can do this if you set things up in a "can't lose" fashion.

How can you benefit from an actual failure? It's simple! You remind yourself that your life won't be destroyed. Getting a B, in fact, is one of the best things that can happen to you if you are a straight A student because it will force you to confront and accept your humanness. This will lead to personal growth. The real tragedy occurs when a student is so bright and compulsive that he or she successfully wards off any chance of failure through overwhelming personal effort, and ends up graduating with a perfect straight A average. The paradox in this situation is that success has a dangerous effect of turning these students into cripples or slaves whose lives become obsessively rigid attempts to ward off the fear of being less than perfect. Their careers are rich in achievement but frequently impoverished in joy.

7. Another method for overcoming perfectionism involves developing a process orientation. This means you focus on processes rather than outcomes as a basis for evaluating things. When I first opened my practice, I had the feeling I had to do outstanding work with each patient every session. I thought my patients and peers expected this of me, and so I worked my tail off all day long. When a patient

indicated he benefited from a session, I'd tell myself I was successful and I'd feel on top of the world. In contrast, when a patient gave me the runaround or responded negatively to that day's session, I'd feel miserable and tell myself I had failed.

I got tired of the roller-coaster effect and reviewed the problem with my colleague, Dr. Beck. His comments were extremely helpful, so I'll pass them on to you. He suggested I imagine I had a job driving a car to City Hall each day. Some days I'd hit mostly green lights and I'd make fast time. Other days I'd hit a lot of red lights and traffic jams, and the trip would take much longer. My driving skill would be the same each day, so why not feel equally satisfied with the job I did?

He proposed I could facilitate this new way of looking at things by refusing to try to do an excellent job with any patient. Instead, I could aim for a good, consistent effort at each session regardless of how the patient responded, and in this way I could guarantee 100 percent success forever.

How could you set up process goals as a student? You could make it your aim to (1) attend lectures; (2) pay attention and take notes; (3) ask appropriate questions; (4) study each course between classes a certain amount each day; (5) review class study notes every two or three weeks. All these processes are within your control, so you can *guarantee* success. In contrast, your final grade is not under your control. It depends on how the professor feels that day, how well the other students did, where he sets the curve, etc.

How could you set up process goals if you were applying for a job? You could (1) dress in a confident, appealing manner; (2) have your résumé edited by a knowledgeable friend and typed professionally; (3) give the prospective employer one or more compliments during the interview; (4) express an interest in the company and encourage the interviewer to talk about himself; (5) when the prospective employer tells you about his work, say something positive, using an upbeat approach; (6) if the interviewer makes a critical or negative comment about you, immediately *agree*, using the disarming technique introduced in Chapter 6.

For example, in my negotiations with a prospective publisher about this book, I noticed the editor expressed a number of negative reactions in addition to a few positive ones. I found the use of the disarming technique worked extremely well in keeping the waters flowing nonturbulently during potentially difficult discussions. For example,

EDITOR X: One of my concerns, Dr. Burns, involves the emphasis on symptomatic improvement in the here and now. Aren't you overlooking the causes and origins of depressions?

(In the first draft of this book, I had written several chapters on the silent assumptions that give rise to depression, but apparently the editor was not adequately impressed with this material or had not read it. I had the option of counterattacking in a defensive manner—which would have only polarized the editor and made her feel defensive. Instead, I chose to disarm her in the following way.)

DAVID: That's an excellent suggestion, and you're absolutely right. I can see you've been doing your homework on the manuscript, and I appreciate hearing about your ideas. The readers obviously would want to learn more about *why* they get depressed. This might help them avoid future depressions. What would you think about expanding the section on silent assumptions and introducing it with a new chapter we could call "Getting Down to Root Causes"?

EDITOR: That sounds great!

DAVID: What other negative reactions do you have to the book? I'd like to learn as much as I can from you.

I then continued to find a way to *agree* with each criticism and to praise Editor X for each and every suggestion. This was not insincere because I was a greenhorn in popular writing, and Editor X was a very talented, well-established individual who was in a position to give me some much-needed guidance. My negotiating style made it clear to her that I respected her, and let her know that we would be able to have a productive working relationship.

Suppose instead that I had been fixed on the *outcome* rather than on the negotiating process when the editor interviewed me. I would have been tense and preoccupied with only one thing—would she or would she not make an offer for the book? Then I would have seen her every criticism as a danger, and the whole interpersonal process would have fallen into unpleasant focus.

Thus, when you are applying for work, do *not* make it your aim to *get* the job! Especially if you *want* the job! The outcome depends on numerous factors that are ultimately out of your control, including the number of applicants, their qualifications, who knows the boss's daughter, etc. In fact, you would do better to try to get as many rejections as possible for the following reason: Suppose on the average it takes about ten to fifteen interviews for each acceptable job offer you receive in your profession (a typical batting average for people I know who have been recently looking for work). This means you've got to go out and get those nine to fourteen rejections over with in order to get the job you want! So each morning say, "I'll try to get as many rejections as possible today." And each time you *do* get rejected you can say, "I was successfully rejected. This brings me one important step closer to my goal."

8. Another way to overcome perfectionism involves assuming responsibility for your life by setting strict time limits on all your activities for one week. This will help you change your perspective so you can focus on the flow of life and enjoy it.

If you are a perfectionist, you are probably a real procrastinator because you insist on doing things so thoroughly. The secret to happiness is to set modest goals to accomplish them. If you want misery, then by all means cling to your perfectionism and procrastination. If you would like to change, then as you schedule your day in the morning, decide on the amount of time you will budget on each activity. Quit at the end of the time you have set aside whether or not you have completed it, and go onto the next project. If you play the piano and tend to play for many hours or not at all, decide instead to play only an hour a day. I think you'll enhance your satisfaction and output substantially this way.

9. I'll bet you're afraid of making mistakes! What's so terrible about making mistakes? Will the world come to an end if you're wrong? Show me a man who can't stand to be wrong, and I'll show you a man who is afraid to take *risks* and has given up the capacity for growth. A particularly powerful method for defeating perfectionism involves learning to make mistakes.

Here's how you can do this. Write an essay in which you spell out why it is both *irrational* and *self-defeating* to try to be perfect or to fear making mistakes. The following was written by Jennifer, the student mentioned earlier:

Why It's Great to Be Able to Make Mistakes

1. I fear making mistakes because I see everything in absolutist, perfectionistic terms—*one mistake and the whole is ruined*. This is erroneous. A small mistake certainly doesn't ruin an otherwise fine whole.

2. It's good to make mistakes because then we learn—in fact, we won't learn *unless* we make mistakes. No one can avoid making mistakes—and since it's going to happen in any case, we may as well accept it and learn from it.

3. Recognizing our mistakes helps us to adjust our behavior so that we can get results we're more pleased with—so we might say that mistakes ultimately operate *to make us happier and make things better*.

4. *If we fear making mistakes, we become paralyzed*—we're afraid to do or try anything, since we might (in fact, probably will) make some mistakes. If we restrict our activities so that we won't make mistakes, then we are really defeating ourselves. The more we try and the more mistakes we make, the faster we'll learn and the happier we'll be ultimately.

5. Most people aren't going to be mad at us or dislike us because we make mistakes—they all make mistakes, and most people feel uncomfortable around "perfect" people.

6. We don't die if we make mistakes.

Although such an essay does not *guarantee* that you will change, it can help get you started in the right direction. Jennifer reported an enormous improvement the week after she wrote the essay. She found it useful in her studies to focus on learning rather than obsessing constantly about whether or not she was great. As a result, her anxiety decreased and her ability to get things done increased. This relaxed, confident mood persisted through the final examination period at the end of the first semester—a time of extreme anxiety for the majority of her classmates. As she explained, "I realized I didn't *have* to be perfect. I'm going to make my share of mistakes. So what? I can learn from my mistakes, so there's *nothing* to worry about." And she was right!

Write a memo to yourself along these lines. Remind yourself that the world won't come to an end if you make a mistake, and point out the potential benefits. Then read the memo every morning for two weeks. I think this will go a long way toward helping you join the human race!

10. In your perfectionism you are undoubtedly great at focusing on all the ways you fall short. You have the bad habit of picking out the things you haven't done and ignoring those you have. You spend your life cataloging every mistake and shortcoming. No wonder you feel inadequate! Is somebody forcing you to do this? Do you *like* feeling that way?

Here's a simple method of reversing this absurd and painful tendency. Use your wrist counter to click off the things you do *right* each day. See how many points you can accumulate. This may sound so unsophisticated that you are convinced it couldn't help you. If so, experiment with it for two weeks. I predict you'll discover that you will begin to focus more on the positives in your life and will consequently feel better about yourself. It sounds simplistic because it is! But who cares, if it works?

11. Another helpful method involves exposing the absurdity in the all-or-nothing thinking that gives rise to your perfectionism. Look around you and ask yourself how many things in the world can be broken down into all-or-nothing categories. Are the walls around you totally clean? Or do they have at least *some* dirt? Am I totally effective with all of my writing? Or partially effective? Certainly every single paragraph of this book isn't polished to perfection and breathtakingly helpful. Do you know anyone who is *totally* calm and confident *all* the time? Is your favorite movie star perfectly beautiful?

Once you recognize that all-or-nothing thinking doesn't fit reality very often, then look out for your all-or-nothing thoughts throughout the day, and when you notice one, talk back to it and shoot it down. You'll feel better. Some examples of how a number of different individuals combat all-or-nothing thoughts appear in Figure 14–5.

12. The next method to combat perfectionism involves personal disclosure. If you feel nervous or inadequate in a situation, then share it with people. Point out the things you feel you've done inadequately instead of covering them up.

Figure 14–5. How to replace all-or-nothing thoughts with others that are more in tune with reality. These examples were contributed by a variety of individuals.

All-or-Nothing Thinking	*Realistic Thoughts*
1. What a lousy day!	1. A couple of bad things have happened, but everything hasn't been a disaster.
2. This meal I cooked really turned out terrible.	2. It's not the best meal I ever cooked, but it's okay.
3. I'm too old.	3. Too old for what? Too old to have fun? No. Too old for occasional sex? No. Too old to enjoy friends? No. Too old to love or be loved? No. Too old to enjoy music? No. Too old to do some productive work? No. So what am I "too old" for? It really has no meaning!
4. Nobody loves me.	4. Nonsense. I have many friends and family. I may not get *as much* love as I want when I want it, but I can work on this.
5. I'm a failure.	5. I've succeeded at some things and failed at others, just like everybody.
6. My career is over the hill.	6. I can't do as much as when I was younger, but I can still work and produce and create, so why not enjoy it?
7. My lecture was a flop!	7. It wasn't the best lecture I ever gave. In fact, it was below my average. But I did get some points across, and I can work to improve my next lectures. Remember—half my lectures will be below my average, and half will be above!
8. My boyfriend doesn't like me!	8. He doesn't like me enough for what? He may not want to marry me, but he takes me out on dates, so he *must* like me partially.

Ask people for suggestions on how to improve, and if they're going to reject you for being imperfect, let them do it and get it over with. If in doubt as to where you stand, ask if they think less of you when you make a mistake.

If you do this, you must of course be prepared to handle the possibility that people *will* look down on you because of your imperfections. This actually happened to me during a teaching session I was conducting for a group of therapists. I pointed out an error I felt I had made in reacting angrily to a difficult, manipulative patient. I then asked if any of the therapists present thought less of me after hearing about my foible. I was taken aback when one replied in the affirmative, and the following conversation took place:

THERAPIST (in the audience): I have two thoughts. One thought is a positive one. I appreciate your taking that risk to point out your error in front of the group because I would have been scared to do it. I think it takes great courage on your part to do this. But I have to admit I'm ambivalent about you now. Now I know that you *do* make mistakes, which is realistic, but . . . I feel disappointed in you. In all honesty, I do.

DAVID: Well, I *knew* how to handle the patient, but I was so overcome with my anger that I just got caught up in the moment and retaliated. I was overly abrupt in the way I reacted to her. I admit I handled it quite poorly.

THERAPIST: I guess in the context that you see so many patients each week for so many years, if you make one blunder like that it's definitely not earthshattering. It's not going to kill her or anything. But I do feel let down, I have to admit.

DAVID: But it *isn't* just one rare error. I believe that all therapists make many blunders every single day. Either obvious ones or subtle ones. At least I do. How will you come to terms with that? It seems you're quite disappointed in me because I didn't handle that patient effectively.

THERAPIST: Well, I am. I thought you had a sufficiently wide behavioral repertoire that you could easily handle nearly *anything* a patient said to you.

DAVID: Well, that's untrue. I *sometimes* come up with very helpful things to say in difficult situations, but sometimes I'm not as effective as I'd like to be. I still have a lot to learn. Now with that knowledge, do you think less of me?

THERAPIST: Yeah. I really do. I have to say that. Because now I see that there's a reasonably easy kind of conflict that can upset you. You were unable to handle it without showing your vulnerabilities.

DAVID: That's true. At least *that time* I didn't handle it well. It's an area where I need to focus my efforts and grow as a therapist.

THERAPIST: Well, it shows that at least in that case, and I assume in others, that you don't handle things as well as I thought you did.

DAVID: I think that's correct. But the question is, why do you think less of me because I am imperfect? Why are you looking down on me? Does it make me less a person to you?

THERAPIST: You're exaggerating the whole thing now, and I don't feel that you are necessarily of less value as a human or anything like that. But on the other hand, I think you're not as good as a therapist as I thought you were.

DAVID: That's true. Do you think less of me because of that?

THERAPIST: As a therapist?

DAVID: As a therapist *or* as a person. Do you think less of me?

THERAPIST: Yes, I suppose I do.

DAVID: Why?

THERAPIST: Well, I don't know how to say this. I think "therapist" is the primary role that I know you in. I'm disappointed to find you're so imperfect. I had a higher expectation of you. But perhaps you're better in other areas of your life.

DAVID: I hate to disappoint you, but you'll discover that in many other aspects of my life I'm even *more* imperfect. So if you're looking down on me as a therapist, I presume you'll look down on me more as a person.

THERAPIST: Well, I do think less of you as a person. I think that's an accurate description of how I'm feeling about you.

DAVID: Why do you think less of me because I don't measure up to your standard of perfection? I'm a human and not a robot.

THERAPIST: I'm not sure I understand that question. I judge people in terms of their performance. You goofed up, so you have to face the fact I'll judge you negatively. It's tough, but it's reality. I thought you should perform better because you're our preceptor and our teacher. I expected *more* of you. Now it sounds like I could have handled that patient better than you did!

DAVID: Well, I think you *could* have done better than I did with that patient that day, and this is

> an area where I think I can learn from you. But why do you look down on me for this? If you get disappointed and lose respect each time you notice I've made a mistake, pretty soon you'll be totally miserable, and you'll have no respect for me at all because I've been making errors every day since I was born. Do you want all that discomfort? If you want to continue and enjoy our friendship, and I hope you will, you'll just have to accept the fact that I'm not perfect. Maybe you'd be willing to look for mistakes I make and point them out to me so I can learn from you while I'm teaching you. When I stop making mistakes, I'll lose much of my capacity to grow. Recognizing and correcting my errors and learning from them is one of my greatest assets. And if you can accept my humanity and imperfection, maybe you can also accept your own. Maybe you'll want to feel that it's okay for you to make mistakes too.

This kind of dialogue transcends the possibility you will feel put down. Asserting your right to make mistakes will paradoxically make you a greater human being. If the other person feels disappointed, the fault is really his for having set up the unrealistic expectation you are more than human. If you don't buy into that foolish expectation, you won't have to become angry or defensive when you do goof up—nor will you have to feel any sense of shame or embarrassment. The choice is clear-cut: You can either try to be perfect and end up miserable, or you can aim to be human and imperfect and feel enhanced. Which do you choose?

13. The next method is to focus mentally on a time in your life when you were really happy. What image comes to mind? For me the image is of climbing down into Havasupai Canyon one summer vacation when I was a college student. This canyon is an isolated part of the Grand Canyon, and you have to hike into it or arrange for horses. I went with a friend. Havasupai, an Indian word meaning "blue-green water people," is the name of a turquoise river that bubbles out of the desert floor and turns the narrow canyon into a lush paradise many miles long. Ultimately, the Havasupai River empties into the Colorado River. There are a number of waterfalls several hundred feet high, and at the bottom of each, a green chemical in the water precipitates out and makes the river's bottom and edges smooth and polished, just like a turquoise swimming pool. Cottonwood trees and Jimsonweed with purple flowers like trumpets line the river in abundance. The Indians who live there are easygoing and friendly. It is a blissful memory. Perhaps you have a similar happy memory. Now ask yourself—what was *perfect* about that experience? In my case, *nothing*! There were no toilet facilities, and we slept in sleeping bags outdoors. I didn't hike perfectly or swim perfectly, and nothing was perfect. There was no electricity available in most of the village because of its remoteness, and the only available food in the store was canned beans and fruit cocktail—no meat or vegetables. But the food tasted darn good after a day of hiking and swimming. So who needs perfection?

How can you use such a happy memory? When you are having a presumably pleasurable experience—eating out, taking a trip, going to a movie, etc.—you may unnecessarily sour the experience by making an inventory of all the ways it falls short and telling yourself you can't possibly enjoy it. But this is hogwash—it's your *expectation* that upsets you. Suppose the motel bed is too lumpy and you paid fifty-six dollars for the room. You called the front desk, and they have no other beds or rooms available. Tough! Now you can double your trouble by demanding perfection, or you can conjure up your "happy, imperfect" memory. Remember the time you camped out and slept on the ground and loved it? So you can certainly enjoy yourself in this motel room if you choose! Again, it's up to you.

14. Another method for overcoming perfectionism is the "greed technique." This is based on the simple fact that most of us try to be perfect so we can get ahead in life. It may not have occurred to you that you might end up much more successful if your standards were lower. For example, when I started my academic career, I spent over two years writing the first research paper I published. It was an excellent product, and I'm still quite proud of it. But I noticed that in the same time period, many of my peers who were of equal intelligence wrote and published numerous papers. So I asked myself—am I better off with one publication that contains ninety-eight "units of excellence," or ten papers that are each worth only eighty "units of excellence"? In the latter case, I would actually end up with 800 "excellence units," and I would be way ahead of the game. This realization was a strong personal persuader, and I decided to lower my standards a bit. My productivity then became dramatically enhanced, as well as my levels of satisfaction.

How can this work for you? Suppose you have a task and you notice you're moving slowly. You may find that you've already reached the point of diminishing returns, and you'd do better by moving on to the next task. I'm not advocating that you slough off, but you may find that you as well as others will be equally if not more pleased with many good, solid performances than with one stress-producing masterpiece.

15. Here's the last approach. It involves simple logic. Premise one: All human beings make mistakes. Do you agree? Okay, now tell me: What are you? A human being, you say? Okay. Now, what follows? Of course—you *will* and *should* make mistakes! Now tell yourself this every time you persecute yourself because you made an error. Just say, "I was *supposed* to make that mistake because I'm human!" or "How human of me to have made that mistake."

In addition, ask yourself, "What can I learn from my mistake? Is there some good that could come from this?" As an experiment, think about some error you've made and write down everything you learned from it. Some of the best things can be learned only through making mistakes and learning from them. After all, this is how you learned to talk and walk and do just about everything. Would you be willing to give up that kind of growth? You may even go so far as to say your imperfections and goof-ups are some of your greatest assets. Cherish them! Never give up your capacity for being wrong because then you lose the ability to move forward. In fact, just think what it would be like if you *were* perfect. There'd be *nothing* to learn, *no way* to improve, and life would be completely void of challenge and the satisfaction that comes from mastering something that takes effort. It would be like going to kindergarten for the rest of your life. You'd know all the answers and win every game. Every project would be a guaranteed success because you would do everything correctly. People's conversations would offer you nothing because you'd already know it all. And most important, nobody could love or relate to you. It would be impossible to feel any love for someone who was flawless and knew it all. Doesn't that sound lonely, boring, and miserable? Are you so sure you still want perfection?

Part V

Defeating Hopelessness and Suicide

Chapter 15

The Ultimate Victory: Choosing to Live

Dr. Aaron T. Beck reported in a study that suicidal wishes were present in approximately one-third of individuals with a mild case of depression, and in nearly three-quarters of people who were severely depressed.* It has been estimated that as many as 5 percent of depressed patients do actually die as a result of suicide. This is approximately twenty-five times the suicide rate within the general population. In fact, when a person with a depressive illness dies, the chances are one in six that suicide was the cause of death.

No age group or social or professional class is exempt from suicide; think of the famous people you know of who have killed themselves. Particularly shocking and grotesque—but by no means rare—is suicide among the very young. In a study of seventh- and eighth-grade students in a suburban Philadelphia parochial school, nearly one third of the youngsters were significantly depressed and had suicidal thoughts. Even *infants* who undergo maternal separation can develop a depressive syndrome in which failure

*Beck, Aaron T. *Depression: Causes and Treatment*. Philadelphia: University of Pennsylvania Press, 1972, pp. 30–31.

to thrive and even self-imposed death from starvation can result.

Before you get overwhelmed, let me emphasize the positive side of the coin. First, suicide is unnecessary, and the impulse can be rapidly overcome and eliminated with cognitive techniques. In our study, suicidal urges were reduced substantially in patients treated with cognitive therapy *or* with antidepressant drugs. The improved outlook on life occurred within the first week or two of treatment in many cognitively treated patients. The current intensive emphasis on the prevention of depressive episodes in individuals prone to mood swings should also result in a long-term reduction in suicidal impulses.

Why do depressed individuals so frequently think of suicide, and what can be done to prevent these impulses? You will understand this if you examine the thinking of people who are actively suicidal. A pervasive, pessimistic vision dominates their thoughts. Life seems to be nothing but a hellish nightmare. As they look into the past, all they can remember are moments of depression and suffering.

When you feel down in the dumps, you may also feel so low at times that you get the feeling you were never really happy and never will be. If a friend or relative points out to you that, except for such periods of depression, you were quite happy, you may conclude they're mistaken or only trying to cheer you up. This is because while you are depressed you actually distort your memories of the past. You just can't conjure up any memories of periods of satisfaction or joy, so you erroneously conclude they did not exist. Thus, you mistakenly conclude that you always have been and always will be miserable. If someone insists that you have been happy, you may respond as a young patient recently did in my office, "Well, that period of time doesn't count. Happiness is an illusion of some kind. The real me is depressed and inadequate. I was just fooling myself if I thought I was happy."

No matter how bad you feel, it would be bearable if you

had the conviction that things would eventually improve. The critical decision to commit suicide results from your illogical conviction that your mood *can't* improve. You feel certain that the future holds only more pain and turmoil! Like some depressed patients, you may be able to support your pessimistic prediction with a wealth of data which seems to you to be overwhelmingly convincing.

A depressed forty-nine-year-old stockbroker recently told me, "Doctor, I have already been treated by six psychiatrists over a ten-year-period. I have had shock treatments and all types of antidepressants, tranquilizers, and other drugs. But in spite of it all, this depression won't let up for one minute. I have spent over eighty thousand dollars trying to get well. Now I am emotionally and financially depleted. Every doctor has said to me. 'You'll beat this thing. Keep your chin up.' But now I realize it wasn't true. They were all lying to me. I'm a fighter, so I fought hard. You'd better realize when you are defeated. I've got to admit I'd be better off dead."

Research studies have shown that your unrealistic sense of hopelessness is one of the most crucial factors in the development of a serious suicidal wish. Because of your twisted thinking, you see yourself in a trap from which there seems to be no escape. You jump to the conclusion that your problems are insoluble. Because your suffering feels unbearable and appears unending, you may erroneously conclude that suicide is your only way of escape.

If you have had such thoughts in the past, or if you are seriously thinking this way at present, let me state the message of this chapter loud and clear:

> You Are Wrong in Your Belief That Suicide Is the Only Solution or the Best Solution to Your Problem.

Let me repeat that. *You Are Wrong!* When you think that you are trapped and hopeless, your thinking is illogical, distorted, and skewed. No matter how thoroughly you have convinced yourself, and even if you get other people to agree with you, you are just plain *mistaken* in your belief that it is ever advisable to commit suicide because of depressive illness. This is not the most rational solution to your misery. I will explain this position and help point the way out of the suicide trap.

Assessing Your Suicidal Impulses

Although suicidal thoughts are common even in individuals who are not depressed, the occurrence of a suicidal impulse if you *are* depressed is always to be regarded as a dangerous symptom. It is important for you to know how to pinpoint those suicidal impulses which are the most threatening. In the Burns Depression Checklist in Chapter 2, questions 23, 24, and 25 refer to your suicidal thoughts and impulses. If you have checked a one, two, three, or four on these questions, suicidal fantasies are present, and it is important to evaluate their seriousness and to intervene if necessary (see page 21).

The most serious error you could make with regard to your suicidal impulses is to be overly inhibited in talking them over with a counselor. Many people are afraid to talk about suicidal fantasies and urges for fear of disapproval or because they believe that even talking about them will bring on a suicide attempt. This point of view is unwarranted. You are more likely to feel a great sense of relief in discussing suicidal thoughts with a professional therapist, and consequently you have a much better chance of defusing them.

If you do have suicidal thoughts, ask yourself if you are taking such thoughts seriously. Are there times when you wish you were dead? If the answer is yes, is your death wish active or passive? A passive death wish exists if you would prefer to be dead, but you are unwilling to take active steps to bring this about. One young man confessed to me, "Doctor, every night when I go to bed I pray to God to let me wake up with cancer. Then I could die in peace, and my family would understand."

An *active* death wish is more dangerous. If you are seriously planning an actual suicide attempt, then it's important to know the following: Have you thought about a method? What is your method? Have you made plans? What specific preparations have you made? As a general rule, the more concrete and well-formulated your plans are, the more likely you may actually make a suicide attempt. The time to seek professional help is now!

Have you ever made a suicide attempt in the past? If so, you should view any suicidal impulse as a danger signal to seek help immediately. For many people these previous attempts seem to be "warm-ups," in which they flirt with suicide but have not mastered the particular method they have selected. The fact that an individual has made this attempt unsuccessfully on several occasions in the past indicates an increased risk of success in the future. It is a dangerous myth that unsuccessful suicide attempts are simply gestures or attention-getting devices and are therefore not to be taken seriously. Current thinking suggests that all suicidal thoughts or actions are to be taken seriously. It can be highly misleading to view suicidal thoughts and actions as a "plea for help." Many suicidal patients want help *least* of all because they are 100 percent convinced they are hopeless and beyond help. Because of this illogical belief, what they really want is death.

Your degree of hopelessness is of the greatest importance in assessing whether or not you are at risk for making an active suicide attempt at any time. This one factor seems more closely linked with actual suicide attempts than any other. You must ask yourself, "Do I believe that I have absolutely no chance of getting better? Do I feel that I have exhausted all treatment possibilities and that nothing could possibly help? Do I feel convinced beyond all doubt that my suffering is unbearable and could never come to an end?" If you answer yes to these questions, then your degree of hopelessness is high, and professional treatment is indicated *now*! I would like to emphasize that hopelessness is as much a symptom of depression as a cough is a symptom of pneumonia. The feeling of hopelessness does *not* in fact prove that you are hopeless, any more than a cough proves you are doomed to succumb to pneumonia. It just proves that you are suffering from an illness, in this case, depression. This sense of hopelessness is *not* a reason to make a suicide attempt, but gives you a clear signal to seek competent treatment. So, if you feel hopeless, seek help! Do not consider suicide for one more minute!

The last important factor concerns deterrents. Ask yourself, "Is there anything that is preventing me from committing suicide? Would I hold back because of my family, friends, or religious beliefs?" If you have no deterrents, the possibility is greater that you would consider an actual suicide attempt.

SUMMARY: If you are suicidal, it is of great importance for you to evaluate these impulses in a matter-of-fact manner, using your common sense. The following factors put you in a high-risk group:

1. If you are severely depressed and feel hopeless;
2. If you have a past history of suicide attempts;
3. If you have made concrete plans and preparations for suicide; and
4. If no deterrents are holding you back.

If one or more of these factors apply to you, then it is vital to get professional intervention and treatment immediately. While I firmly believe that the attitude of self-help is important for all people with depression, you clearly must seek professional guidance right away.

The Illogic of Suicide

Do you think depressed people have the "right" to commit suicide? Some misguided individuals and novice therapists are unduly concerned with this issue. If you are counseling or trying to help a chronically depressed individual who is hopeless and threatening self-destruction, you may ask yourself, "Should I intervene aggressively, or should I let him go ahead? What are his rights as a human being in this regard? Am I responsible for preventing this attempt, or should I tell him to go ahead and exercise his freedom of choice?"

I regard this as an absurd and cruel issue that misses the point entirely. The real question is not whether a depressed individual has the right to commit suicide, but whether he is *realistic* in his thoughts when he is considering it. When I talk to a suicidal person, I try to find out why he is feeling that way. I might ask, "What is your motive for wanting to kill yourself? What problem in your life is so terrible that there is no solution?" Then I would help that person expose the illogical thinking that lurks behind the suicidal impulse as quickly as possible. When you begin to think more realistically, your sense of hopelessness and the desire to end your life will fade away and you will have the urge to live. Thus, I recommend joy rather than death to suicidal individuals, and I try to show them how to achieve it as fast as possible! Let's see how this can be done.

Holly was a nineteen-year-old woman who was referred to me for treatment by a child psychoanalyst in New York City. He had treated her unsuccessfully with analytic therapy for many years since the onset of a severe unremitting depression in her early teens. Other doctors had also been unable to help her. Her depression originated during a period of family turbulence that led to her parents' separation and divorce.

Holly's chronic blue mood was punctuated by numerous wrist-slashing episodes. She said that when periods of frustration and hopelessness would build up, she would be overcome by the urge to rip into her flesh and would experience relief only when she saw the blood flowing across her skin. When I first met Holly, I noticed a mass of white scar tissue across her wrists that attested to this behavior. In addition to these episodes of self-mutilation, which were not suicide attempts, she had tried to kill herself on a number of occasions.

In spite of all the treatment she had received, her depression would not let up. At times it became so severe that she had to be hospitalized. Holly had been confined to a closed ward of a New York hospital for several months at the time she was referred to me. The referring doctor recommended a minimum of three years of additional continuous hospitalization, and appeared to agree with Holly that her prognosis for substantial improvement, at least in the near future, was poor.

Ironically, she was bright, articulate, and personable. She had done well in high school, in spite of being unable to go to classes during the times she was confined to hospitals. She had to take some courses with the help of tutors. Like a number of adolescent patients, Holly's dream was to become a mental-health professional, but she had been told by her previous therapist that this was unrealistic because of the nature of her own explosive, intractable emotional problems. This opinion was just one more crushing blow for Holly.

After graduation from high school, she spent the majority of her time in inpatient mental-hospital facilities because she was considered too ill and uncontrollable for outpatient therapy. In a desperate attempt to find help, her father contacted the University of Pennsylvania because he had read about our work in depression. He requested a consultation to determine whether any promising treatment alternatives existed for his daughter.

After speaking to me by phone, Holly's father obtained custody of her and drove to Philadelphia so that I could talk to her and review the possibilities for treatment. When I met them, their personalities contrasted with my expectations. He proved to be a relaxed, mild-mannered individual; she was strikingly attractive, pleasant, and cooperative.

I administered several psychological tests to Holly. The Beck Depression Inventory indicated severe depression, and other tests confirmed a high degree of hopelessness and serious suicidal intent. Holly put it to me bluntly, "I want to kill myself." The family history indicated that several relatives had attempted suicide—two of them successfully. When I asked Holly why she wanted to kill herself, she told me that she was a lazy human being. She explained that because she was lazy, she was worthless and so deserved to die.

I wanted to find out if she would react favorably to cognitive therapy, so I used a technique that I hoped would capture her attention. I proposed we do some role-playing, and she was to imagine that two attorneys were arguing her case in court. Her father, by the way, happened to be an attorney who specialized in medical malpractice suits! Because I was a novice therapist at the time, this intensified my own anxious, insecure feelings about tackling such a tough case. I told Holly to play the role of the prosecutor, and she was to try to convince the jury that she deserved a death sentence. I told her I would play the role of the defense attorney, and that I would challenge the validity of every accusation she made. I told her that this way we could review her reasons for living and her reasons for dying, and see where the truth lay:

HOLLY: For this individual, suicide would be an escape from life.

DAVID: That argument could apply to anyone in the world. By itself, it is not a convincing reason to die.

HOLLY: The prosecutor replies that the patient's life is so miserable, she cannot stand it one minute longer.

DAVID: She has been able to stand it up until now, so maybe she can stand it a while longer. She was not always miserable in the past, and there is no proof that she will always be miserable in the future.

HOLLY: The prosecutor points out that her life is a burden to her family.

DAVID: The defense emphasizes that suicide will not solve this problem, since her death by suicide may prove to be an even more crushing blow to her family.

HOLLY: But she is self-centered and lazy and worthless, and deserves to die!

DAVID: What percentage of the population is lazy?

HOLLY: Probably twenty percent . . . no, I'd say only ten percent.

DAVID: That means twenty million Americans are lazy. The defense points out that they don't have to die for this, so there is no reason the patient should be singled out for death. Do you think laziness and apathy are symptoms of depression?

HOLLY: Probably.

DAVID: The defense points out that individuals in our culture are not sentenced to death for the symptoms of illness, whether it be pneumonia, depression, or any other disease. Furthermore, the laziness may disappear when the depression goes away.

Holly appeared to be involved in this repartee and amused by it. After a series of such accusations and defenses, she conceded that there was no convincing reason she should have to die, and that any reasonable jury would have to rule in favor of the defense. What was more important was that Holly was learning to challenge and answer her negative thoughts about herself. This process brought her partial but

immediate emotional relief, the first she had experienced in many years. At the end of the consultation session, she said to me, "This is the best that I have felt in as long as I can remember. But now the negative thought crosses my mind, 'This new therapy may not prove to be as good as it seems.'" In response to this she felt a sudden surge of depression again. I assured her, "Holly, the defense attorney points out that this is no real problem. If the therapy isn't as good as it seems to be, you'll find out in a few weeks, and you'll still have the alternative of a long-term hospitalization. You'll have lost nothing. Furthermore, the therapy may be partially as good as it seems, or conceivably even better. Perhaps you would be willing to give it a try." In response to this proposal, she decided to come to Philadelphia for treatment.

Holly's urge to commit suicide was simply the result of cognitive distortions. She confused the symptoms of her illness, such as lethargy and loss of interest in life, with her true identity and labeled herself as a "lazy person." Because Holly equated her worth as a human being with her achievement, she concluded she was worthless and deserved to die. She jumped to the conclusion that she could never recover, and that her family would be better off without her. She magnified her discomfort by saying, "I can't stand it." Her sense of hopelessness was the result of the fortune-telling error—she illogically jumped to the conclusion that she could not improve. When Holly saw that she was simply trapping herself with unrealistic thoughts, she felt a sudden relief. In order to maintain such improvement, Holly had to learn to correct her negative thinking on an ongoing basis and that took hard work! She wasn't going to give in that easily!

Following our initial consultation, Holly was transferred to a hospital in Philadelphia, where I visited her twice a week to initiate cognitive therapy. She had a stormy course in the hospital with dramatic mood swings, but was able to be discharged after a five-week period, and I persuaded her to enroll as a part-time summer-school student. For a while her moods continued to oscillate like a yo-yo, but she showed an overall improvement. At times Holly would report feeling very good for several days. This constituted a real breakthrough, since these were the first happy periods she had experienced since the age of thirteen. Then she would suddenly relapse into a severe depressive state. At these times she would again become actively suicidal, and would try her best to convince me that life was not worth living. Like many adolescents, she seemed to carry a grudge against all mankind, and insisted there was no point in living any longer.

In addition to feeling negative about her own sense of worth, Holly had developed an intensely negative and disillusioned view of the entire world. Not only did she see herself as trapped by an endless, untreatable depression, but like many of today's adolescents, she had adopted a personal theory of nihilism. This is the most extreme form of pessimism. Nihilism is the belief that there is no truth or meaning to anything, and that *all* of life involves suffering and agony. To a nihilist like Holly, the world offers *nothing* but misery. She had become convinced that the very essence of every person and object in the universe was evil and horrible. Her depression was thus the experience of hell on earth. Holly envisioned death as the *only* possible surcease, and she longed for death. She constantly complained and harangued cynically about the cruelties and miseries of living. She insisted that life was totally unbearable at all times, and that all human beings were totally lacking in redeeming qualities.

The task of getting such an intelligent and persistent young woman to see and admit how distorted her thinking was provided a real challenge to this therapist! The following lengthy dialogue illustrates her intensely negative attitudes as well as my struggles to help her penetrate the illogic in her thinking:

HOLLY: Life is not worth living because there is more bad than good in the world.

DAVID: Suppose I was the depressed patient and you were my therapist and I told you that, what would you say?

(I used this maneuver with Holly because I knew her goal in life was to be a therapist. I figured she'd say something reasonable and upbeat, but she outfoxed me in her next statement.)

HOLLY: I'd say that I can't argue with you!

DAVID: So, if I were your depressed patient and told you that life is not worth living, you'd advise me to jump out the window?

HOLLY (laughing): Yes. When I think about it, that's the best thing to do. If you think about all the bad things that are going on in the world, the right thing to do is to get really upset about them and be depressed.

DAVID: And what are the advantages to that? Does that help you correct the bad things in the world or what?

HOLLY: No. But you *can't* correct them.

DAVID: You can't correct *all* the bad things in the world, or you can't correct *some* of them?

HOLLY: You can't correct anything of importance. I guess you can correct small things. You can't really make a dent in the badness of this universe.

DAVID: Now, at the end of each day if I said that to myself when I went home, I could really become upset. In other words, I could either think about the people that I did help during the day and feel good, or I could think of all the thousands of people that I will never get a chance to see and work with, and I could feel hopeless and helpless. That would incapacitate me, and I don't think that it is to my advantage to be incapacitated. Is it to your advantage to be incapacitated?

HOLLY: Not really. Well, I don't know.

DAVID: You *like* being incapacitated?

HOLLY: No. Not unless I were completely incapacitated.

DAVID: What would that be like?

HOLLY: I would be dead, and I think I would be better off being that way.

DAVID: Do you think being dead is enjoyable?

HOLLY: Well, I don't even know what it's like. I suppose it might be horrible to be dead and to experience nothing. Who knows?

DAVID: So it might be horrible, or it might be nothing. Now the closest thing to nothing is when you are being anesthetized. Is that enjoyable?

HOLLY: It's not enjoyable, but it's not unenjoyable either.

DAVID: I'm glad you admit that it's not enjoyable. And you're right, there's really nothing enjoyable about nothing. But there are some things enjoyable about life.

(At this point I thought I had really made a mark. But again, in her adolescent insistence that things were no good, she continued to outmaneuver me and contradict everything I said. Her contrariness made my work with her challenging and more than a bit frustrating at times.)

HOLLY: But you see, there are so *few* things that are enjoyable about life, and there is so much other stuff that you have to go through to get those few enjoyable things that it seems to me it just doesn't weigh out.

DAVID: How do you feel when you're feeling good? Do you feel that it doesn't weigh out then, or do you just feel this way when you're feeling bad?

HOLLY: It all depends on what I want to focus on, right? The only way I get myself not to be depressed is if I don't think about all the lousy things in this universe that make me depressed. Right? So when I am feeling good, that means I'm focusing on the good things. But all the bad things are still there. Since there is so much more bad than good, it is dishonest and phony to look only at the good and feel good or feel happy, and that's why suicide is the best thing to do.

DAVID: Well, there are two kinds of bad things in this universe. One is the pseudo-bad. This is the unreal bad that we create as a figment of our imagination by the way we think about things.

HOLLY (interrupting): Well, when I read the newspapers, I see rapes and murders. That seems to me to be the *real* bad.

DAVID: Right. That's what I call the real bad. But let's look at the pseudo-bad first.

HOLLY: Like what? What do you mean by pseudo-bad?

DAVID: Well, take your statement that life is no good. That statement is an inaccurate exaggeration. As you pointed out, life has its good elements, its bad elements, and its neutral elements. So the statement that life is no good or that everything is hopeless is just exaggerated and unrealistic. This is what I mean by the pseudo-bad. On the other hand, there are the real problems in life. It's true that people do get murdered and that people do get cancer, but in my experience these unpleasant things can be coped with. In fact, in your life you will probably make the decision to commit yourself to some aspect of the world's problems where you think you can make a contribution to a solution. But even there, the meaningful approach involves interaction with the problem in a positive way rather than getting overwhelmed by it and sitting back and moping.

HOLLY: Well, see, that's what I do. I just get immediately overwhelmed with the bad things I encounter, and then I feel like I ought to kill myself.

DAVID: Right. Well, it might be nice if there were a universe where there were no problems and no suffering, but then there would be no opportunity for people to grow or solve these problems either. One of these days you'll probably take one of the problems in the world, and contributing to its solution will become a source of satisfaction to you.

HOLLY: Well, that's not fair to use problems in that way.

DAVID: Why don't you test it out? I wouldn't want you to believe anything that I say unless you test it out for yourself and find out if it's true. The way to test it out is to begin getting involved in things, to go to classes, do your work, and establish relationships with people.

HOLLY: That's what I am beginning to do.

DAVID: Well, you can see how it works out over a period of time, and you may find that going to summer school and making a contribution to this world, and meeting with friends and getting involved with activities, and doing your work and getting adequate grades, and experiencing a sense of achievement and pleasure in doing what you can—all of this might not be satisfying to you, and you might conclude, "Hey, depression was better than this." And "I don't like being happy." You might say, "Hey, I don't like being involved in life." If that's true, you can always go back to being depressed and hopeless. I'm not going to take anything away from you. But don't knock happiness until you've tried it.

> Check it out. See what life is like when you get involved and make an effort. Then we'll see where the chips fall at that time.

Holly again experienced a substantial emotional relief as she realized, at least in part, that her intense conviction that the world was no good and life was not worth living was simply the result of her illogical way of looking at things. She was making the mistake of focusing only on negatives (the mental filter) and arbitrarily insisting that the positive things in the world didn't count (disqualifying the positive). Consequently, she got the impression that everything was negative and that life was not worth living. As she learned to correct this error in her thinking, she began to experience some improvement. Although she continued to have a number of ups and downs, the frequency and severity of her mood swings diminished with time. She was so successful in her summer-school work that she was accepted in the fall as a full-time student at a top Ivy League college. Although she made many pessimistic predictions that she would flunk out because she didn't have the brains to make it in academics, to her great surprise she did outstandingly well in her classes. As she learned to transform her intense negativity into productive activity, she became a top-notch student.

Holly and I had a parting of the ways after less than a year of weekly sessions. In the middle of an argument, she fled from the office, slammed the door, and vowed never to return. Maybe she didn't know any other way to say goodbye. I believe she felt she was ready to try and make it on her own. Perhaps she finally got tired of trying to batter me down; after all, I was just as stubborn as she was! She called me recently to let me know how things turned out. Although she still struggles with her moods at times, she is now a senior and at the top of her class. Her dream of going to graduate school to pursue a professional career appears to be a certainty. God bless you, Holly!

Holly's thinking represents many of the mental traps that

can lead to a suicidal impulse. Nearly all suicidal patients have in common an illogical sense of hopelessness and the conviction they are facing an insoluble dilemma. Once you expose the distortions in your thinking, you will experience considerable emotional relief. This can give you a basis for hope and can help you avert a dangerous suicide attempt. In addition, the emotional relief can give you some breathing room so you can continue to make more substantive changes in your life.

You may find it difficult to identify with a turbulent adolescent like Holly, so let's take a brief look at another more common cause for suicidal thoughts and attempts—the sense of disillusionment and despair that sometimes hits us in middle age or in our senior years. As you review the past, you may conclude that your life hasn't really amounted to much in comparison with the starry-eyed expectations of your youth. This has been called the mid-life crisis—that stage in which you review what you have actually done with your life compared with your hopes and plans. If you cannot resolve this crisis successfully, you may experience such intense bitterness and such profound disappointment that you may attempt suicide. Once again, the problem turns out to have little, if anything, to do with reality. Instead, your turmoil is based on twisted thinking.

Louise was a married woman in her fifties who had emigrated from Europe to the United States during World War II. Her family brought her to my office one day after she had been discharged from an intensive care unit, where she had been treated for an almost successful and totally unexpected suicide attempt. The family was unaware she had been experiencing serious depression, so her sudden suicide attempt was a complete surprise. As I spoke with Louise, she told me bitterly that her life had not measured up. She had never experienced the joy and fulfillment that she dreamed of as a girl: she complained of a sense of inadequacy and was convinced she was a failure as a human being. She told me that she had accomplished nothing worthwhile and concluded her life was not worth living.

Because I felt a rapid intervention was necessary in order to prevent a second suicide attempt, I used cognitive techniques to demonstrate to her as fast as possible the illogic of what she was saying to herself. I first asked her to give me a list of things she *had* accomplished in life as a way of testing her belief that she hadn't succeeded at anything worthwhile.

LOUISE: Well, I helped my family escape from the Nazi terrorism and relocate in this country during World War II. In addition, I learned to speak many languages fluently—five of them—when I was growing up. When we came to the United States, I worked at an unpleasant job so that enough money would be available for my family. My husband and I raised a fine young son, who went on to college and is now a highly successful businessman. I'm a good cook; and in addition to perhaps being a good mother, my grandchildren seem to think I'm a good grandmother. These would be the things which I feel I have accomplished during my life.

DAVID: In light of all these accomplishments, how can you tell me you have accomplished nothing?

LOUISE: You see, *everyone* in my family spoke five languages. Getting out of Europe was just a matter of survival. My job was ordinary and required no special talent. It is a mother's duty to raise her family, and any good housewife should learn to cook. Because these are all the things I was supposed to do, or that anyone could have done, they are not real accomplishments. They are just ordinary, and this is why I have decided to commit suicide. My life is not worthwhile.

I realized that Louise was upsetting herself unnecessarily by saying, "It doesn't count" with regard to anything good about herself. This common cognitive distortion, called "disqualifying the positive," was her main enemy. Louise focused *only* on her inadequacies or errors, and insisted that her successes weren't worth anything. If you discount your achievements in this way, you will create the mental illusion that you are a worthless zero.

In order to demonstrate her mental error in a dramatic fashion, I proposed that Louise and I do some role-playing. I told her that I would play the role of a depressed psychiatrist, and she was to be my therapist, who would try to find out why I have been feeling so depressed.

LOUISE (as therapist): Why is it you feel depressed, Dr. Burns?

DAVID (as depressed psychiatrist): Well, I realize that I've accomplished nothing with my life.

LOUISE: So you feel you've accomplished nothing? But that doesn't make sense. You must have accomplished something. For example, you care for many sick depressed patients, and I understand you publish articles about your research and give lectures. It sounds like you have accomplished a great deal at such a young age.

DAVID: No. None of those things count. You see, it is every doctor's obligation to care for his patients. So that doesn't count. I'm just doing what I'm supposed to do. Furthermore, it is my duty at the university to do research and publish the results. So these are not *real* accomplishments. All the faculty members do this, and my research is not very important, at any rate. My ideas are just ordinary. My life is basically a failure.

LOUISE (laughing at herself—no longer being the therapist): I can see that I have been criticizing myself like that for the past ten years.

DAVID (as therapist again): Now, how does it feel when you continually say to yourself, "It doesn't count" whenever you think about the things you have accomplished?

LOUISE: I feel depressed when I say this to myself.

DAVID: And how much sense does it make to think of the things that you haven't done that you might have liked to do, and to overlook the things that you have done which turned out well and were the result of substantial effort and determination?

LOUISE: It doesn't make any sense at all.

As a result of this intervention, Louise was able to see she had been arbitrarily upsetting herself by saying over and over, "What I have done isn't good enough." When she recognized how arbitrary it was to do this to herself, she experienced immediate emotional relief, and her urge to commit suicide disappeared. Louise realized that no matter how much she had accomplished in her life, if she wanted to upset herself she would always be able to look back and say, "It wasn't enough." This indicated to her that her problem was not *realistic* but simply a mental trap she had fallen into. The role-reversal seemed to evoke a sense of amusement and laughter in her. This stimulation of her sense of humor appeared to help her recognize the absurdity of her self-criticism, and she achieved a much needed sense of compassion for herself.

Let's review why your conviction that you are "hopeless" is both irrational and self-defeating. First, remember that depressive illness is usually, if not always, self-limiting, and in most cases eventually disappears even without treatment. The purpose of treatment is to speed the recovery process. Many effective methods of drug therapy and psychotherapy now exist, and others are being rapidly developed. Medical science is in a constant state of evolution. We are currently experiencing a renaissance in our approaches to depressive illness. Because we cannot predict yet with complete certainty which psychological intervention or medication will be most helpful for a particular patient, a number of techniques must sometimes be applied until the right key to the locked-up potential for happiness is found. Although this does require patience and hard work, it is crucial to keep in mind that nonresponse to one or even to several techniques does not indicate that all methods will fail. In fact, the opposite is more often true. For example, recent drug research has shown that patients who do not respond to one antidepressant medication often have a better than average chance of responding to another. This means if you fail to respond to one of the agents, your chances for improvement when you are given another may actually be enhanced. When you consider that there are large numbers of effective antidepressants, psychotherapeutic interventions, and self-help techniques, the probability for eventual recovery becomes tremendously high.

When you are depressed, you may have a tendency to confuse feeling with facts. Your feelings of hopelessness and total despair are just *symptoms* of depressive illness, not facts. If you think you are hopeless, you will naturally feel this way. Your feelings only trace the illogical pattern of your thinking. Only an expert, who has treated hundreds of depressed individuals, would be in a position to give a meaningful prognosis for recovery. Your suicidal urge merely indicates the need for treatment. Thus, your conviction that you are "hopeless" nearly always proves you are not. Therapy, not suicide, is indicated. Although generalizations can be misleading, I let the following rule of thumb guide me: Patients who *feel* hopeless *never actually are* hopeless.

The conviction of hopelessness is one of the most curious aspects of depressive illness. In fact, the degree of hopelessness experienced by seriously depressed patients who have an excellent prognosis is usually greater than in terminal malignancy patients with a poor prognosis. It is of great importance to expose the illogic that lurks behind your hopelessness as soon as possible in order to prevent an actual suicide attempt. You may feel convinced that you have an insoluble problem in your life. You may feel that you are caught in a trap from which there is no exit. This may lead to extreme frustration and even to the urge to kill yourself as the only escape. However, when I confront a depressed patient with respect to precisely what kind of trap he or she is in, and I zero in on the person's "insoluble problem," I invariably find that the patient is deluded. In this situation, you are like an evil magician, and you create a hellish illusion with mental magic. Your suicidal thoughts are illogical, distorted, and erroneous. Your twisted thoughts and faulty assumptions, not reality, create your suffering. When you learn to look behind the mirrors, you will see that you are fooling yourself, and your suicidal urge will disappear.

It would be naïve to say that depressed and suicidal individuals never have "real" problems. We *all* have real problems, including finances, interpersonal relationships, health, etc. But such difficulties can nearly always be coped with in a reasonable manner without suicide. In fact, meeting such challenges can be a source of mood elevation and personal growth. Furthermore, as pointed out in Chapter 9, real problems can never depress you even to a small extent. Only distorted thoughts can rob you of valid hopes or self-esteem. I have never seen a "real" problem in a depressed patient which was so "totally insoluble" that suicide was indicated.

Part VI

Coping with the Stresses and Strains of Daily Living

Chapter 16

How I Practice What I Preach

"Physician, heal thyself."—Luke 4:23

A recent study of stress has indicated that one of the world's most demanding jobs—in terms of the emotional tension and the incidence of heart attacks—is that of an air-traffic controller in an airport tower. The work involves precision, and the traffic controller must be constantly alert—a blunder could result in tragedy. I wonder however if that job is more taxing than mine. After all, the pilots are cooperative and intend to take off or land safely. But the ships I guide are sometimes on an intentional crash course.

Here's what happened during one thirty-minute period last Thursday morning. At 10:25 I received the mail, and skimmed a long, rambling, angry letter from a patient named Felix just prior to the beginning of my 10:30 session. Felix announced his plans to carry out a "blood bath," in which he would murder three doctors, including two psychiatrists who had treated him in the past! In his letter Felix stated, "I'm just waiting until I get enough energy to drive to the store and purchase the pistol and the bullets." I was unable to reach Felix by phone, so I began my 10:30 session with Harry. Harry was emaciated and looked like a concentration camp victim. He was unwilling to eat because of a delusion that his bowels had "closed off," and he had lost seventy pounds. As I was discussing the unwelcome option of hospitalizing Harry for forced tube feeding to prevent his death from starvation, I received an emergency telephone call from a patient named Jerome, which interrupted the session. Jerome informed me he had placed a noose around his neck and was seriously considering hanging himself before his wife came home from work. He announced his unwillingness to continue outpatient treatment and insisted that hospitalization would be pointless.

I straightened out these three emergencies by the end of the day, and went home to unwind. At just about bedtime I received a call from a new referral—a well-known woman VIP referred by another patient of mine. She indicated she'd been depressed for several months, and that earlier in the evening she'd been standing in front of a mirror practicing slitting her throat with a razor blade. She explained she was calling me only to pacify the friend who referred her to me, but was unwilling to schedule an appointment because she was convinced her case was "hopeless."

Every day is not as nerve-racking as that one! But at times it does seem like I'm living in a pressure cooker. This gives me a wealth of opportunities to learn to cope with intense uncertainty, worry, frustration, irritation, disappointment, and guilt. It affords me the chance to put my cognitive techniques to work on myself and see firsthand if they're actually effective. There are many sublime and joyous moments too.

If you have ever gone to a psychotherapist or counselor, the chances are that the therapist did nearly all the listening and expected you to do most of the talking. This is because many therapists are trained to be relatively passive and nondirective—a kind of "human mirror" who simply reflects what you are saying.* This one-way style of communication

*Some of the newer forms of psychiatric treatment, such as cognitive therapy, allow for a natural fifty-fifty dialogue between the client and therapist, who work together as equal members of a team.

may have seemed unproductive and frustrating to you. You may have wondered—"What is my psychiatrist really like? What kinds of feelings does he have? How does he deal with them? What pressures does he feel in dealing with me or with other patients?"

Many patients have asked me directly, "Dr. Burns, do you actually practice what you preach?" The fact is, I often do pull out a sheet of paper on the train ride home in the evening, and draw a line down the center from top to bottom so I can utilize the double-column technique to cope with any nagging emotional hangovers from the day. If you are curious to take a look behind the scenes, I'll be glad to share some of my self-help homework with you. This is your chance to sit back and listen while the *psychiatrist* does the talking! At the same time, you can get an idea of how the cognitive techniques you have mastered to overcome clinical depression can be applied to all sorts of daily frustrations and tensions that are an inevitable part of living for all of us.

Coping With Hostility: The Man Who Fired Twenty Doctors

One high-pressure situation I often face involves dealing with angry, demanding, unreasonable individuals. I suspect I have treated a few of the East Coast's top anger champions. These people often take their resentment out on the people who care the most about them, and sometimes this includes me.

Hank was an angry young man. He had fired twenty doctors before he was referred to me. Hank complained of episodic back pain, and was convinced he suffered from some severe medical disorder. Because no evidence for any physical abnormality had ever surfaced, in spite of lengthy, elaborate medical evaluations, numerous physicians told him that his aches and pains were in all likelihood the result of emotional tension, much like a headache. Hank had difficulty accepting this, and he felt his doctors were writing him off and just didn't give a damn about him. Over and over he'd explode in a fury, fire his doctor, and seek out someone new. Finally, he consented to see a psychiatrist. He resented this referral, and after making no progress for about a year, he fired his psychiatrist and sought treatment at our Mood Clinic.

Hank was quite depressed, and I began to train him in cognitive techniques. At night when his back pain flared up, Hank would work himself up into a frustrated rage and impulsively call me at home (he had persuaded me to give him my home number so he wouldn't have to go through the answering service). He would begin by swearing and accusing me of misdiagnosing his illness. He'd insist he had a medical, not a psychiatric, problem. Then he'd deliver some unreasonable demand in the form of an ultimatum: "Dr. Burns, either you arrange for me to get shock treatments tomorrow or I'll go out and commit suicide tonight." It was usually difficult, if not impossible, for me to comply with most of his demands. For example, I don't give shock treatments, and furthermore I didn't feel this type of treatment was indicated for Hank. When I would try to explain this diplomatically, he would explode and threaten some impulsive destructive action.

During our psychotherapy sessions Hank had the habit of pointing out each of my imperfections (which are real enough). He'd often storm around the office, pound on the furniture, heaping insults and abuse on me. What used to get me in particular was Hank's accusation that I didn't care about him. He said that all I cared about was money and maintaining a high therapy success rate. This put me in a dilemma, because there was a grain of truth in his criticisms—he was often several months behind in making payments for his therapy, and I was concerned that he might drop out of treatment prematurely and end up even more disillusioned. Furthermore, I *was* eager to add him to my list of successfully treated individuals. Because there was some truth in Hank's haranguing attacks, I felt guilty and defensive when he would zero in on me. He, of course, would sense this, and consequently the volume of his criticism would increase.

I sought some guidance from my associates at the Mood Clinic as to how I might handle Hank's outbursts and my own feelings of frustration more effectively. The advice I received from Dr. Beck was especially useful. First, he emphasized that I was "unusually fortunate" because Hank was giving me a golden opportunity to learn to cope with criticism and anger effectively. This came as a complete surprise to me; I hadn't realized what good fortune I had. In addition to urging me to use cognitive techniques to reduce and eliminate my own sense of irritation, Dr. Beck proposed I try out an unusual strategy for interacting with Hank when he was in an angry mood. The essence of this method was: (1) Don't turn Hank off by defending yourself. Instead, do the opposite—urge him to say all the worst things he can say about you. (2) Try to find a grain of truth in all his criticisms and then agree with him. (3) After this, point out any areas of disagreement in a straightforward, tactful, nonargumentative manner. (4) Emphasize the importance of sticking together, in spite of these occasional disagreements. I could remind Hank that frustration and fighting might slow down our therapy at times, but this need not destroy the relationship or prevent our work from ultimately becoming fruitful.

I applied this strategy the next time Hank started storming around the office screaming at me. Just as I had planned, I urged Hank to keep it up and say all the worst things he could think of about me. The result was immediate and dramatic. Within a few moments, all the wind went out of his sails—all his vengeance seemed to melt away. He began communicating sensibly and calmly, and sat down. In fact, when I agreed with some of his criticisms, he suddenly began to defend me and say some nice things about me! I was so impressed with this result that I began using the same approach with other angry, explosive individuals, and I actually did begin to enjoy his hostile outbursts because I had an effective way to handle them.

I also used the double-column technique for recording and talking back to my automatic thoughts after one of Hank's midnight calls (see Figure 16–1, page 415). As my associates suggested, I tried to see the world through Hank's eyes in order to gain a certain degree of empathy. This was a specific antidote that in part dissolved my own frustration and anger, and I felt much less defensive and upset. It helped me to see his outbursts more as a defense of his own self-esteem than as an attack on me, and I was able to comprehend his feelings of futility and desperation. I reminded myself that much of the time he was damn hardworking and cooperative, and how foolish it was for me to demand he be totally cooperative at all times. As I began to feel more calm and confident in my work with Hank, our relationship continually improved.

Eventually, Hank's depression and pain subsided, and he terminated his work with me. I hadn't seen him for many months when I received a message from my answering service that Hank wanted me to call him. I suddenly felt apprehensive; memories of his turbulent tirades flooded my mind, and my stomach muscles tensed up. With some hesitation and mixed feelings, I dialed his number. It was a sunny Saturday afternoon, and I'd been looking forward to a much needed rest after an especially taxing week. Hank answered the phone: "Dr. Burns, this is Hank. Do you remember me? There's something I've been meaning to tell you for some time . . ." He paused, and I braced for the impending explosion. "I've been essentially free of pain and depression since we finished up a year ago. I went off disability and I've gotten a job. I'm also the leader of a self-help group in my own hometown."

This wasn't the Hank I remembered! I felt a wave of relief and delight as he went on to explain, "But that's not why I'm calling. What I want to say to you is . . ." There was another moment of silence—"I'm grateful for your

Figure 16–1. Coping with Hostility.

Automatic Thoughts	*Rational Responses*
1. I've put more energy into working with Hank than nearly anyone, and this is what I get—abuse!	1. Stop complaining. You sound like Hank! He's frightened and frustrated, and he's trapped in his resentment. Just because you work hard for someone, it doesn't necessarily follow that they'll feel appreciative. Maybe he will some day.
2. Why doesn't he trust me about his diagnosis and treatment?	2. Because he's in a panic, he's extremely uncomfortable and in pain, and he hasn't yet gotten any substantial results. He'll believe you once he starts getting well.
3. But in the meantime, he should at least treat me with respect!	3. Do you expect him to show respect *all* the time or *part* of the time? In general, he exerts tremendous effort in his self-help program and does treat you with respect. He's determined to get well—if you don't expect perfection, you won't have to feel frustrated.
4. But is it fair for him to call me so often at home at night? And does he have to be so abusive?	4. Talk it over with him when you're both feeling more relaxed. Suggest that he supplement his individual therapy by joining a self-help group in which the various patients call each other for moral support. This will make it easier for him to cut down on calls to you. But for now, remember that he doesn't *plan* these emergencies, and they are very terrifying and real to him.

efforts, and I now know you were right all along. There was nothing dreadfully wrong with me, I was just upsetting myself with my irrational thinking. I just couldn't admit it until I knew for sure. Now, I feel like a whole man, and I had to call you up and let you know where I stood . . . It was hard for me to do this, and I'm sorry it took so long for me to get around to telling you."

Thank you, Hank! I want you to know that some tears of joy and pride in you come to my eyes as I write this. It was worth the anguish we both went through a hundred times over!

Coping With Ingratitude: The Woman Who Couldn't Say Thank You

Did you ever go out of your way to do a favor for someone only to have the person respond to your efforts with indifference or nastiness? People *shouldn't* be so unappreciative, right? If you tell yourself this, you will probably stew for days as you mull the incident over and over. The more inflammatory your thoughts and fantasies become, the more disturbed and angry you will feel.

Let me tell you about Susan. After high-school graduation, Susan sought treatment for a recurrent depression. She was very skeptical that I could help her and continually reminded me that she was hopeless. She had been in a hysterical state for several weeks because she couldn't decide which of two colleges to attend. She acted as though the world would come to an end if she didn't make the "right" decision, and yet the choice was simply not clear-cut. Her insistence on eliminating all uncertainty was bound to cause her endless frustration because it simply couldn't be done.

She cried and sobbed excessively. She was insulting and abusive to her boyfriend and her family. One day she called me on the phone, pleading for help. She just had to make up her mind. She rejected every suggestion I made, and angrily demanded I come up with some better approach. She kept insisting, "Since I can't make this decision, it proves your cognitive therapy won't work for me. Your methods are no damn good. I'll never be able to decide, and I can't get better." Because she was so upset, I arranged my afternoon schedule so that I could have an emergency consultation with a colleague. He offered several outstanding suggestions; I called her right back and gave her some tips on how to resolve her indeciveness. She was then able to come to a satisfactory decision within fifteen minutes, and felt an instantaneous wave of relief.

When she came in for her next regularly scheduled session, she reported she had been feeling relaxed since our talk, and had finalized the arrangements to attend the college that she chose. I anticipated waves of gratitude because of my strenuous efforts on her behalf, and I asked her if she was still convinced that cognitive techniques would be ineffective for her. She reported, "Yes, indeed! This just proves my point. My back was up against the wall, and I *had* to make a decision. The fact that I'm feeling good now doesn't count because it can't last. This stupid therapy can't help me. I'll be depressed for the rest of my life." My thought: "My God! How illogical can you get? I could turn mud into gold, and she wouldn't even notice!" My blood was boiling, so I decided to use the double-column technique later that day to try and calm my troubled and insulted spirits (see Figure 16–2, page 418).

After writing down my automatic thoughts, I was able to pinpoint the irrational assumption that caused me to get upset over her ingratitude. It was, "If I do something to help someone, they are duty-bound to feel grateful and reward me for it." It would be nice if things worked like this, but it's simply not the case. No one has a moral or legal obligation to credit me for my cleverness or praise my good efforts on their behalf. So why expect it or demand it? I decided to tune in to reality and adopt a more realistic attitude: "If I do something to help someone, the chances are the person *will* be appreciative, and that will feel good.

Figure 16–2. Coping with Ingratitude.

Automatic Thoughts	*Rational Responses*
1. How can such a brilliant girl be so illogical?	1. Easily! Her illogical thinking is the cause of her depression. If she didn't continually focus on negatives and disqualify positives, she wouldn't be depressed so often. It's your job to train her in how to get over this.
2. But I can't. She's determined to beat me down. She won't give me an ounce of satisfaction.	2. She doesn't have to give you any satisfaction. Only you can do this. Don't you recall that only *your* thoughts affect your moods? Why not credit yourself for what you did? Don't wait around for her. You just learned some exciting things about how to guide people in making decisions. Doesn't that count?
3. But she should admit I helped her! She should be grateful!	3. Why "should" she? That's a fairy tale. If she could she probably would, but she can't yet. In time she'll come around, but she'll have to reverse an ingrained pattern of illogical thinking that's been dominating her mind for over a decade. She may be *afraid* to admit she's getting help so she won't end up disillusioned again. Or she might be afraid you'll say, "I told you so." Be like Sherlock Holmes and see if you can figure out this puzzle. It's pointless to demand that she be different from the way she is.

But every now and then, someone will not respond the way I want. If the response is unreasonable, this is a reflection on that person, not me, so why get upset over it?" This attitude has made life much sweeter for me, and overall I have been blessed with as much gratitude from patients as I could desire. Incidentally, Susan gave me a call just the other day. She'd done well at college and was about to graduate. Her father had been depressed, and she wanted a referral to a good cognitive therapist! Maybe that was her way of saying thank you!

Coping With Uncertainty and Helplessness: The Woman Who Decided to Commit Suicide

On my way to the office on Monday, I always wonder what the week will hold in store. One Monday morning I was in for an abrupt shock. As I unlocked the office, I found some papers had been slipped under the door over the weekend—a twenty-page letter from a patient named Annie. Annie had been referred to me several months earlier on her twentieth birthday, after having received eight years of completely successful treatment from several therapists for a horrible, grotesque mood disorder. From age twelve on, Annie's life had deteriorated into a nightmarish pattern of depression and self-mutilation. She loved to slash her arms to shreds with sharp objects, one time requiring 200 stitches. She also made a number of nearly successful suicide attempts.

I tensed as I picked up her note. Annie had recently expressed a deep sense of despair. In addition to depression, she suffered from a severe eating disorder, and the previous week had engaged in a bizarre three-day spree of compulsive, uncontrollable binge-eating. Going from restaurant to restaurant, she would stuff herself for hours nonstop. Then she'd vomit it all up and eat some more. In her note she described herself as a "human garbage disposal," and explained that she was beyond hope. She indicated that she had decided to give up trying because she realized she was basically "a nothing."

Without reading further, I called her apartment. Her roommates told me that she had packed up and "left town" for three days without giving any indication of where or why. Alarms sounded in my head! This is exactly what she had done on her last several suicide attempts prior to treatment—she'd drive to a motel, sign in under an assumed name, and overdose. I continued to read her letter. In it she stated, "I'm drained, I'm like a burnt-out light bulb. You can pipe electricity into it, but it just won't light up. I'm sorry but I guess it's just too late. I'm not going to feel false hope any longer . . . During the last few moments I do not feel particularly sad. Once every so often I try to grasp onto life, hoping to clench my hands around something, anything—but I keep grasping nothing, empty."

It sounded like a bona fide suicide note, although no explicit intention was announced. I suddenly became submerged by a massive uncertainty and helplessness—she had disappeared and left no traces. I felt angry and anxious. Because I could do nothing for her, I decided to write down the automatic thoughts that flowed through my mind. I hoped some rational responses would help me cope with the intense uncertainty I was facing (see Figure 16–3, page 421).

After recording my thoughts, I decided to call my associate, Dr. Beck, for a consultation. He agreed that I should assume she was alive unless it was proved otherwise. He suggested that if she were found dead, I could then learn to cope with one of the professional hazards of working with depression. If she was alive, as we assumed, he emphasized the importance of persisting with treatment until her depression finally broke.

The effect of this conversation and the written exercise was magnificent. I realized I was under no obligation to assume "the worst," and that it was my right to choose not to make myself miserable over her possible suicide attempt.

Figure 16–3. Coping with Uncertainty.

Automatic Thoughts	*Rational Responses*
1. She's probably made a suicide attempt—and succeeded.	1. There's no proof she's dead. Why not assume she's alive until proven otherwise? Then you won't have to worry and obsess in the meantime.
2. If she's dead, it means I killed her.	2. No, you're not a killer. You're trying to help.
3. If I'd done something different last week, I could have prevented this. It's my fault.	3. You're not a fortune teller—you can't predict the future. You do the best you can based on what you know—draw the line there and respect yourself on this basis.
4. This shouldn't have happened—I tried so hard.	4. Whatever happened did happen. Just because you make maximal efforts, there's no guarantee about the results. You can't control her, only your efforts.
5. This means my approach is second-rate.	5. Your approach is one of the finest ever developed, and you apply yourself with great effort and commitment, and get outstanding results. You are *not* second-rate.
6. Her parents will be angry with me.	6. They may and they may not. They know how you've knocked yourself out for her
7. Dr. Beck and my associates will be angry with me—they'll know I'm incompetent, and they'll look down on me.	7. Extremely unlikely. We'll all be disappointed to lose a patient we've gone to such extreme lengths to help, but your peers won't feel you've let them down. If you're at all concerned, call them! Practice what you preach, Burns.

Figure 16–3. cont.

Automatic Thoughts	*Rational Responses*
8. I'll feel miserable and guilty until I find out what happened. I'm expected to feel that way.	8. You'll only feel miserable if you make a negative assumption. Odds are (a) she's alive and (b) she'll get better. Assume this and you'll feel good! You have no obligation to feel bad—you have the *right* to refuse to get upset.

I decided I couldn't take on responsibility for her actions, only for mine, and that I had done well with her and would stubbornly continue to do so until she and I had finally defeated her depression and tasted victory.

My anxiety and anger disappeared completely, and I felt relaxed and peaceful until I received the news by telephone on Wednesday morning. She had been found unconscious in a motel room fifty miles from Philadelphia. This was her eighth suicide attempt, but she was alive and complaining as usual in the Intensive Care Unit of an outlying hospital. She would survive, but would require plastic surgery to replace the skin over her elbows and ankles because of sores which had developed during the long period of unconsciousness. I arranged for her transfer to the University of Pennsylvania, where she would be back in my relentless, cognitive clutches again!

When I spoke with her, she was enormously bitter and hopeless. The next couple of months of therapy were especially turbulent. But the depression finally began to lift in her eleventh month, and exactly one year to the day of her referral, her twenty-first birthday, the symptoms of depression disappeared.

The Payoff. **My joy was enormous. Women must have this feeling when they first see their child after delivery—all**

the discomfort of pregnancy and the pain of delivery are forgotten. It's the celebration of life—quite a heady experience. I find that the more chronic and severe the depression, the more intense the therapeutic struggle becomes. But when the patient and I at last discover the combination that unlocks the door to their inner peace, the riches inside far exceed any effort or frustration that occurred along the way.

Part VII

The Chemistry of Mood

NOTE: Numbered Notes and References for Chapters 17–20 can be found on pages 682–687. Because some References are cited more than once, the superscript numbers assigned to those References will appear in these chapters more than once.

Chapter 17

The Search for "Black Bile"

(Notes and References appear on pages 682–687.)

Some day, scientists may provide us with frightening technology that will allow us to change our moods at will. This technology may be in the form of a safe, fast-acting medication that relieves depression in a matter of hours with few or no side effects. This breakthrough will represent one of the most extraordinary and philosophically confusing developments in human history. In a sense, it will almost be like discovering the Garden of Eden again—and we may face new ethical dilemmas. People will probably ask questions like these: When should we use this pill? Are we entitled to be happy all the time? Is sadness sometimes a normal and healthy emotion, or should it always be considered an abnormality that needs treatment? Where do we draw the line?

Some people think such technology has already arrived in the form of a pill called Prozac. When you read the next few chapters, you will see that this is not really the case. Although we have large numbers of antidepressant medications that work for some people, many people do not respond to antidepressant medications in a satisfactory way, and when they do improve, the improvement is often incomplete. Clearly, we are still a long way from our goal.

In addition, we still do not really know how the brain

creates emotions. We do not know why some people are more prone to negative thinking and gloomy moods throughout their lives, whereas others seem to be eternal optimists who always have a positive outlook and a cheerful disposition. Is depression partially genetic? Is it due to some type of chemical or hormonal imbalance? Is it something we're born with, or something we learn? The answers to these questions still elude us. Many people wrongly believe we already have the answers.

The answers to questions about treatment are equally unclear. Which patients should be treated with medications? Which patients need psychotherapy? Is the combination better than either type of treatment alone? You will see that the answers to questions as basic as these are more controversial than you might expect.

In this chapter, I address these issues. I discuss whether depression is caused more by biology (nature) or the environment (nurture). I explain how the brain works, and review evidence that depression might be caused by a chemical imbalance in the brain. I also describe how antidepressant drugs attempt to correct this imbalance.

In Chapter 18, I discuss the "mind-body problem" and address the current controversies about treatments that affect the "mind" (for instance, cognitive therapy) versus treatments that affect the "body" (for instance, antidepressants.) In Chapters 19 and 20, I will give you practical information about all the antidepressant drugs that are currently prescribed for mood problems.

Do Genetic or Environmental Influences Play a Greater Role in Depression?

Although much research is being conducted to try to tease out the relative strengths of the genetic and environmental influences on depression, scientists do not yet know which influences are the most important. With regard to bipolar (manic-depressive) illness, the evidence is quite strong: genetic factors seem to play a strong role. For example, if one identical twin develops bipolar manic-depressive illness, the odds are high that the other twin will also develop this disorder (50 percent to 75 percent). In contrast, when one of two nonidentical twins develops bipolar (manic-depressive) illness, the odds that the other twin will develop the same illness are lower (15 percent to 25 percent). The odds of developing bipolar illness if a parent or nontwin sibling has this disorder are around 10 percent. All these odds are considerably higher than the odds that someone in the general population will develop bipolar illness—the lifetime risk is estimated at less than 1 percent.

Keep in mind that identical twins have identical genes, whereas nonidentical twins share only half their genes. This is probably why the likelihood of bipolar (manic-depressive) illness is so much higher if you have an identical twin than if you have a nonidentical twin with this disorder, and why these rates are so much higher than the rates for bipolar illness in the general population. The increased risk for bipolar illness among identical twins is even true if the identical twins are separated at birth and raised by different families. Although the adoption of identical twins by separate families is rare, it does happen on occasion. In some cases, scientists have been able to locate the twins later in life to determine how similar or different they are. These "natural" experiments can tell us a great deal about the relative importance of genes versus environment because the separately raised identical twins have identical genes but their environments are different. Such studies highlight the importance of strong genetic influences in bipolar disorder.

With regard to the far more common garden-variety depression without episodes of uncontrollable mania, the evidence for genetic factors is still quite fuzzy. Part of the problem facing genetic researchers is that the diagnosis of depression is much less clear-cut than the diagnosis of bipolar (manic-depressive) illness. Bipolar manic-depressive illness is such an unusual disorder, at least in its more severe forms, that the diagnosis is often obvious. The patient has a sudden and alarming change in personality that comes on without drugs or alcohol, along with symptoms such as:

- intense euphoria, often with irritability;
- incredible energy with constant exercising or restless, agitated body movements;
- very little need for sleep;
- nonstop, pressured talking;
- racing thoughts that skip from subject to subject;
- grandiose delusions (for example, the sudden belief that one has a plan for world peace);
- impulsive, reckless, and inappropriate behaviors (such as spending money foolishly);
- inappropriate, excessive flirtatiousness and sexual activity;
- hallucinations (in severe cases).

These symptoms are usually unmistakable and often so uncontrollable that the patient may require hospitalization with medication treatment. Following recovery, the individual usually returns to absolutely normal functioning again. These distinct features of bipolar illness make genetic research relatively straightforward, since it is usually not difficult to determine when individuals have the disorder and when they do not. In addition, this disorder usually begins fairly early in life, with the first episode often occurring by the age of twenty to twenty-five.

In contrast, the diagnosis of depression is much less obvious. Where does normal sadness end and clinical depression begin? The answer is somewhat arbitrary, but the decision will have a big impact on the results of research. Another difficult question genetic researchers face is this: How long should we wait before we decide whether or not a person has developed a clinical depression during his or her life? Suppose, for example, that an individual with a strong family history of depression dies in an auto accident at the age of twenty-one without ever having had an episode of clinical depression. We might conclude that she or he did not inherit the tendency for depression. But if that individual had not died, she or he might have developed an episode of depression later on in life, since a first episode of depression can often occur when you are older than twenty-one.

Problems like this are not insurmountable, but they do make genetic research on depression difficult. In fact, many previously published studies on the genetics of depression are quite flawed and do not permit us to make any unambiguous conclusions about the importance of heredity versus environment in this disorder. Fortunately, more sophisticated studies are now under way, and we may have better answers to these questions during the next five to ten years.

Is Depression Caused by a "Chemical Imbalance" in the Brain?

Throughout the ages, humans have searched for the causes of depression. Even in ancient times there was some suspicion that blue moods were due to an imbalance in body chemistry. Hippocrates (460–377 B.C.) thought that "black bile" was the culprit. In recent years scientists have spearheaded an intensive search for the elusive black bile. They have tried to pinpoint the imbalances in brain chemistry that might cause depression. There are hints about the answer, but in spite of increasingly sophisticated research tools, scientists have not yet discovered the causes of depression.

At least two major arguments have been advanced to

support the notion that some type of chemical imbalance or brain abnormality may play a role in clinical depression. First, the physical (somatic) symptoms of severe depression support the notion that organic changes might be involved. These physical symptoms include agitation (increased nervous activity such as pacing or hand-wringing) or enormous fatigue (motionless apathy—you feel like a ton of bricks and do nothing). You also may experience a "diurnal" variation in your mood. This refers to a worsening of the symptoms of depression in the morning and an improvement toward the end of the day. Other physical symptoms of depression include disturbed sleep patterns (insomnia is the most common), constipation, changes in appetite (usually decreased, sometimes increased), trouble concentrating, and a loss of interest in sex. Because these symptoms of depression "feel" quite physical, there is a tendency to think that the causes of depression are physical.

A second argument for a physiologic cause for depression is that at least some mood disorders seem to run in families, suggesting a role for genetic factors. If there is an inherited abnormality that predisposes some individuals to depression, it could be in the form of a disturbance in body chemistry, as with so many genetic diseases.

The genetic argument is interesting but the data are inconclusive. The evidence for genetic influences in bipolar manic-depressive illness is much stronger than the evidence for genetic influences in the more common forms of depression that afflict most people. In addition, lots of things that do not have genetic causes run in families. For example, families in the United States nearly always speak English, and families in Mexico nearly always speak Spanish. We can say that the tendency to speak a certain language also runs in families, but the language you speak is learned and not inherited.

I don't mean to discount the importance of genetic factors. Recent studies of identical twins who were separated at birth and raised in different families show that many traits we think of as being learned are actually inherited.

Even such personality traits as a tendency toward shyness or sociability appear to be partly inherited. Personal preferences, such as liking a particular flavor of ice cream, may also be strongly influenced by our genes. It seems plausible that we may also inherit a tendency to look at things either in a positive, optimistic way or in a negative, gloomy way. Much more research will be needed to sort out this possibility.

How Does the Brain Work?

The brain is essentially an electrical system that is similar in some ways to a computer. Different portions of the brain are specialized for different kinds of functions. For example, the surface of the brain toward the back of your head is called the "occipital cortex." This is where vision takes place. If you had a stroke that affected this region of the brain, you would have trouble with your vision. A small region on the surface of the left half of your brain is called "Broca's area." This is the part of your brain that allows you to talk to other people. If this part of your brain were injured by a stroke, you would have difficulty talking. You might be able to think of what you wanted to say, but find that you had "forgotten" how to speak the words. A primitive part of your brain called the "limbic system" is thought to be involved in the control of emotions such as joy, sadness, fear, or anger. However, our knowledge of where and how the brain creates positive and negative emotions is still very limited.

We do know that nerves are the "wires" that make up the electrical circuits in the brain. The long thin part of a nerve is called the "axon." When a nerve is stimulated, it sends an electrical signal along the axon to the end of the nerve. A nerve is much more complex than a simple wire, however. For example, a nerve may receive input from tens of thousands of other nerves. Once it is stimulated, its axon may send out signals to tens of thousands of other nerves.

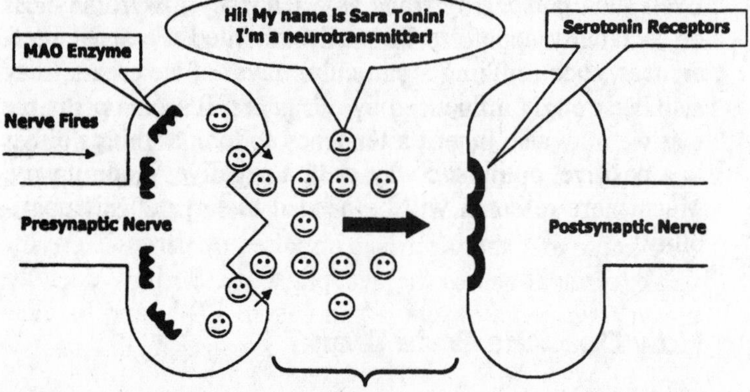

Figure 17–1. When the presynaptic nerve fires, packets of serotonin molecules (neurotransmitters) are released into the synapse. They swim over to the receptors on the surface of the postsynaptic nerve.

This is because the axon can divide and send out many branches. Each of these branches also divides into even more branches, in much the same way that the trunk of a tree divides into more and more branches. Because of this branching tendency, a single nerve in the brain may send out signals to as many as 25,000 other nerves that are located throughout the entire brain.

How do the nerves in your brain communicate their electrical signals to other nerves? To understand this, take a look at Figure 17–1 above. You can see a simplified diagram of two nerves. The region where they meet is called the "synapse." You may not be familiar with that term, but don't feel intimidated by it. It just means the space between two nerves. The left-hand nerve is called the "presynaptic nerve" and the right-hand nerve is called the "postsynaptic nerve." Again, these terms do not have any other fancy or special meanings. They merely refer to the nerve that ends (presynaptic nerve) or begins (postsynaptic nerve) on the left or right edge of the synapse in the figure.

The communication of the electrical signal across this

synapse is important to our understanding of how the brain works. The synaptic region between the presynaptic nerve on the left and the postsynaptic nerve on the right is filled with fluid. This discovery was a major breakthrough in the history of neuroscience. When you think of it, this discovery is not so surprising since our bodies are made up primarily of water. However, scientists were puzzled because they knew that the electrical impulses of nerves were too weak to travel across the synaptic fluid. So how does the presynaptic nerve on the left in Figure 17–1 send its electrical signal across the fluid-filled synapse to the postsynaptic nerve?

As an analogy, imagine that you are hiking and you come to a river. You really need to get to the other side, but the water is too deep. Furthermore, there's no bridge and it's too far to jump. How do you get to the other side? You might need a canoe, or you might have to swim for it.

Nerves face a similar problem. Because their electrical impulses are too weak to jump across synapses, the nerves send little swimmers across with their messages. These little swimmers are chemicals called "neurotransmitters." The nerve in Figure 17–1 uses a neurotransmitter called serotonin.

You can see in Figure 17–1 that when the presynaptic nerve fires, it releases many tiny packets of serotonin into the synapse. Once released, these chemical messengers migrate or "swim" through a process called diffusion across the fluid-filled synapse. At the other side of the synapse, the serotonin molecules become attached to receptors on the surface of the postsynaptic nerve. This signal tells the postsynaptic nerve to fire, as illustrated in Figure 17–2 on page 436.

Different kinds of nerves use different kinds of neurotransmitters. There are a great many of these neurotransmitters in the brain. Chemically, many of them are categorized as "biogenic amines" because they are manufactured from amino acids in the foods we eat. These amine transmitters are the brain's biochemical messengers.

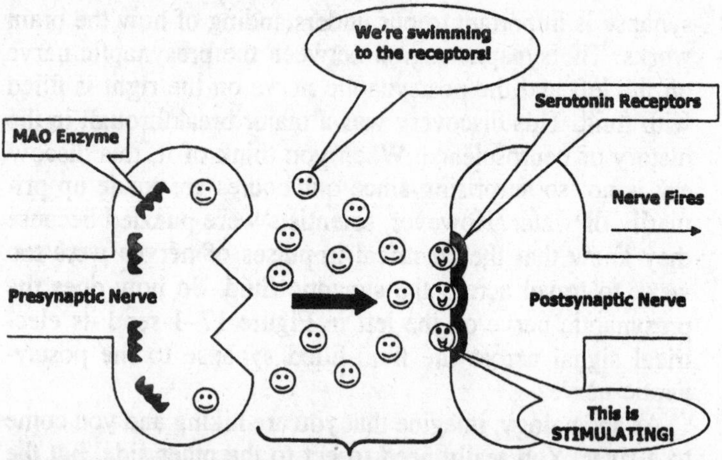

Figure 17–2. The serotonin molecules become attached to the receptors on the postsynaptic nerve. This stimulates the nerve to fire.

Three of the amine transmitters in the limbic (emotional) regions of the brain are called serotonin, norepinephrine, and dopamine. These three transmitters have been theorized to play a role in many psychiatric disorders and have been intensively studied by psychiatric researchers. Because these chemical messengers are called biogenic amines, the theories linking them to depression or mania are sometimes referred to as the biogenic amine theories. But we are getting ahead of ourselves.

How does a chemical messenger cause the postsynaptic nerve to fire once it becomes attached to the nerve? Let's imagine for a moment that the chemical transmitter in the presynaptic nerve is serotonin. (I could have chosen any of them, since they all work in a similar manner.) On the surface of the postsynaptic nerve there are tiny areas called "serotonin receptors." You can think of these receptors as locks because they cannot be opened up without the right key. These receptors are on the membranes that form the outer surface of nerves. These nerve membranes are something like the skin that covers your body.

Now, think of the serotonin as the key to the lock on the

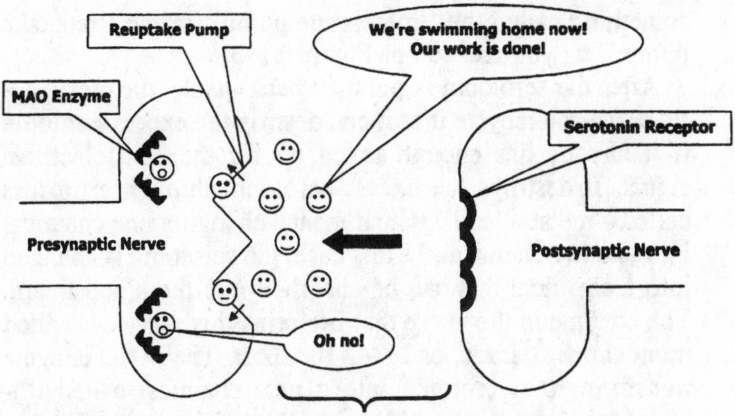

Figure 17–3. The serotonin molecules swim back to the presynaptic nerve where they are pumped back inside. Once inside, MAO destroys them.

postsynaptic nerve. Just like a real key, the serotonin works only because it has a specific shape. There are many other chemicals floating around in the synaptic region, but they will not open the serotonin lock because they do not have the right molecular shape. Once the key fits into the lock, the lock opens up. This triggers additional chemical reactions that cause the postsynaptic nerve to fire electrically. When the nerve fires, the serotonin (the key) is released from the receptor (the lock) on the postsynaptic nerve and ends in the synaptic fluid again. Finally, it "swims" back to the presynaptic nerve (again, through a process called diffusion), as illustrated in Figure 17–3 above.

The serotonin has done its job, and the presynaptic nerve needs to get rid of it; otherwise it will hang around in the synapse and it might swim back to the postsynaptic nerve again. This could create confusion, because the postsynaptic nerve may think there is a new signal and it may get stimulated to fire again.

To solve this problem, the presynaptic nerve has a pump on its surface. Once the serotonin swims back, it attaches to a receptor (another "lock") on the surface of the presynaptic nerve and it is pumped back into the nerve by something called the "membrane pump" or the "reuptake pump," as you can see in Figure 17–3.

After the serotonin is pumped back inside, the presynaptic nerve can recycle it or it can destroy the excess serotonin if it already has enough saved up for the next electrical signal. It destroys the excess serotonin through a process called "metabolism," which means changing one chemical into another chemical. In this case, the serotonin is changed into a chemical that can be absorbed into the bloodstream. The enzyme in the nerve that performs this service is called monoamine oxidase, or MAO for short. The MAO enzyme transforms the serotonin into a new chemical called "5-hydroxyindoleacetic acid," or 5-HIAA. That is another big name, but you can simply think of 5-HIAA as the waste product of the serotonin. The 5-HIAA leaves your brain, enters your bloodstream, and is carried to your kidneys. Your kidneys remove the 5-HIAA from your blood and send it to your bladder. Finally, you get rid of the 5-HIAA when you urinate.

That's the end of the serotonin cycle. Of course, the presynaptic nerve must continually manufacture a new supply of serotonin to use in nerve-firing so that the total amount of serotonin does not get depleted.

What Goes Wrong in Depression?

First of all, let me reemphasize that scientists do not yet know the cause of depression or any other psychiatric disorder. There are lots of interesting theories, but none of them has yet been proven. One day, we may have the answer and look back on the thinking of this era as a quaint historical curiosity. However, science has to start somewhere, and research on the brain is moving forward at an explosive rate. New and very different theories will undoubtedly emerge in the next decade.

The explanations in this section will be very simplified. The brain is enormously complex and our knowledge about how it works is still extremely primitive. There is a vast amount we do not know about the brain's hardware and software. How does the firing of a nerve or a series of nerves get translated into a thought or a feeling? This is one of the deepest mysteries of science, as amazing to me as questions about the origin of the universe.

We won't even attempt to answer those questions here; for the moment, our goals are much more humble. If you understood Figures 17–1 to 17–3, it should be pretty easy for you to understand current theories about what goes wrong in depression.

You have already learned that nerves in the brain send messages to each other with chemical messengers called neurotransmitters. You also know that some of the nerves in the limbic system of the brain use serotonin, norepinephrine, and dopamine as their chemical messengers. Some scientists have hypothesized that depression may result from a deficiency of one or more of these biogenic amine transmitter substances in the brain, while mania (states of extreme euphoria or elation) may result from an excess of one or more of them. Some researchers believe that serotonin plays the most important role in depression and mania; others believe that abnormalities in norepinephrine or dopamine also play a role.

A corollary of these biogenic amine theories is that antidepressant drugs may work by boosting the levels or activity of serotonin, norepinephrine, or dopamine in depressed patients. We will talk some more about how these drugs work in a little while.

What would happen if a chemical messenger such as serotonin became depleted from the presynaptic nerve in Figure 17–1? Then this nerve could not send its nerve signals properly across the synapses to the postsynaptic nerve. The wiring in the brain would develop faulty connections, and the result would be mental and emotional static, much like the music that comes out of a radio with a loose wire in the tuner. One type of emotional static (serotonin deficiency) would cause depression, and another type of static (serotonin excess) would cause mania.

Recently, these amine theories have been modified quite a bit. Some scientists no longer believe that a deficiency or excess of serotonin causes depression or mania. Instead, they postulate that abnormalities in one or more of the receptors on the nerve membranes may lead to mood abnormalities. Examine Figure 17–2 again, and imagine that there is something wrong with the serotonin receptors on the postsynaptic nerve. For example, there might not be enough of them. What would happen to the communication between the nerves? Although there might be plenty of serotonin molecules in the synapse, the postsynaptic nerves might not fire consistently when the presynaptic nerves fired. And if there were too many serotonin receptors, this could have the opposite effect of causing overactivity in the serotonin system.

To date, at least fifteen different kinds of serotonin receptors have been identified throughout the brain and more are being identified all the time. All these receptors probably have different effects on hormones, feelings, and behavior. Scientists do not have a very clear picture of what any of these different receptors do, nor do they know if abnormalities in any of them play a causal role in depression or mania. Research in this area is evolving at an extremely rapid pace, and we will have better information about the physiologic and psychological effects of these many serotonin receptors in the near future.

Although our knowledge about the role of serotonin receptors in brain function is still quite limited, there is evidence that the number of receptors on the postsynaptic nerves may change in response to antidepressant drug therapy. For example, if you give a drug that boosts the levels of serotonin in the synapses between the nerves, the number of serotonin receptors on the postsynaptic nerve membranes will decrease after a few weeks. This might be a way that the nerves attempt to compensate for the excess stimulation—the nerves are trying to turn down the volume of the signal, so to speak. This kind of reaction is called "down-regulation." In contrast, if you deplete the serotonin from the presynaptic nerve in Figure 17–1, much less serotonin will be released into the synapse. After several weeks, the postsynaptic nerves may compensate by increasing the number of serotonin receptors. The nerves are trying to turn up the volume of the signal. This kind of reaction is called "up-regulation."

Again, these are big words with simple meanings. "Up-regulation" means "more receptors," and "down-regulation" means "fewer receptors." We could also say that up-regulation means turning the system up, and down-regulation means turning the system down—just like a radio.

It is known that antidepressant drugs usually require several weeks or more to become effective. Researchers have been trying to figure out why. Some researchers have speculated that down-regulation may account for the antidepressant effects of these drugs. In other words, antidepressants may work not because they boost the serotonin system, as originally proposed, but because they turn the serotonin system down after several weeks. This would imply that decreased serotonin levels might not be the cause of depression after all. Depression might instead be due to *increased* serotonin activity in the brain. Antidepressant drugs may correct this after several weeks because they turn the serotonin system down.

How well established and proven are these theories? Not at all. As I have suggested, it is awfully easy to make up a theory, but much harder to prove it. To date, it has not been possible to validate or disprove any of these theories in a convincing way. In addition, there are no clinical or laboratory tests we could give to groups of patients or to individual patients that will reliably detect any chemical imbalance that causes depression.

The main value of the current theories is to stimulate research so that our knowledge of brain function will become more sophisticated over time. Eventually, I believe we will develop much more refined theories and far better tools for testing them.

Now you may be thinking, "Is that all there is to it?" Do scientists just sit around and say, "Depression could be due to an excess or a deficiency of this or that transmitter or receptor in the brain?" On some level, that really is all there is to it. Part of the problem is that our models of the brain are still very primitive, and so our theories of depression are not yet very sophisticated either.

It may turn out that depression is not due to problems with any transmitter chemical or receptor. We may one day discover that depression is actually more of a "software" problem, and not a "hardware" problem. In other words, if you have a computer, you know that computers crash all the time. Sometimes this results from a problem with the hardware. For example, your hard drive may become defective. But more often, there's a problem with the software—a bug that makes the program work poorly in certain situations. So with regard to brain research on depression, we may be looking for a problem in the "hardware" (for example, a chemical imbalance we are born with) whereas the real problem is in the "software" (for example, a negative thinking pattern based on learning). Both kinds of problems would be "organic," since brain tissue is involved, but the solutions to them would be radically different.

Another major problem facing depression researchers is the chicken-versus-the-egg dilemma. Are changes we measure in the brain the cause of the depression or the result? To illustrate this problem, let's conduct a thought experiment involving a deer in a forest. The deer is happy and contented. Imagine that we have a special machine that allows us to visualize the chemical and electrical activity in the deer's brain. We might have, for example, a futuristic portable brain imaging machine that can work from a distance, like the laser guns the police use to see how fast you're driving. However, the deer does not know we are monitoring its brain activity. Suddenly, the deer spots a pack of hungry wolves approaching. Panic strikes! Our brain imaging machine detects instantaneous massive changes in the electrical and chemical activity in the deer's brain. Are these chemical and electrical changes the cause of the fear or the result of the fear? Would we say the deer is afraid because it has developed a sudden "chemical imbalance" in its brain?

Similarly, there are all kinds of chemical and electrical changes in the brains of depressed patients. Our brains change quite dramatically when we feel happy, angry, or frightened. Which brain changes result from the strong emotions we feel, and which brain changes are the causes? Separating cause from effect is one of the thorniest challenges facing depression researchers. This problem is not impossible to solve, but it is not easy, and those eager to endorse the current theories about depression do not always acknowledge it.

Clearly, the research necessary to test any of these theories can be daunting. One significant problem is that it is still very difficult to get accurate information about the chemical and electrical process in the human brain. We can't just open up the brain of a depressed individual and look inside! And even if we could, we really wouldn't know where or how to look. But new tools, such as PET (positron emission tomography) scanning and MRI (magnetic resonance imaging), do make such research possible. For the first time, scientists can begin to "see" the activity of nerves and chemical processes inside the brains of human beings. This research is still in its infancy, and we can look forward to a great deal of progress in the next decade.

How Do Antidepressant Drugs Work?

The modern era of research on the chemistry of depression got a big boost accidentally in the early 1950s when researchers were testing a new drug for tuberculosis called iproniazid.[1] As it turned out, iproniazid was not an effective treatment for tuberculosis. However, the investigators noticed pronounced mood elevations in a number of patients who received this drug, and hypothesized that iproniazid might have antidepressant properties. This led to an explosion of research by drug companies who wanted to be the first to develop and market antidepressant drugs.

Researchers knew that iproniazid was an inhibitor of the MAO enzyme discussed previously. The drug was therefore categorized as an MAO inhibitor, or MAOI for short. Several new MAOI drugs that were similar in chemical structure to iproniazid were developed. Two of them, phenylzine (Nardil) and tranylcypromine (Parnate), are still in use today. A third MAOI called selegiline (trade name Eldepryl) has been approved for the treatment of Parkinson's disease. This drug is also occasionally used in the treatment of mood disorders. Other new MAOIs in use abroad may eventually be marketed in the United States.

The MAOIs are no longer prescribed nearly as frequently as they used to be. This is because they can cause dangerous elevations of blood pressure if the patient combines them with certain foods such as cheese. The MAOIs can also cause toxic reactions when combined with certain drugs. Because of these hazards, newer and safer antidepressants have been developed. These new drugs work quite differently from the MAOIs. Nevertheless, the MAOIs can be extremely helpful for some depressed patients who do not respond to other medications, and they can be used safely if the patient and doctor follow a number of guidelines that I will spell out in Chapter 20.

The iproniazid discovery helped to usher in a new era of biological research on depression. Scientists were eager to find out how the MAOIs worked. It was known that the MAOIs prevented the breakdown of serotonin, norepinephrine, and dopamine, the three chemical messengers that are concentrated in the limbic regions of the brain. Scientists hypothesized that a deficiency in one or more of these substances might cause depression and that antidepressant drugs might work by increasing the levels of these substances. This is how the biogenic amine theories actually originated.

Now let's see how much you've learned about how the brain works. Look at Figures 17–1 to 17–3 again. When the presynaptic nerve fires, serotonin is released into the synapse. After it attaches to a receptor on the postsynaptic nerve, it swims back to the presynaptic nerve, where it is pumped back inside this nerve and destroyed by the MAO enzyme. Now ask yourself this question: What would happen if we prevented the MAO enzyme from destroying the serotonin?

As you have probably guessed, the serotonin would accumulate in the presynaptic nerve, because this nerve is always manufacturing new serotonin. If this nerve could not get rid of its serotonin, the concentration of serotonin in the nerve would continue to increase. Whenever the presynaptic nerve fired, it would release much more serotonin than usual into the fluid-filled synaptic region. The excess serotonin in the synapse would cause a greater-than-expected stimulation of the postsynaptic nerve. This would be the chemical equivalent of turning up the volume on the radio. These effects of the MAOI antidepressants are illustrated in Figure 17–4 on page 446.

Could this be the reason the MAOI drugs cause a mood elevation? This is possible, and scientists have hypothesized that this is exactly how these MAOI drugs work. Research studies have confirmed that when these MAOI drugs are given to humans or animals, brain levels of serotonin, norepinephrine, and dopamine do increase. However, it is not known for certain if the antidepressant effects result from an increase in one of these biogenic amines, or from some other effect of these drugs on the brain.

Can you think of another theory about why or how these MAOI drugs might work? Does the increase in mood have to result from the extra stimulation of the postsynaptic nerve, or could there be another possible explanation? Think about what you read about down-regulation in the

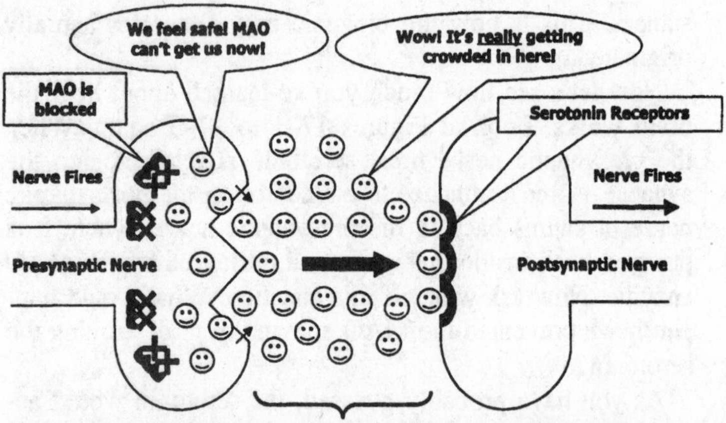

Figure 17–4. MAOIs block the MAO enzyme inside the presynaptic nerve, so serotonin levels increase. The excess serotonin is released into the synaptic region whenever the nerve fires. This provides a stronger stimulation of the postsynaptic nerve.

previous section and see if you can come up with an answer before you read any further.

You probably recall that the effects on the postsynaptic nerves after several weeks can be the opposite of the effects on these nerves when you first take a drug. All the extra serotonin in the synapse may cause a down-regulation of the postsynaptic serotonin receptors after several weeks, and this down-regulation may correspond to the antidepressant effects. (Remember that although some scientists think depression results from a serotonin deficiency, others believe depression results from increased brain serotonin activity.) If you thought of this, it shows you are really learning your neurochemistry. You get an A-plus on this pop quiz!

If you said that the antidepressant effects of the MAOI drug could result from effects on some other system in the brain, you also get an A-plus. These theories about how the antidepressant drugs relieve depression are not proven facts. The effects of the MAOIs on the brain are vastly more complex than the simple model depicted in Figure 17–4. The effects of any antidepressant are probably not limited to one specific region or one specific type of nerve in the brain. Remember that each nerve in the brain connects with many thousands of other nerves, and all of them in turn connect with thousands of others. When you take an antidepressant, there are massive changes in numerous chemical and electrical systems throughout your brain. Any of these changes could be responsible for the improvement in your mood. Trying to figure out exactly how these drugs work is still a little like looking for a needle in a haystack. But the important thing for the moment is that these drugs do seem to help some depressed patients, regardless of how or why they work.

As I have mentioned, many new and different kinds of antidepressant drugs have been developed and marketed since the 1950s. Unlike the MAOIs, the newer antidepressants do not cause a buildup of transmitters like serotonin in the presynaptic nerve depicted in Figure 17–4. Instead, they mimic the effects of the brain's natural transmitter substances by attaching to receptors on the surfaces of the presynaptic or postsynaptic nerves.

To understand how these newer antidepressants can do this, remember our analogy of the lock and the key. A natural transmitter substance is like a key, and the receptor on the surface of the nerve is like a lock. The key is able to unlock the lock only because it has a certain shape. But if you were a magician, like the famous Harry Houdini, you could easily pick the lock and open it without the key.

An antidepressant medication is like a counterfeit key that a drug company has manufactured. Because the chemists know the three-dimensional shape of a natural transmitter like serotonin, norepinephrine, or dopamine, they can create new drugs that have a very similar shape. These drugs will fit into the receptors on the surfaces of nerves and mimic the effects of the natural transmitters. The brain does not know that an antidepressant is in the lock—the brain has been tricked into thinking that the natural transmitter chemical is attached to the receptor on the surface of the nerve.

In theory, the artificial key (the antidepressant) can do one of two things when it becomes attached to the receptor. It can either open the lock, or it can jam the lock without actually opening it. Drugs that open the locks are called "agonists." Agonists are simply drugs that mimic the effects of the natural transmitters. Drugs that jam up these locks are called "antagonists." Antagonists block the effects of the natural transmitters and prevent them from being effective.

We can imagine several different ways that antidepressant drugs could influence the receptors on the presynaptic and postsynaptic nerves. For the purpose of this discussion, imagine that the transmitter used by the presynaptic nerve is serotonin, but the same considerations apply to any transmitter. What would happen if we blocked the receptors on the reuptake pump? The presynaptic nerve could no longer pump the serotonin from the synapse back inside. Each time the nerve fired, more and more serotonin would be released into the synaptic region. As a result, the synapse would get flooded with serotonin.

This is precisely how most of the currently prescribed antidepressants work. As you can see in Figure 17–5 on page 449, they block the receptors for the reuptake pumps on presynaptic nerves, and so the transmitters build up in the synaptic region. The end result of this process is similar to the effects of giving the MAOI drugs discussed above. In both instances, the levels of serotonin build up in the synaptic region. When the presynaptic nerve fires, more serotonin than normal will "swim" to the postsynaptic nerve and stimulate it to fire. Once again, we have "turned up" the serotonin system, so to speak.

Is this good? Is this why these antidepressant drugs can improve our moods? That's the current theory, but no one really knows the answers to this question yet.

Different antidepressants block different amine pumps and some of them have more specific effects than others. The older "tricyclic" antidepressants, such as amitriptyline (Elavil) or imipramine (Tofranil) and others, block the

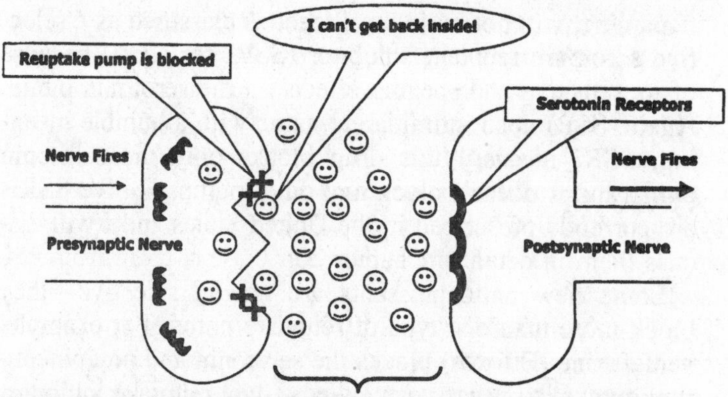

Figure 17–5. Most antidepressants block the reuptake pumps, so serotonin remains in the synapse after the nerve fires. Because serotonin builds up in the synaptic region, the stimulation of the postsynaptic nerve is stronger.

reuptake pumps for serotonin and norepinephrine. (Tricyclic means "three wheels," like a tricycle, because the chemical structure of these drugs resembles three linked rings.) Therefore, these transmitters build up in the brain if you take one of these drugs. Some tricyclic antidepressants have relatively stronger effects on the serotonin pump, and some of them have relatively stronger effects on the norepinephrine pump. Drugs with stronger effects on the serotonin pump are called "serotonergic" and drugs with relatively stronger effects on the norepinephrine pump are called "noradrenergic." What do you think we would call a drug with a strong effect on the dopamine pump? If you guessed "dopaminergic," you would be correct!

Some of the newer antidepressants, such as fluoxetine (Prozac), differ from the older tricyclic compounds in that they have highly selective and specific effects on the serotonin pump. If we want to use one of our new words, we can say that Prozac is highly "serotonergic" because levels of serotonin will build up in the brain when you take it. However, because Prozac blocks only the serotonin pump, the levels of other transmitters, such as norepinephrine and dopamine, will not build up. Prozac is classified as a selective serotonin reuptake inhibitor (SSRI for short) because of its selective and specific effects on the serotonin pump. Again, SSRI is an intimidating name with a humble meaning. SSRI means, "this drug blocks only the serotonin pump and it doesn't block any other pumps." Five SSRIs are currently prescribed in the United States and I will discuss them in detail in Chapter 20.

Some new antidepressants are not so selective—they block more than one type of reuptake pump. For example, venlafaxine (Effexor) blocks the serotonin and norepinephrine pumps, so it has been called a dual reuptake inhibitor. The drug company that manufactures venlafaxine promotes the idea that this drug may be more effective because the levels of two transmitters (serotonin and norepinephrine) increase, rather than just one. Actually, this is not such a novel feature. As you just learned, most of the older (and much cheaper) antidepressants do exactly the same thing. In addition, there is no evidence that venlafaxine works any better or any faster than the older drugs. However, venlafaxine has fewer side effects than some of the older tricyclic antidepressants. This might justify the increased cost of venlafaxine in some instances.

So far you have learned about the MAOIs and the pump inhibitors, such as the tricyclics and the SSRIs. Are there any other ways that antidepressant drugs might work? If you were a chemist working for a drug company and you wanted to create a completely novel antidepressant, what kinds of effects would your new drug have? One possibility would be to create a drug that directly stimulated the serotonin receptors on the postsynaptic nerves. A drug like this would mimic the effect of the natural serotonin. It would be a kind of counterfeit serotonin. Buspirone (BuSpar) works like this. This drug directly stimulates serotonin receptors on postsynaptic nerves. Buspirone was marketed a number of years ago as the first nonaddictive drug for anxiety, but it also has some mild antidepressant effects. However, its antidepressant and antianxiety prop-

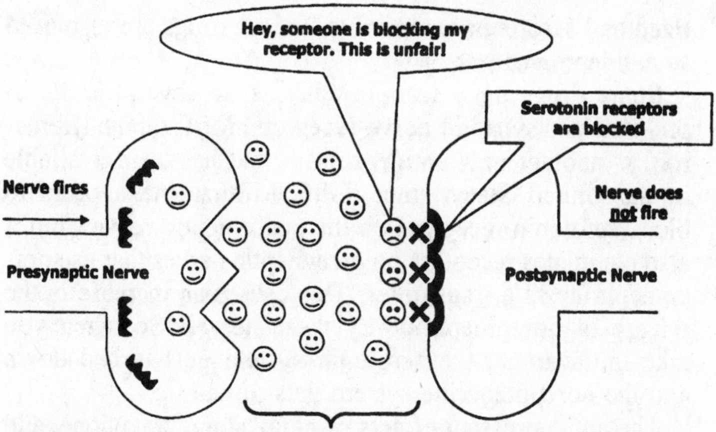

Figure 17–6. Serotonin antagonists block the serotonin receptors on the postsynaptic nerve, so the serotonin cannot stimulate the postsynaptic nerve after the presynaptic nerve fires.

erties are not especially strong. As a result, buspirone has not emerged as a particularly popular drug for anxiety or depression.

Why is it that buspirone is not more effective for depression? Scientists don't actually know the answer. Remember, though, that there are at least fifteen different kinds of serotonin receptors throughout the brain. All of these receptors have different functions that are not yet fully understood. Perhaps drugs that stimulated different kinds of serotonin receptors would have stronger antidepressant effects. As you might have gathered, things get complicated fairly quickly as we learn more and more about how the brain works.

If you were a drug company chemist, you could also create drugs that blocked the serotonin receptors on the postsynaptic nerves, as illustrated in Figure 17–6 above. Because such drugs would prevent the natural serotonin from having its effects, they would theoretically make depression worse. In fact, drugs that block serotonin receptors have been created. Two of them are called nefazodone (Serzone) and trazodone (Desyrel). Although they are categorized as "serotonin antagonists," these drugs are also used as antidepressants.

Some drugs have complex effects on several kinds of pre- and postsynaptic nerve receptors. Mirtazapine (Remeron) is another new antidepressant that has been available in the United States since 1996. Mirtazapine appears to block serotonin receptors on the postsynaptic nerves, but it also stimulates receptors on presynaptic nerves that use norepinephrine as a transmitter. This causes an increase in the release of norepinephrine by these nerves. So when you take mirtazapine, the serotonin system gets turned down and the norepinephrine system gets turned up.

The antidepressant effects of nefazodone, trazodone, and mirtazapine are exactly the opposite of what you might predict from the serotonin theory. Although they turn the serotonin system off, they are antidepressants. How can this be possible? If you are starting to get confused, join the club! Remember that there are many types of serotonin receptors in the brain and they all have different kinds of effects. Remember, too, that there are many high-speed and complex interactions among the different circuits in the brain. When we perturb one system of nerves in one region of the brain, we almost instantly create changes in thousands or millions of other nerves in other regions of the brain. In the final analysis, even the world's top neuroscientists do not have a very clear understanding of why or how these drugs relieve depression.

In summary, most of the currently prescribed antidepressants have effects on the serotonin, norepinephrine, or dopamine systems. Some of them are highly selective for one transmitter system, and others have effects on many transmitter systems. However, the effects of the currently prescribed antidepressants on these three systems do not really account for their beneficial effects in a very consistent or convincing way. For example, you have learned that some antidepressants stimulate serotonin levels, some of them block serotonin receptors, and some of them seem to have no effects at all on serotonin. And yet they all work about equally well. Clearly, the models I have drawn in Figures 17–4 to 17–6 are overly simplified, and current theories about how antidepressant medications work appear to be incomplete at best.

I do not mean to sound overly negative. Keep in mind that I am not challenging the effectiveness of the currently prescribed antidepressant drugs; I am simply saying that our theories about how these drugs work do not account for all the facts.

Fortunately, most neuroscience researchers now acknowledge this. The focus of research has expanded greatly. Instead of focusing narrowly on levels of one or another biogenic amine, researchers are pursuing a wide variety of strategies which focus on regulatory mechanisms throughout the brain, and new theories have been proposed. These theories deal with other transmitters in the brain, or with a variety of pre- or postsynaptic receptors, or with "second messenger" systems within the nerves, or with ion flux across nerve membranes, as well as with neuroendocrine systems, immune systems, and biological rhythm abnormalities. I believe the wider net that has now been cast will eventually lead to much better understanding of how the brain regulates moods.

Sophistication in brain research has accelerated tremendously and will accelerate even more rapidly in the next decade. This research will hopefully lead to improvements such as these:

- clinical tests for the chemical imbalance that causes depression (if, indeed, such an imbalance actually exists);
- tests to detect the genetic abnormalities that make certain individuals more vulnerable to depression as well as manic-depressive illness;
- safer medications with fewer side effects—as you will learn in Chapter 20, significant advances have already been made in this area;

- drugs and psychotherapeutic treatments that are more effective and faster-acting;
- drugs and psychotherapeutic treatments that minimize or entirely prevent relapses of depression following recovery.

Although our current level of understanding is still primitive, an important scientific effort has been launched. One day this effort may even lead us to the identification of the mysterious "black bile."

Chapter 18

The Mind-Body Problem

(Notes and References appear on pages 682–687.)

Ever since the time of the French philosopher, René Descartes, scholars have been puzzled by the "mind-body problem." This is the idea that as human beings we have at least two separate levels of existence—our minds and our bodies. Our minds consist of our thoughts and our feelings, which are invisible or ethereal. We know they are there because we can experience them, but we do not know why or how they exist.

In contrast, our bodies consist of tissue—blood, bones, muscle, fat, and so forth. The tissue ultimately consists of molecules, and the molecules are ultimately made up of atoms. These building blocks are inert—presumably, atoms have no consciousness. So how can the inert tissue in our brains give rise to our conscious minds, which can see, feel, hear, love, and hate?

According to Descartes, our minds and bodies must be connected in some manner. Descartes called the portion of the brain that links these two separate entities the "seat of the soul." For centuries, philosophers have tried to locate the "seat of the soul." In the modern era, neuroscientists continue this search as they attempt to figure out how our brains create emotions and conscious thoughts.

The belief that our minds and bodies are separate is reflected in our treatments for problems such as depression. We have biological treatments, which work on the "body," and psychological treatments, which work on the "mind." Biological treatments usually involve medications, and psychological treatments usually involve some type of talking therapy.

There is often intense competition between the "drug therapy" camp and the "talking therapy" camp. On the average, psychiatrists are more likely to be in the drug therapy camp. This is because psychiatrists are first trained as physicians (M.D.s). They can prescribe medications, and they are more likely to be influenced by the medical model of diagnosis and treatment. If you are depressed and you go to a psychiatrist, there's a good chance that she or he will tell you that your depression is caused by a chemical imbalance in your brain, and will recommend treatment with an antidepressant medication. If your family physician treats your depression, drug treatment is also very likely. This is because many family physicians have little training in psychotherapy and very little time to talk to patients about the problems in their lives.

In contrast, psychologists, clinical social workers, and other types of counselors are more likely to be in the talking therapy camp. They do not have medical training and cannot prescribe medications.[2] Their education usually focuses more on the psychological and social factors that may cause depression. If you are depressed and you go to a therapist in the talking therapy camp, she or he is more likely to focus on your upbringing, your attitudes, or stressful events such as the loss of love or the loss of your job. Your therapist will probably also recommend psychotherapeutic treatment, such as cognitive behavioral therapy. However, there are many exceptions to this generalization. Many nonmedical therapists believe that biological factors do play a role in depression, and many psychiatrists are gifted psychotherapists. Psychiatrists and nonmedical therapists sometimes work together in teams so that their patients can benefit from both types of treatment.

Nevertheless, the split between the mind (psychological) and body (biological) schools is sharp, and the dialogue between them is often intense, combative, and bitter. Political and financial considerations sometimes seem to influence the tone of these discussions more than scientific findings. Some recent studies suggest that these arguments may amount to much ado about nothing and that the dichotomy between the mind and the brain may be illusory. These studies indicate that antidepressant drugs and psychotherapy may have similar effects on our minds and on our brains—in other words, they might work in the same way.

For example, in a classic study published in the *Archives of General Psychiatry* in 1992, Drs. Lewis R. Baxter, Jr., Jeffrey M. Schwartz, Kenneth S. Bergman, and their colleagues at UCLA School of Medicine studied changes in the brain chemistry of eighteen patients with obsessive-compulsive disorder (OCD). Half of these patients were treated with cognitive behavioral therapy (and no drugs) and half were treated with antidepressant drugs (and no psychotherapy).[3] The patients in the no-drug group received individual and group psychotherapy that had two main components. The first component was exposure and response prevention. This is a behavior therapy technique which involves encouraging patients not to give in to their compulsive urges to check locks, to wash their hands repeatedly, and so forth. The second component was cognitive therapy along the lines described in this book. Remember that patients in this group did not receive any medications at all.

These investigators used positron emission tomography (PET scanning) to study the metabolic rate for sugar (glucose) in various brain regions before and after ten weeks of treatment with either drugs or psychotherapy. This method of brain scanning assesses the activity of the nerves in different areas of the brain. One brain region they were particularly interested in was the caudate nucleus on the right half of the brain.

Both treatments were effective: the majority of patients in both groups improved, and there were no significant differences in the two treatments. This was not surprising; previous researchers have also reported that drugs and cognitive behavioral psychotherapy have similar effects in the treatment of OCD. However, the results of the PET study were quite surprising. The investigators reported comparable reductions in the activity in the right caudate nucleus in the successfully treated patients regardless of whether they were treated with drugs and no psychotherapy, or psychotherapy and no drugs. In addition, the symptoms and thinking patterns of the two groups improved to a similar degree—neither treatment was superior. Finally, the amount of improvement in symptoms was significantly correlated with the degree of change in the right caudate nucleus. In other words, patients who improved the most had, on average, the greatest reductions in brain activity in the right caudate nucleus. The reduced activity meant that the nerves in this region of the brain had calmed down, regardless of whether they were treated with drugs or psychotherapy.

One implication of this study is that excessive activity in the right caudate nucleus might play a role in the development or maintenance of the symptoms of obsessive-compulsive disorder. A second important implication is that antidepressant medications and cognitive behavioral therapy might be equally effective in restoring the structure and function of the brain back to normal.

Like most published studies, this one had some fairly significant flaws. One problem is that any brain changes you observe in a particular psychiatric disorder might simply represent "downstream" effects rather than true causal effects. In other words, the increased neural activity in the right caudate nuclei of patients with obsessive-compulsive disorder might simply reflect a more general pattern of distress throughout the brain and may not be the cause of the symptoms, as we have discussed above.

Another problem was that the number of patients studied

was extremely small, and the number of brain regions the investigators studied was fairly large, so it is possible—even likely—that these findings were the result of chance. This possibility is consistent with the fact that other investigators have reported different patterns of brain activity in patients treated with antidepressant medications. This is why replications with more patients conducted by independent investigators are needed before the results of any study can be accepted. In spite of these limitations, the report by Dr. Baxter and his colleagues was the first of its kind and may open the door to an important new type of integrated research on the ways that drugs and psychotherapy can influence brain function and emotions.

Other studies have shown that antidepressants may actually work by helping depressed patients change their negative thinking patterns. Indeed, in an investigation conducted at Washington University School of Medicine in St. Louis, Drs. Anne D. Simons, Sol L. Garfield, and George E. Murphy randomly assigned depressed patients to treatment with either antidepressants alone or cognitive therapy alone. They studied changes in the negative thinking patterns of both groups of patients. They discovered that the negative thinking of patients who responded to the antidepressants improved as much as the negative thinking of depressed patients who responded to the cognitive therapy.[4] Remember that the drug patients received no psychotherapy and the cognitive therapy patients received no medications. Thus this study indicated that antidepressant drugs change negative thinking patterns in much the same way that cognitive therapy does. The effect of antidepressant drugs on attitudes and thoughts may explain their antidepressant effects just as well or even better than more biological explanations of their effects on different transmitter systems in the brain.

These remarkable studies suggest that we might do better to let go of this "mind-body" split and begin to think about how these different treatments may be working in tandem on the mind and on the brain. This combined approach could foster a greater sense of teamwork among therapists and researchers approaching the problem from different angles and may lead to more rapid advances in our understanding of emotional disorders. Even if there is some type of genetic or biological disorder in at least some depressions, psychotherapy can often help to correct these problems, even without medications. Many research studies, as well as my own clinical experience, have confirmed that severely depressed patients who appear very "biologically" depressed with lots of physical symptoms often respond rapidly to cognitive therapy alone without any drugs.[5]

It can work the other way as well. I have worked with many depressed patients who were still stuck after I had tried numerous psychotherapeutic interventions. When I prescribed an antidepressant medication, many of these patients started to turn the corner, and the psychotherapy began to work better. It seemed as if the medication really did help them change their negative thinking patterns as they recovered from the depression.

If Depression Is Inherited, Doesn't It Mean We Should Treat It with Drugs?

In Chapter 17 we talked about the fact that we don't yet know how strong the genetic influences are in the more common forms of depression that do not involve mania. But suppose scientists eventually discover that nearly all forms of depression are inherited, at least in part. Would it mean we should treat depression with drugs?

The answer is: not necessarily. For example, a blood phobia is thought to be at least partially genetic, but it can nearly always be treated quickly and easily with behavior therapy. The treatment of choice for most phobias is to expose the person to the frightening situation and to urge them to face it and endure the anxiety until the fear diminishes and disappears. Most patients are so frightened that they resist the treatment at first, but if they can be persuaded to hang in there, the success rate is extraordinarily high.

I can attest to this personally. While growing up, I was terrified of blood. When, in medical school, it was time to draw blood from each other's arms, I felt so unenthusiastic that I dropped out of medical school. For the next year, I decided to work in the clinical laboratory of the Stanford University Hospital so I could try to get over my fear. They gave me a job doing nothing but drawing blood out of people's arms and I had to do this all day long. The first few times I had to draw blood, it made me very anxious, but after those initial anxious moments, I got used to it. Pretty soon, I *loved* my new job. This shows that at least some genetic tendencies can respond to a behavioral treatment without drugs.

To state an even more commonplace example, we all inherit a tendency to have a particular type of body. Some of us are genetically taller or shorter than others. Some of us have larger frames, others have smaller frames. But our diets and habits hugely influence the types of bodies we have as adults. Many professional bodybuilders were skinny and embarrassed about their looks when growing up. This motivated them to go to the gym and work out. This intense effort transformed many of them into champions. Their genes may have greatly influenced what they were born with, but their behaviors and determination dominated where they ended up.

The opposite is also true. If it turned out that depression was entirely caused by the environment and that there were no genetic influences, this would not minimize the potential value of antidepressant drugs. For example, if you are exposed to someone with a strep throat, you may get a strep throat because streptococcal bacteria are so infectious. We can say that the causes of your strep throat are almost entirely environmental and not genetic. Nevertheless, we would still treat your strep throat with an antibiotic, and not with behavior therapy!

With regard to bipolar manic-depressive illness, the answer is clear. This disorder appears to have an extremely strong biological cause, and although we don't yet know exactly what this cause may be, treatment with a mood stabilizer such as lithium or valproic acid (Depakene) is usually a must. Other medications will also be used during episodes of depression or severe mania. However, good psychotherapy can also make a big contribution in the treatment of bipolar illness. In my experience, the combination of a drug like lithium or valproic acid along with cognitive therapy has been far more effective than treatment with medications alone.

From a practical point of view, the question I face as a clinician is this: How can I best treat each particular patient who is suffering from depression, regardless of the cause? Whether or not genes play a role, drugs can sometimes help and psychotherapy can sometimes help. Sometimes, a combination of psychotherapy and antidepressant medications seems to be the best approach.

Is It Better to Be Treated with Drugs or Psychotherapy?

A number of studies have compared the effectiveness of antidepressant drug treatment with cognitive therapy.[5–8] On the whole, these studies have indicated that during the acute phase of treatment, when patients first seek treatment for their depressions, both treatments seem to work reasonably well. Following recovery, the picture is a little different. Several long-term studies indicate that patients who receive cognitive therapy, alone or in combination with antidepressant medications, appear to stay undepressed longer than patients who receive only antidepressant medication therapy and no psychotherapy.[5] This is probably because cognitively treated patients have learned many coping tools to help them to deal with any mood problems they might experience in the future.

If you would like to learn more about recent research on

the effectiveness of drugs versus psychotherapy, you can read an excellent article on this topic by Drs. David O. Antonuccio and William G. Danton from the University of Nevada and Dr. Gurland Y. DeNelsky from the Cleveland Clinic.[5] These authors reviewed the world research literature on the effectiveness of psychotherapy versus medications for depression and came up with some rather startling conclusions that are quite different from the popular perceptions about these treatments. They argue that cognitive therapy appears to be at least as effective, if not more effective, than medications in the treatment of depression. They conclude that this is even true for severe depressions that appear to be "biological" because they have many physical side effects such as fatigue or a loss of interest in sex. The authors also question the methods used by drug companies to test new antidepressants. This scholarly and provocative article is clearly written, so look it up if you are curious.

My own clinical experience has convinced me that pure "test-tube treatment" with drugs alone is not the answer for most patients. There appears to be a definite role for effective psychological interventions, even if you have had the good fortune to respond to an antidepressant medication. If you learn cognitive therapy self-help techniques like those described in this book, I believe you will be better prepared to cope with any mood problems that develop again in the future.

My clinical practice has always been predicated on an integrated approach. At my clinic in Philadelphia, approximately 60 percent of our patients received cognitive therapy with no drugs, and approximately 40 percent of our patients received a combination of cognitive therapy along with antidepressants. Patients in both groups did well, and we found both types of treatment tools to be valuable. We did not treat patients with drugs alone and no psychotherapy because in my experience this approach has not been satisfactory.

It may be that for certain types of depression, the addition of the proper antidepressant to help your treatment program might make you more amenable to a rational self-help program and greatly speed up the therapy. As I have mentioned earlier above, I can think of many depressed individuals who seemed to "see the light" with regard to their illogical, twisted, negative thoughts more rapidly once they began taking an antidepressant. My own philosophy is this: I'm in favor of any reasonably safe tool that will help you!

I believe that your feelings about the type of treatment that you receive may be important to the outcome. If you are more biologically oriented, you may do better with drug treatment. In contrast, if you are more psychologically oriented, you may do better with psychotherapy. If you and your therapist do not see eye to eye, you may lose confidence and resist the treatment, and this can reduce the chances for a successful result. In contrast, if the treatment makes sense to you, you will feel more hope, trust and confidence in your doctor. Consequently your chances for a positive outcome will be increased.

I have also seen that certain negative attitudes and irrational thoughts can interfere with proper drug treatment or with psychotherapeutic treatment. I would like to expose twelve hurtful myths at this time. The first eight myths concern medication treatment and the last four myths concern psychotherapy. With regard to medications, I believe that enlightened caution in taking any drug is well advised, but an excessively conservative attitude based on half-truths can be equally destructive. I also believe that one should be appropriately skeptical and cautious about psychotherapy, but that too much pessimism can also interfere with effective treatment.

Myth Number 1. "If I take this drug, I won't be my true self. I'll act strange and feel unusual." Nothing could be further from the truth. Although these drugs can sometimes eliminate depression, they do not usually create abnormal mood elevations and, except in rare cases, they will not make you feel abnormal, strange, or "high." In fact, many patients report that they feel much *more* like themselves after they take an antidepressant medication.

Myth Number 2. "These drugs are extremely dangerous." Wrong. If you are receiving medical supervision and cooperate with your doctor, you will have no reason to fear most antidepressant drugs. Adverse reactions are rare and can usually be safely and effectively managed when you and your doctor work together as a team. The antidepressants are far safer than the depression itself. After all, depression, if left untreated, can kill—through suicide!

This does not mean you should be complacent about antidepressant drugs—or any drug you take, for that matter, including aspirin. In the following chapters, you will learn about the side effects and toxic effects of all the different antidepressants and mood-stabilizing agents. If you are taking one or more of these drugs, educate yourself and read about them in Chapter 20. This should not be difficult, and the information will enhance your chances of having a safe and effective experience with the antidepressant your doctor has prescribed.

Myth Number 3. "But the side effects will be intolerable." No, the side effects are mild and can usually be made barely noticeable by adjusting the dose properly. If in spite of this you find the medication uncomfortable, you can usually switch to another medication that will be equally effective with fewer side effects.

Remember, too, that untreated depression also has many "side effects." These include feelings of tiredness, increases or decreases in appetite, difficulties sleeping, a loss of motivation and energy, a loss of interest in sex, and so forth. And if you respond favorably to an antidepressant, these "side effects" will usually disappear.

Myth Number 4. "But I'm bound to get out of control and use these drugs to commit suicide." Some of the antidepressant drugs *do* have a lethal potential if you take them in overdose or combine them with certain other drugs, but this need not be a problem if you discuss your concerns with your physician. If you feel actively suicidal, it might be helpful to obtain only a few days' or one week's supply at a time. Then you will not be likely to have a lethal supply on hand. Your doctor may also decide to treat you with one of the newer antidepressant drugs that are much safer than the older antidepressants if taken in accidental or intentional overdose. Remember that as the drug begins to work, you will feel less suicidal. You should also see your therapist frequently and receive intensive therapy, either as an outpatient or as an inpatient, until any suicidal urges have passed.

Myth Number 5. "I'll become hooked and addicted, like the junkies on the street. If I ever try to go off the drug, I'll fall apart again. I'll be stuck with this crutch forever." Wrong again. Unlike sleeping pills, opiates, barbiturates, and minor tranquilizers (benzodiazepines), the addictive potential of antidepressants is extremely low. Once the drug is working, you will *not* need to take larger doses to maintain the antidepressant effect. As noted above, if you are learning cognitive therapy techniques and focusing on relapse prevention, in most instances your depression will not return when you discontinue the drug.

When it is time to go off the medicine, it would be advisable to do this gradually, tapering off over a week or two. This will minimize any discomfort that might occur from abruptly stopping the medicine, and will help you nip any relapse in the bud before it becomes full blown.

Many doctors now advocate long-term maintenance therapy for patients with severe depressions that return on many occasions. A prophylactic effect can sometimes be achieved if you take the antidepressant over a period of a year or two after you have recovered. That can minimize the probability of your depression returning. If you have had a significant problem with recurrences of depression over a period of years, this might be a wise step for you. But you should be reassured that antidepressant drugs are definitely *not* addictive. In my practice through the years, I have had very few patients who had to remain on antidepressant drugs for more than a year, and almost no patients who stayed on antidepressants indefinitely.

Myth Number 6. "I won't take any psychiatric drug because that would mean I was crazy." This is quite misleading. Antidepressants are given for depression, not for "craziness." If your doctor recommends an antidepressant, this would indicate he or she is convinced you have a mood problem. It does *not* mean that she or he thinks you are crazy. However, it *is* crazy to refuse an antidepressant on this basis because you may bring about greater misery and suffering for yourself. Paradoxically, you may feel normal more quickly with the help of the medicine.

Myth Number 7. "But other people are bound to look down on me if I take an antidepressant. They'll think I'm inferior." This fear is unrealistic. Other people will not know you're taking an antidepressant unless you tell them—there's no other way they could know. If you do tell someone, they're more likely to feel relieved. If they care about you, they'll probably think *more* of you because you're doing something to help eliminate your painful mood disorder.

Of course, it is possible that someone might question you about the advisability of taking a drug, or even criticize your decision. This will give you the golden opportunity to learn to cope with disapproval and criticism along the lines discussed in Chapter 6. Sooner or later, you're going to have to decide to believe in yourself and stop giving in to the disabling terror that someone might or might not agree with something you do.

Myth Number 8. "It is shameful to have to take a pill. I should be able to eliminate the depression on my own."

Research on mood disorders conducted throughout the world has clearly shown that many individuals *can* recover without pills if they engage in an active, structured, self-help program of the type outlined in this book.[5, 9–13]

However, it is also clear that psychotherapy does not work for everyone, and that some depressed patients recover faster with the help of an antidepressant. In addition, in many cases an antidepressant can facilitate your efforts to help yourself, as described above.

Does it really make sense to mope and suffer endlessly, stubbornly insisting you must "do it on your own" without a medication? Obviously, you must do it yourself—with or without a pharmacological boost. An antidepressant may give you that little edge you need to begin to cope in a more productive manner. This can accelerate the natural healing process.

Myth Number 9. "I feel so severely depressed and overwhelmed that only a drug could help me." Drugs and psychotherapy both have a lot to offer in the treatment of severe depression. I believe that the passive attitude of letting a drug do it for you is unwise. My own research has indicated that the willingness to do something to help yourself can have powerful antidepressant effects, whether or not you are also taking a medication. The self-help work patients complete between sessions also seems to speed recovery.[14, 15] So if you combine a medication with a good form of psychotherapy, you will have more weapons in your arsenal.

As I have already stated, many patients I have treated with drugs alone did not recover completely. When I added the cognitive therapy, many of them improved. I believe that the combination of drugs and psychotherapy can work better and quicker than drugs alone and frequently leads to better long-term results. This seems to be true for mildly depressed patients and for severely depressed patients as well. For example, we treat many severely depressed inpatients at the Stanford University Hospital with group cognitive therapy techniques. These techniques are similar to the ones you have learned about in this book. We have found that the group format can be especially helpful. I have seen many of these patients improve significantly during these therapy groups. The improvement often occurs within the actual therapy group. At the moment the patient sees how to talk back to his or her negative thoughts in a convincing manner, there is often a strong, immediate uplift in mood and outlook. Keep in mind that these inpatients also receive antidepressant drugs that their attending psychiatrists prescribe for them. So nearly all of them receive a combination of drugs and psychotherapy—we are not purists devoted only to one approach or the other.

I can recall one woman who was so severely depressed that she would burst into tears almost every time she tried to speak. If you even looked at her, it seemed it was enough to trigger an outburst of uncontrollable sobbing. I asked what she was thinking about when she was sobbing. She said she was thinking about something that her psychiatrist told her. He said her depression was "biological" and the causes were genetic. She concluded that if the depression was genetic, it meant she must have passed it down to her children and her grandchildren. One of her sons was, in fact, having a hard time. She attributed this to his "depression gene" and blamed herself for ruining his life. She castigated herself for even having gotten married and given birth to children in the first place and felt certain they would all endure horrible suffering forever. As she explained this, she began sobbing again.

Now from your perspective, her self-blame may seem incredibly unrealistic. Her insistence that all her children and grandchildren would lead lives of endless and irreversible suffering may seem equally unrealistic. But from her perspective, all her self-criticisms seemed entirely justified and negative predictions seemed completely valid. Her self-loathing and suffering were incredibly intense.

After she stopped crying, I asked what she would say to another depressed woman with children. Would she be so hard on her? This intervention did not work. She did not even seem to comprehend what I said. Instead of answering my question, she sobbed so uncontrollably that her entire body shook as the tears streamed down her cheeks.

After a while she stopped crying again. I asked if two other patients would volunteer to do a role-play to help her out. I call this exercise "externalization of voices" because you verbalize the negative thoughts in your mind and learn to talk back to them. I wanted the other patients to illustrate how she might talk back to her own negative thoughts so that all she would have to do was watch. I told her to imagine that these other women were very similar to her. They were depressed and had children and grandchildren.

The first volunteer played the role of the negative part of her mind and said out loud the sort of things the depressed woman had been thinking: "If my depression is partly genetic, then it means I am to blame for my son's depression." The second volunteer played the role of the more positive, realistic, self-loving part of her mind. This volunteer talked back to the negative thought along these lines: "I certainly wouldn't blame another depressed woman for her son's depression, so it makes no sense to blame myself, either. If there is a conflict with my son, or if he is having problems, I can try to be helpful to him. That's what any loving mother would try to do." Then they continued with this dialogue and modeled ways she could talk back to her other self-critical thoughts. The two volunteers took turns in the roles of the negative thoughts and the positive thoughts.

After the role-play was over, I asked the tearful patient which voice was winning and which voice was losing. Was it the negative voice or the positive voice? Which voice was more realistic, more believable? She said that the negative voice was unrealistic, and that the positive voice was winning. I pointed out that the volunteers were actually verbalizing her own self-criticisms.

Although her depression did not improve dramatically by the end of that group, it seemed that the clouds lifted just a little bit. The next time I saw her in a group, her mood had brightened up considerably. She was quite personable and could talk without crying for the first time since admission. She said she wanted to practice the role-playing in the group so she could learn how to do it. She said she was also intent on getting a referral to a cognitive therapist near her home after discharge so she could continue the work that was proving to be so helpful to her.

The method that helped this patient is also called the "double-standard technique." It is based on the idea that many of us operate on a double standard. We may judge ourselves in a harsh, critical, demanding way, and yet we judge others in a more compassionate and reasonable manner. The idea is to give up this double standard and agree to judge all human beings, including ourselves, by one set of standards that is based on truth and compassion instead of using a separate standard that is distorted and mean when we judge ourselves.

Myth Number 10. "It is shameful to receive psychotherapy because it means I am weak or neurotic. It is more acceptable to be treated with a drug because it means I have a medical illness, like diabetes." Actually, the sense of shame is common in depressed patients who are treated with drugs *or* psychotherapy. Often, the double-standard technique just described above can be helpful. Imagine, for example, that you've just discovered that a dear friend of yours received psychotherapy for depression and found that the treatment was helpful. Ask yourself what you would say to your friend. Would you say: "Oh, the psychotherapy just shows what a weak and defective neurotic you are. You should have taken a drug instead. What you did was shameful." If you would not say this to a friend, then why give yourself these messages? That's the essence of the double-standard technique.

Myth Number 11. "My problems are real, so psychotherapy couldn't possibly help me." Actually, cognitive therapy seems to work the best with depressed individuals with real problems in their lives, including catastrophic medical problems such as terminal cancer or an amputation, bankruptcy, or severe personal relationship problems. In many cases, I have seen individuals with problems like this who improved in a handful of cognitive therapy sessions. In contrast, chronically depressed individuals without any obvious problems that triggered their depressions are often more difficult to treat. Although the prognosis is excellent, they may require more intensive and prolonged treatment.

Myth Number 12. "My problems are hopeless, so no psychotherapy or drug could possibly help me." This is your depression talking, and not reality. Hopelessness is a common but horrible symptom of depression that is based on twisted thinking, just as the other symptoms are. One of the distortions is called "emotional reasoning." The depressed individual may reason: I *feel* hopeless, therefore I must *be* hopeless. Another cognitive distortion that leads to feelings of hopeless is fortune-telling—you are making a negative prediction that you will never improve, and assuming this prediction is really a fact. Other distortions can lead to feelings of hopelessness as well. These include the following:

- all-or-nothing thinking—you think of yourself as completely happy or completely depressed; shades of gray do not count, so if you are not completely happy or completely recovered, you assume you are completely depressed and hopeless;
- overgeneralization—you see your current feelings of depression as a never-ending pattern of defeat and suffering;
- mental filter—you selectively think of all the times you have been depressed, and end up thinking your whole life will be bad forever;

- discounting the positive—you insist that the times you were not depressed don't count;
- "should" statements—you use up all your energy telling yourself you "shouldn't" be depressed (or you "shouldn't" have gotten depressed again) instead of systematically working to overcome the feelings;
- labeling—you tell yourself you are hopelessly and irreversibly defective and conclude that you could never really feel whole, or happy, or worthwhile.

Other cognitive distortions, such as magnification or personalization, can also lead to feelings of hopelessness. Although these feelings are not realistic, they can act like self-fulfilling prophecies. If you give up, nothing will change and you will conclude that you really were hopeless.

Patients who feel hopeless usually cannot see that they are deceiving themselves. They are nearly always convinced these feelings are entirely valid. If I can persuade them to challenge these hopeless feelings and try to get better—even though they feel in their hearts that this is impossible—they usually do begin to improve, slowly at first and then more rapidly, until they feel a whole lot better.

One of the most important tasks of any therapist is to help depressed patients find the courage and determination to resist and fight these hopeless feelings. This battle is often fierce and rarely easy, but nearly always rewarding in the long run.

Chapter 19

What You Need to Know about Commonly Prescribed Antidepressants

(Notes and References appear on pages 682–687.)

This chapter contains practical general information about the use of antidepressants. You will learn who is the most—and least—likely to benefit from an antidepressant, how you can tell whether an antidepressant drug is really working, how much mood elevation you can anticipate, how long you should stay on it, and what you can do if it doesn't work. You will also learn how to monitor and minimize side effects and prevent potentially dangerous interactions between antidepressants and other drugs you may take, including prescription drugs as well as nonprescription (over-the-counter) drugs you can obtain at the drug store or grocery store. In the next chapter, I will provide specific information about each antidepressant and mood-stabilizing drug currently in use.

When you read this chapter, keep in mind that the use of antidepressants is still a blend of art and science. Each practitioner has a slightly different philosophy, and your doctor's approach may differ from mine. I will state my own biases up front.

First, I am quite demanding in terms of what I expect

from an antidepressant. I believe that any antidepressant medication should have a pretty profound and dramatic effect in order to justify its continued use. In addition, I firmly believe that every patient taking antidepressants should take a mood test like the one in Chapter 2 at least once a week. Your score on this test (or any other good depression test) is a highly reliable measure of how well your antidepressant is working. I do not encourage patients to continue taking drugs that have only modest or questionable beneficial effects on mood. When the score on the test goes down only a little bit (for example, a 30 percent or 40 percent improvement), I would be inclined to call this a placebo effect and not a real drug effect. This amount of improvement could be due to the passage of time, the psychotherapy, or the belief that the drug will work. If the improvement in mood is minimal, and assuming the patient has had a sufficient dose of the medication for a sufficient period of time, I would probably take the patient off the drug and try another medication, a combination of medication and psychotherapy, or psychotherapy alone.

Now some readers may think, "but a 40-percent improvement in my mood sounds pretty good. This sounds like *real* improvement. I'm almost half better." Certainly, any improvement is desirable, but research studies indicate that inactive placebos can also have large antidepressant effects. A 40-percent improvement has been shown to be a typical placebo response. The only justification for taking any antidepressant drug is this: Is the drug doing its job? To my way of thinking, the goal of treatment is to recover from depression. Most patients want complete recovery, not just a slight or moderate improvement in their mood. If an antidepressant is not accomplishing this goal after a reasonable trial, then I would recommend switching to another drug or treatment approach.

Second, I never treat patients with medications alone. If I prescribe an antidepressant for a patient, I always combine the medication treatment with psychotherapy as well. Although I tried the medication-only approach with large numbers of patients early in my career, I almost never found this approach to be satisfactory.

For example, when I was a postdoctoral fellow following my residency training at the University of Pennsylvania, I ran the lithium clinic at the Philadelphia VA Hospital. I treated many depressed veterans suffering from bipolar manic-depressive illness with a combination of lithium and other antidepressant drugs. Although the medications appeared to be helpful, the results were not very encouraging. Most of these poor veterans were going in and out of the hospital almost constantly, and few were leading productive, joyous, stable lives. Later in my career, when I learned cognitive therapy, I treated all my manic-depressive patients with a combination of medications plus psychotherapy. The results were much better. From that point on, I can recall only one manic-depressive patient I treated who required hospitalization for an episode of mania.

The results with depressed patients were similar. Early in my career, I treated depressed patients with the drugs alone or drugs combined with traditional supportive psychotherapy. I administered a depression test like the one in Chapter 2 to every patient every session. I could see very clearly that while some patients were helped a lot by antidepressants, many were not. A lot of patients improved only slightly, and some did not improve at all. Later in my career, I began to combine antidepressant drugs with the new cognitive therapy techniques I was learning, and saw much better results. Eventually, I gave up treating patients with drugs alone.

Third, I usually use one medication at a time, rather than a combination of many different kinds of drugs, although there are certainly many exceptions to this or any rule. The idea behind polypharmacy is that if one drug is good, two, three, or more will be even better. Some doctors also use additional drugs to try to combat the side effects of other drugs the patient is taking. The potential drawbacks to polypharmacy are many, including more side effects and more possible adverse drug interactions. I discuss polypharmacy in detail at the end of Chapter 20 and describe a number of specific situations in which the use of more than one drug may be justified.

Finally, I have usually not kept patients on antidepressant drugs indefinitely following recovery. Instead, I slowly taper patients off their antidepressants after they have been feeling really good for several months. I have found that in most cases, patients who have recovered can continue to remain undepressed without medications. Keep in mind that all my patients have received cognitive therapy, whether or not they also received an antidepressant. The cognitive therapy is probably responsible for the good long-term results, because patients learn tools they can use for the rest of their lives whenever they are feeling upset.

Many doctors practice very differently. They tell their patients that they must continue taking their antidepressants indefinitely to correct a "chemical imbalance in the brain" and to prevent relapses into depression. While relapse is an important issue, I have found that training patients to use their cognitive therapy tools whenever they need them seems to maintain improvement following recovery. In fact, a number of well-controlled long-term follow-up studies have confirmed that this works better than drugs to prevent relapses.

While this is my philosophy in a nutshell, remember that there is no single "correct" approach, and your doctor's philosophy might differ from mine. In addition, there are many exceptions to any rule, and your own diagnosis or personal history may mandate a different approach from the one I have just outlined. If you have questions about your treatment, discuss your concerns with your physician. In my experience, the sense of teamwork and mutual respect is still the most important ingredient in any successful treatment.

If I Am Depressed, Does It Mean that I Have a "Chemical Imbalance" in My Brain?

There is an almost superstitious belief in our culture that depression results from a chemical or hormonal imbalance of some type in the brain. But this is an unproven theory and not a fact. As discussed in Chapter 17, we still do not know the cause of depression and we do not know how or why antidepressant drugs work. The theory that depression results from a chemical imbalance has been around for at least two thousand years, but there is still no proof of this, so we really do not know for sure. Furthermore, there is no test or clinical symptom that could demonstrate that a particular patient or group of patients has a ''chemical imbalance'' that is causing the depression.

If I Am Depressed, Does It Mean that I Should Take an Antidepressant?

Many people also believe that if you are depressed you should be on an antidepressant. However, I do not insist that every depressed patient must take an antidepressant. Large numbers of well-controlled studies published in respected scientific journals indicate that the newer forms of psychotherapy can be just as effective as, and sometimes more effective than, antidepressants.

Certainly many depressed people have been treated successfully with antidepressants and swear by them. They are valuable tools and I am glad to have them available in my treatment arsenal. Sometimes antidepressants are helpful, but they are rarely total answers, and often they are not necessary.

How Can I Decide Whether or Not to Take an Antidepressant?

I always ask my patients during their initial evaluations whether or not they would prefer to take an antidepressant. If a patient strongly feels that she or he would prefer to be treated without an antidepressant, I treat with cognitive therapy alone, and this is usually successful. However, if the patient has been working hard in therapy for six to ten weeks without any improvement, I sometimes suggest we try to add an antidepressant to put some "high octane" in the tank, so to speak. In some cases, this makes the psychotherapy more effective.

If a patient feels strongly that she or he would like to receive an antidepressant at the initial evaluation, I treat with a combination of an antidepressant medication and psychotherapy right away. However, I almost never treat patients with antidepressant medications alone, as noted previously. In my experience, the drugs-only approach has not been satisfactory. The combination of medications with psychotherapy seems to produce better results in the short term and in the long term than treating patients with drugs alone.

It may sound unscientific to base the medication decision on the patient's preferences, and certainly there are exceptional cases where I feel I have to make a recommendation that differs from my patient's wishes. But the majority of time, I have found that patients do well when treated with the approach they are most comfortable with.

So if you are depressed and you have strong positive feelings that an antidepressant drug will help you, this increases the likelihood that you will be helped by one of these medications. And if you feel strongly that you would prefer to be treated with a drug-free form of therapy, the likelihood of a successful outcome is also good. But I would urge flexibility in your thinking. If you are receiving a medication, I strongly believe that cognitive or interpersonal psychotherapy can enhance your recovery. If you are receiving psychotherapy and your progress is slow, an antidepressant might accelerate your recovery.

Can Anyone Take an Antidepressant?

Most people can, but competent medical supervision is a must. For example, special precautions are indicated if you have a history of epilepsy, heart, liver, or kidney disease, high blood pressure, or certain other disorders. For the very young and elderly, some medications should be avoided, and smaller dosages may be indicated. And, as noted above, if you are taking medicines in addition to an antidepressant, special precautions are sometimes required. Properly administered, an antidepressant is safe and may be lifesaving. But don't try to regulate it or administer it on your own. Medical supervision is a must.

Should a pregnant woman use an antidepressant? This sensitive question often requires consultation between the psychiatrist and the obstetrician. Since fetal abnormalities might occur, the potential benefit, the severity of the depression, and the stage of pregnancy must all be taken into account. Other treatment approaches should usually be employed first, and an active self-help program of the type described in this book might eliminate the need for a medication. This would give optimal protection to the developing child, of course. On the other hand, if the depression is very severe, there may be cases where it makes sense to use an antidepressant.

Who Is Most—and Least—Likely to Benefit from an Antidepressant Drug?

Your chance of responding to an appropriate drug may be enhanced:

1. If you are unable to carry on with your day-to-day activities because of your depression.
2. If your depression is characterized by many organic symptoms, such as insomnia, agitation, retardation, a worsening of symptoms in the morning, or an inability to feel cheered up by positive events.
3. If your depression is severe.
4. If your depression had a reasonably clear-cut beginning.
5. If your symptoms are substantially different from the way you normally feel.
6. If you have a family history of depression.
7. If you have had a beneficial response to antidepressant drugs in the past.
8. If you strongly feel that you would like to take an antidepressant drug.
9. If you are strongly motivated to recover.
10. If you are married.

Your chance of responding to an appropriate drug may be diminished:

1. If you are very angry.
2. If you have a tendency to complain and to blame others.
3. If you have a history of an exaggerated sensitivity to drug side effects.
4. If you have a history of multiple physical complaints that your doctor has been unable to diagnose, such as tiredness, stomach ache, headache, or pains in your chest, stomach, arms, or legs.
5. If you have a long history of another psychiatric disorder or hallucinations preceding your depression.

6. If you feel strongly that you do not want to take an antidepressant drug.
7. If you are abusing drugs or alcohol and you are unwilling to go into a recovery program.
8. If you are receiving financial compensation for your depression, or if you hope to receive financial compensation. For example, if you receive disability payments for depression, or if you are involved in a lawsuit and hope to receive financial compensation because of your depression, then any form of treatment is going to be difficult. This is because if you recover, you will lose money. This is a conflict of interest.
9. If you have failed to respond to other antidepressants you have been given.
10. If for any reason you have mixed feelings about getting better.

These guidelines are of a general nature and are not intended to be comprehensive or precise. Our ability to predict who will respond best to a medication or to psychotherapy is still extremely limited. Many people with all the positive indicators may fail to respond to antidepressants, and many people with all the negative indicators may respond beautifully to the first drug they receive. In the future, the use of antidepressant drugs will hopefully become more precise and scientific, just as the use of antibiotics has become.

If you have many of the negative indicators, is this bad? I don't think so. Most patients with all the negative indicators can be treated quite successfully, but it may sometimes take a little longer. In addition, as I have emphasized repeatedly, a combination of medication with good psychotherapy along the lines described in this book is sometimes more effective than treatment with antidepressant drugs alone.

How Fast and How Well Do Antidepressant Drugs Work?

Most studies indicate that approximately 60 percent to 70 percent of depressed patients will respond to an antidepressant medication. Since approximately 30 percent to 50 percent of depressed patients will also respond to a sugar pill (a placebo), these studies indicate that an antidepressant will increase your chances for recovery.

However, remember that the word "respond" is different from the word "recover," and the improvement from an antidepressant is often only partial. In other words, your score on a mood test like the one in Chapter 2 may improve without going into the range considered truly happy (less than 5). This is why I nearly always combine antidepressant medication treatment with cognitive and behavioral techniques like those described in this book. Most people are not interested in just partial improvement. They want the real McCoy. They want to get up in the morning and say, "Hey, it's great to be alive!"

As I have emphasized, most of the depressed and anxious people I have treated have problems in their lives such as a marital conflict or a career difficulty, and nearly all of them beat up on themselves with negative thinking patterns. In my experience, medication therapy is usually more effective—and more satisfying—when it is combined with psychotherapy. Many doctors do prescribe medications alone without psychotherapy, but I have not found this approach to be satisfactory.

Which Antidepressants Are the Most Effective?

All of the currently prescribed antidepressant drugs tend to work about equally well, and equally rapidly, for most patients. So far, no new type of antidepressant medication has been shown to be more effective or faster-acting than the older drugs that have been available for several decades. However, there are dramatic differences in the costs of the different types of antidepressants and in the side effects they have. Essentially, the newer medications are much more expensive because they are still on patent. However, they are far more popular because they usually have fewer side effects than the older, cheaper drugs. If you have certain kinds of medical conditions, some antidepressants will be relatively safer for you than others. I will discuss these issues in greater detail in Chapter 20.

Sometimes a patient will respond particularly well to one antidepressant or kind of antidepressant. Unfortunately, we cannot usually predict this ahead of time for the individual, and so most physicians use a trial-and-error approach. There are, however, a few generalizations about the kinds of antidepressants that work best for certain kinds of problems. For example, drugs that have stronger effects on the serotonin systems in the brain are generally considered to be effective for patients who suffer from obsessive-compulsive disorder (called OCD for short). These patients have recurrent illogical thoughts (like a fear that the stove will catch fire and burn the house down) and perform compulsive rituals over and over (such as checking repeatedly to make sure that the stove is turned off). Drugs often prescribed for OCD include several of the tricyclic antidepressants, including clomipramine (Anafranil), one of the SSRIs, such as fluoxetine (Prozac) or fluvoxamine (Luvox), or one of the MAOIs, such as tranylcypromine (Parnate).

If a depressed patient also has symptoms of anxiety, such as panic attacks or social anxiety, the physician might also choose one of the SSRI or MAOI antidepressants, since these often seem to be quite effective. Or the physician might choose one of the more sedative antidepressants, such as trazodone (Desyrel) or doxepin (Sinequan), thinking that the relaxation might help reduce the anxiety.

In my practice, I have treated many patients with a particularly difficult type of chronic and severe depression known as borderline personality disorder (called BPD for short). Patients with this disorder have intense and constantly fluctuating negative moods such as depression, anxiety, and anger. Patients with BPD also experience lots of turbulence in their personal relationships. In my experience, quite a few BPD patients have responded dramatically to the MAOI antidepressants, and so I might be more inclined to choose an MAOI for patients with these features. Of course, some patients with BPD have poor impulse control, and they may do better with one of the newer and safer antidepressants. This is because the MAOIs can be quite dangerous if patients mix these drugs with certain forbidden foods and medications that I will describe in detail in Chapter 20.

There are a number of other guidelines as well, but they should not be taken too literally because there are so many exceptions to them. The bottom line is this: any depressed patient has a reasonably good chance of having a positive response to almost any antidepressant medication if it is prescribed at the correct dose for a reasonable period of time. You can ask your physician if she or he has a reason for recommending a particular antidepressant. However, most physicians will prescribe antidepressants they are familiar with. This is good practice. Few doctors can master the myriad details about all the currently prescribed antidepressants, and so most doctors try to become familiar with the one or two agents they use most frequently. In this way, they will have the greatest expertise about the medication they are recommending for you.

How Can I Tell if My Antidepressant Is Really Working?

My own philosophy is to use a depression test like the one in Chapter 2 as a guide. Take the test once or twice a week during treatment. This is *really* important. The test will show whether and to what extent you have improved. If you are not getting better, or if you are getting worse, your scores will not improve. If your scores are steadily improving, this indicates the drug is probably helping.

Unfortunately, most doctors do not require their patients to complete a mood test like the one in Chapter 2 between therapy sessions. Instead, they rely on their own clinical judgment to evaluate the effectiveness of the treatment. This is quite unfortunate, because studies have indicated that doctors are often poor judges of how patients feel inside.

How Much Mood Elevation Can I Anticipate?

Your aim should be to reduce the score on the depression test in Chapter 2 until it is in the range considered normal and happy. This is true whether you are being treated with an antidepressant, with psychotherapy, or with a combination of the two. Treatment cannot be considered completely successful if your score remains in the depressed range.

If One Antidepressant Works Somewhat, Will It Be Even Better to Take Two or More Antidepressants at the Same Time?

As a general rule, it is usually not necessary (or even beneficial) to take two or more different antidepressant drugs simultaneously. The two drugs may interact in ways that are unpredictable, and the side effects may increase substantially. There are exceptions to this, of course. For example, if you are restless and having trouble sleeping, your doctor may sometimes add a small dose of a second, more sedating antidepressant at night to help you get a good night's sleep. Or your doctor may add a small dose of a second antidepressant to try to increase the effectiveness of the first antidepressant. This is called an "augmentation" strategy, and I will discuss this approach in greater detail in Chapter 20. But on the average, one drug at a time usually works best.

How Long Will It Take Before I Can Expect to Feel Better?

It typically requires a minimum of two or three weeks before an antidepressant medicine begins to improve your mood. Some drugs may take even longer. For example, Prozac may not become effective for five to eight weeks. It is not known why antidepressants have this delayed reaction (and whoever discovers the reason will probably be a good candidate for a Nobel prize). Many patients have the impulse to discontinue their antidepressants before three weeks have passed because they feel hopeless and believe the medicine is not working. This is illogical, since it is unusual for these agents to become effective right away.

What Can I Do if My Antidepressant Doesn't Work?

I have seen many patients who failed to respond adequately to one or many antidepressants. In fact, at my clinic in Philadelphia, most of the patients were referred to me after unsuccessful treatments with a variety of antidepressant drugs and therapy as well. Most of the time we were eventually able to get an excellent antidepressant effect, often through a combination of cognitive therapy and another medication that the patient had not yet tried. The important thing is to keep persisting in your efforts until you recover. Sometimes this requires enormous dedication and faith. Patients often feel like giving up, but persistence nearly always pays off.

I have stated earlier that the feelings of hopelessness are probably the worst aspect of depression. These feelings sometimes lead to suicide attempts because patients feel so convinced that things will never get any better. They think that things have always been this way and that their feelings of worthlessness and despair will go on forever. In addition, there is a kind of genius about depression. Patients can be so incredibly persuasive about their hopelessness that even their doctors and families may start believing them after a while. Early in my career I grappled with this and often felt tempted to give up on particularly difficult patients. But a trusted colleague urged me never to give in to the belief that any patient was hopeless. Throughout my career, this policy has paid off. No matter what type of treatment you receive, faith and persistence can be the keys to success. I cannot emphasize this enough.

How Long Should I Take an Antidepressant if It Doesn't Seem to Be Working?

Of course, you should always check with your physician before making any changes in your medication, but on average, a trial of four or five weeks should be adequate. If you do not have a clear-cut and fairly dramatic improvement in your mood, then a switch to another drug is probably indicated. It is important, however, that the dose be adjusted correctly during this time, since doses that are too high or too low may not be effective. Sometimes your doctor may order a blood test to make sure the dose you are taking is adequate for you.

One of the commonest errors your doctor may make is to keep you on a particular antidepressant for many months (or even years) when there is no clear-cut evidence that you have improved. This makes absolutely no sense to me! However, I have seen many severely depressed individuals who reported that they had been treated continuously with the same antidepressant for many years but were not aware of any beneficial effects from the medication. Their scores on the mood test in Chapter 2 usually indicated they were still severely depressed. When I asked them why they were taking the drug for such a long time, they usually said that theirs doctors told them that they needed it, or that it was necessary because of their "chemical imbalance." If your mood has not improved, it seems clear that the drug has not worked, so why keep taking it? If a drug does not have fairly substantial beneficial effects, as indicated by a clear and continuing improvement in your score on a depression test like the one in Chapter 2, then it is usually appropriate to switch to another antidepressant medication.

How Long Should I Continue to Take the Antidepressant if It Does Help Me?

You and your doctor will have to make this decision together. If this is your first episode of depression, you can probably go off the medicine after six to twelve months and continue to feel undepressed. In some cases, I have discontinued antidepressants after only three months with good results, and rarely found that treatment for more than six months was necessary. But different doctors have different opinions about this.

One of the strongest predictors of relapse in research studies is the degree of improvement at the end of treatment. In other words, if you are happy and completely free of depression, and this is documented by a score below 5 on the depression test in Chapter 2, the likelihood of a prolonged depression-free period is high. On the other hand, if you are partially improved but your depression score is still somewhat elevated, the likelihood is much greater that the depression will worsen or return in the future, whether or not you continue to take an antidepressant medication.

This is another reason why I like to combine antidepressant medications with cognitive behavioral therapy. The patients usually have a much better response, and very few patients in my private practice appeared to relapse and return for additional treatment following recovery.

What if My Doctor Tells Me I Have to Stay on the Antidepressant Indefinitely?

Patients with certain kinds of depressions will almost definitely need to take medications on a long-term basis. For example, if a patient has bipolar (manic-depressive) illness with uncontrollable highs as well as lows, long-term treatment with a mood-stabilizing medication such as lithium, valproic acid, or carbamazepine may be necessary.

If you have had many years of unremitting depression or if you have been prone to many recurrent attacks of depression, you might want to consider maintenance therapy for a longer period of time. Since doctors are becoming more aware of the relapsing nature of mood disorders, the use of antidepressants on a long-term or prophylactic basis is gaining greater favor.

Some doctors routinely recommend therapy with antidepressants indefinitely, in much the same way they might insist that patients with diabetes must take daily insulin to regulate their blood sugar. Several research studies suggest that such maintenance therapy can reduce the incidence of depressive relapses. However, research studies also indicate that treatment with the cognitive therapy techniques described in this book can also reduce depressive relapses. In addition, these studies suggest that the preventive effect of cognitive therapy may be greater than the preventive effect of antidepressant medications. One important advantage of cognitive behavioral therapy is that you learn new skills to minimize or prevent future depressions. For example, the simple exercise of writing down and challenging your own negative thoughts when you are under stress can be invaluable.

In my private practice, the vast majority of the depressed patients I have treated have not had to stay on antidepressant drugs indefinitely following recovery. Most of them did extremely well with no medications simply by using the cognitive therapy skills they learned whenever they became upset again in the future. This is very encouraging, and it shows there is quite a bit you can do not only to treat your own depression, but also to minimize the probability of severe and prolonged depressions in the future. It also suggests that if you are taking an antidepressant, it might be very helpful for you to study and practice the methods in this book.

Once you discover how to change your own negative thinking patterns using the techniques I describe, you may find that you will be able to remain undepressed without any medications. But certainly, you will want to discuss this with your physician. It is never smart to go off a medicine or to change the dose of a medication unless you talk this over with your doctor first.

What if I Start Getting More Depressed When I Taper Off the Medication?

This is actually pretty common, and I will tell you how I have handled it in my own practice. First, I make sure the patient continues to take the depression test in Chapter 2 at least once or twice a week while she or he is tapering off the medication. Then we develop a plan for slowly reducing the dose of the antidepressant. I tell patients that if they start to feel depressed again while tapering off the drug, and this is reflected by an increased score on the depression test, then they should temporarily raise the dose slightly for a week or two. This usually leads to an improvement in mood again. Then they can slowly continue to taper off the drug again. This approach is reassuring because it puts the patient in control. After a couple tries like this, most patients have been able to taper off their antidepressants without becoming depressed again.

What Should I Do if the Depression Comes Back in the Future?

If your depression returns, the chances are excellent that you will again respond to the same drug that helped you the first time. It may be the proper biological "key" for you. So you can probably use that drug again for any future episode of depression. If any blood relative of yours develops a depression, this drug might also be a good choice for them because a person's response to antidepressants, like the depression itself, appears to be influenced by genetic factors.

The same reasoning applies to the psychotherapy techniques. I have found that for most people, the same kinds of events (for example, being criticized by an authority figure) tend to trigger depression, and the same kinds of cognitive therapy technique usually reverse the depression for a particular patient. In most cases patients have been able to reverse a new episode of depression fairly rapidly without having to take the medication again. I encourage my patients to come in for a little "tune-up" if they become depressed again in the future. Often these "tune-ups" consisted of only one or two therapy sessions, since we were usually able to reapply the same technique that had helped them so much the first time I treated them.

What Are the Most Common Side Effects of the Antidepressants?

As discussed in Chapter 17, all the medications prescribed for depression, anxiety, and other psychiatric problems can cause different kinds of side effects. For example, many of the older antidepressants (such as amitriptyline, trade name Elavil) cause fairly noticeable side effects such as dry mouth, sleepiness, dizziness, and weight gain, among others. Many of the newer antidepressants (such as fluoxetine, trade name Prozac) can cause nervousness, sweating, upset stomach, or a loss of interest in sex as well as difficulties having an orgasm.

I will describe the specific side effects of every antidepressant in Chapter 20. You will see that some medications produce lots of side effects whereas others produce very few.

The Side Effects Checklist on pages 494–496 can provide you and your physician with extremely accurate information about any side effects that you experience while you are taking a medication. If you take this test a couple times per week, this will show how the side effects change over time.

Remember, however, that many of these so-called side effects can occur even if you are not taking any medication, since many side effects are also symptoms of depression. Feeling tired, having trouble sleeping at night, or a loss of interest in sex would be good examples. So it can be very useful to complete the Side Effects Checklist at least once or twice before you start any medication. That way, you can see if a side effect began before or after you started the drug. Obviously, if you had the same side effect before you started taking a drug, then the drug is probably not to blame for it.

It is also good to remember that patients who only take placebo medications (sugar pills) during research studies tend to report lots of side effects. This is because they think they are taking a real drug. So there is no proof that a particular side effect is necessarily caused by the drug you are taking. When in doubt, talk this over with your physician.

Let me give you a particularly vivid example of how the mind can occasionally play tricks on us. I once treated a high school teacher for depression. She was not responding well to the psychotherapy and I had the hunch that she would respond to a particular antidepressant drug called tranylcypromine (Parnate) that is described in Chapter 20. However, she was somewhat stubborn and had a strong fear of taking any medication. She complained that she would

Side Effects Checklist*

Instructions: Put a check (✔) after each item to indicate if you have had this type of side effect during the past several days. **Please answer all the items.**

	0—Not At All	1—Somewhat	2—Moderately	3—A Lot	4—Extremely
Mouth and Stomach					
1. dry mouth					
2. frequently thirsty					
3. loss of appetite					
4. nausea or vomiting					
5. stomach cramps or upset stomach					
6. increase in appetite or eating too much					
7. weight gain or loss					
8. constipation					
9. diarrhea					
Eyes and Ears					
10. blurred vision					
11. overly sensitive to light					
12. changes in vision, such as halos around objects					
13. ringing in your ears					
Skin					
14. sweating too much					
15. rash					
16. excessive sunburn when exposed to sun					
17. change in skin color					
18. bleeding or bruising easily					
Sex					
19. loss of interest in sex					
20. difficulties getting sexually excited					
21. difficulties getting an erection (men)					

Side Effects Checklist continued

	0—Not At All	1—Somewhat	2—Moderately	3—A Lot	4—Extremely
22. difficulties having an orgasm					
23. difficulties with your period (women)					
Stimulation and Nervousness					
24. stimulated					
25. agitated					
26. anxious, worried or nervous					
27. feeling strange or "spaced out"					
28. excess energy					
Sleep Problems					
29. feeling tired or exhausted					
30. loss of energy					
31. sleeping too much					
32. trouble falling asleep					
33. sleep that is restless or disturbed					
34. waking up too early in the morning					
35. nightmares or strange dreams					
Muscles and Coordination					
36. muscle jerks or twitches					
37. slurred speech					
38. tremor					
39. difficulty walking or loss of balance					
40. feeling slowed down					
41. stiffness of the arms, legs, or tongue					
42. feeling restless, like you have to keep moving your arms or legs					
43. hand-wringing					
44. constant, regular, rhythmic leg jiggling					
45. abnormal movements of your face, lips, tongue					

Side Effects Checklist continued

Instructions: Put a check (✔) after each item to indicate if you have had this type of side effect during the past several days. **Please answer all the items.**

	0—Not At All	1—Somewhat	2—Moderately	3—A Lot	4—Extremely
46. abnormal movements of other parts of your body, such as your fingers or shoulders					
47. muscle spasms of your tongue, jaw, or neck					
Other					
48. difficulty remembering things					
49. feeling dizzy, light-headed, or faint					
50. feeling your heart race or pound					
51. swelling in your arms or legs					
52. trouble starting urination					
53. headache					
54. breast swelling or enlargement					
55. milk secretion from the nipples					

Please describe any other side effects: ______________________

____

____

____

____

not be able to tolerate the side effects. I explained that I planned to prescribe a low dose and that in my experience most patients did not have many side effects with this medication, especially when the dose was low. But my efforts were to no avail—she insisted the side effects of the drug would be unbearable, and refused to accept a prescription.

I asked if she would be willing to do a little experiment to check this out. I told her I would give her two weeks' worth of pills in fourteen separate envelopes. Each envelope was labeled with the date and day of the week she was to take the pills inside it. I explained that some envelopes would contain placebo pills that could not have any side effects whatsoever. Half the pills would be yellow and half would be red, but she would not know whether she was taking the real medication or a placebo on any given day. The envelope for the first day contained one yellow pill and the envelope for the second day contained one red pill. The envelopes for the third and fourth days contained two yellow pills each, and the envelopes for the fifth and sixth days contained two red pills each. Finally, each envelope for the second week contained three yellow pills or three red pills.

I asked her to complete the Side Effects Checklist every day and to record the date. I explained how this experiment would help us determine whether any side effects she experienced on a given day were due to the real drug or the placebo effect. She reluctantly agreed, but insisted that her body was very sensitive to drugs and predicted the experiment would prove just how wrong I was.

Shortly after she started taking the pills she started calling me almost daily with alarming reports about severe side effects, especially on the days she was taking the yellow pills. She said these effects also spilled over to the days she took the red pills as well. I explained the side effects usually diminished over time and encouraged her to try to continue.

On Sunday evening she had the answering service page me at home for an emergency. She stated that the side effects did not diminish but were getting worse. In fact, they were so severe that she simply could no longer function. She was dizzy and confused and fatigued. Her mouth was as dry as cotton. She staggered when she tried to walk and could barely get out of bed. She had severe headaches. She said she would not take any more pills and wanted to know why I had put her through such grief.

I apologized, told her to stop the medications immediately and made an appointment to see her the first thing Monday morning for an emergency session. I reassured her that none of her symptoms sounded life-threatening, although she was obviously in great distress. I told her to bring her daily Side Effects Checklists to the session and promised that we would break the code together the next morning so we could find out which days she had taken the placebos and which days she had taken the real pills.

The next morning I explained that *all* the pills she had taken were placebos I had obtained from the hospital pharmacist. They were simply red placebos and yellow placebos—there were no Parnate pills in any of the envelopes.

This information surprised her, and tears began rolling down her cheeks. She acknowledged that she never would have believed that her mind could have such powerful effects on her body. She had been totally convinced the side effects were real. She then went ahead to take the Parnate in small doses and her mood improved substantially over the next month or two. She also started working very hard on her psychotherapy homework between sessions. She continued to fill out the Depression Test and the Side Effects Checklist once a week, but she did not report many side effects.

I do not mean to imply that all side effects are in your mind. On rare occasions this can occur, but most of the time the side effects are quite real, and the vast majority of my patients have reported them accurately. If you use the Side Effects Checklist on a daily basis, it will help you and your doctor assess the specific type and severity of any symptoms you might experience. Then the appropriate medication adjustments can be made if the side effects are excessive or dangerous.

Why Do Antidepressant Drugs Have Side Effects?

You learned in Chapter 17 that antidepressant drugs can stimulate or block the receptors for the neurotransmitter chemicals that nerves use to send messages to each other. In that chapter, we focused on serotonin, since this transmitter is felt to be involved in the regulation of mood. One of the most important and helpful discoveries of the past two decades is that antidepressants can also interact with the receptors for several additional chemical transmitters in the brain. These interactions appear to be responsible for many of the side effects of the antidepressants.

The three brain receptors that have been studied the most intensively are called histamine receptors, alpha-adrenergic receptors, and muscarinic receptors. These are located on nerves that use histamine, norepinephrine, and acetylcholine, respectively, as their chemical transmitters. Drugs that block histamine receptors are called "antihistamines," a term you are probably familiar with. Drugs that block alpha-adrenergic receptors are called "alpha-blockers," and drugs that block muscarinic receptors are called "anticholinergics."

Each type of receptor is responsible for certain kinds of side effects. You can predict the side effects of any drug if you understand how strongly the drug affects each of these three brain systems. Antidepressant medications produce many of their side effects because they block histaminic receptors, alpha-adrenergic receptors, and cholinergic receptors (which are also called "muscarinic" receptors) that are located on the surfaces of nerves inside of your brain and throughout your body as well. In case you do not recall what a "receptor" is, it is simply an area on the surface of a nerve that can turn the nerve on or off. The histamine receptors are located on nerves that use histamine as a chemical transmitter; the alpha-adrenergic receptors are located on nerves that use norepinephrine as a chemical transmitter; and the cholinergic receptors are located on nerves that use acetylcholine as a chemical transmitter. If you block any of these three types of receptors, you will turn the nerves off. The effects of different antidepressant medications on these three receptors help explain many of the side effects of these drugs.

For example, amitriptyline (Elavil) is an older antidepressant that can cause many side effects, including sleepiness, weight gain, dizziness, dry mouth, blurred vision, and forgetfulness, to name just a few of the more common ones. Most of these side effects are not dangerous, but they can be uncomfortable. Let's see if we can understand these side effects a little better by examining the effects of amitriptyline on the three kinds of nerve receptors.

Scientists have learned that amitriptyline blocks the cholinergic, histamine, and alpha-adrenergic receptors in the brain. Let's examine its anti-cholinergic effects first. What do these cholinergic nerves ordinarily do? Among other things, they control the amount of lubrication in your mouth. If you stimulate cholinergic nerves, more fluids will flow into your mouth from glands that are located in your cheeks.

What would happen if you turned off these nerves that normally lubricate your mouth? Your mouth would feel dry. You may have experienced a dry mouth when you were very nervous (cottonmouth) or when you were exercising for a long time in the sunshine without drinking any water. Cholinergic nerves also slow the heart, and so anticholinergic drugs like amitriptyline will cause the heart to speed up. Anticholinergic drugs can also cause forgetfulness, confusion, blurred vision, constipation, and difficulties getting your urine started.

Amitriptyline also blocks alpha-adrenergic receptors on nerves that use norepinephrine as a transmitter substance. If you stimulate these alpha-adrenergic receptors, your blood pressure will usually increase. Conversely, when you block them, your blood pressure will usually fall. This is why amitriptyline can cause a drop in blood pressure in certain individuals. This problem is especially noticeable when you suddenly stand up, because the drop in blood pressure makes you dizzy. Dizziness when standing is a common side effect of amitriptyline and many other antidepressants.

As noted above, amitriptyline also blocks histamine receptors in the brain. Drugs that block these receptors are called "antihistamines." You've probably taken an antihistamine if you've had an allergy or a stuffy nose. Drugs that block histamine receptors can make you sleepy and hungry. This is why amitriptyline, as well as many other antidepressant drugs that block histamine receptors, causes tiredness and weight gain.

Many of the older antidepressant medications are categorized as "tricyclic" antidepressants. The tricyclics have relatively strong effects on these three kinds of brain receptors, and so they tend to cause quite a few side effects. In fact, on pages 530–532 in Chapter 20 you will find a table that lists each tricyclic and shows how strong its effects are on each of these three types of brain receptors. This information indicates how strong the different kinds of side effects will be for each medication.

In contrast, many of the newer antidepressants (such as Prozac and the other SSRIs) generally have only weak effects on the histaminic, alpha-adrenergic, and cholinergic receptors in the brain. Consequently, they usually produce fewer side effects than the older drugs like amitriptyline. For example, the SSRIs are less likely to cause sleepiness, excessive appetite, dizziness, dry mouth, constipation, and so forth. The SSRIs also have little effect on the rate or rhythm of the heart.

However, we are now discovering that the SSRIs such as Prozac have new and different side effects of their own. For example, as many as 30 percent to 40 percent of the patients taking these drugs experience sexual difficulties such as a loss of interest in sex or difficulties having an orgasm. They can also cause upset stomach, loss of appetite, weight gain, nervousness, difficulties sleeping, fatigue, tremor, and excessive sweating, and a number of other side effects.

What Can I Do to Prevent or Minimize these Side Effects?

The likelihood and severity of any side effect usually depends on the dose of the medication you are taking. As a general rule, if you start out with a small dose and increase the dose gradually, the side effects can be minimized. In addition, many side effects tend to diminish over time. Sometimes a reduction in dose will minimize side effects without reducing the effectiveness of an antidepressant; sometimes a change to another type of antidepressant medication will be needed. If you and your doctor work together, you can usually find a medication that will have a beneficial effect on your mood without excessive side effects.

Your doctor might also add a second medication to help combat the side effects of an antidepressant medication or a mood stabilizer. Sometimes this is necessary and justified and sometimes it is not necessary. I will discuss this issue in greater detail in Chapter 20 but I will give you a couple of specific examples here.

Let's assume that you are taking lithium for manic-depressive illness. A common side effect of lithium is a tremor of the hands. You may find it difficult to write your name clearly or your hand may shake while you are attempting to hold a cup of coffee. One of my patients trembled so much that the coffee would actually spill out of the cup. Obviously such a severe side effect is not acceptable.

Your doctor may add one of the drugs called beta-blockers to help combat the tremor. The drug propranolol (Inderal) is often used for this purpose. However, beta-blockers have potent effects on the heart and they can also have a number of side effects of their own. Furthermore, both lithium and beta-blockers have the potential for adverse interactions with other drugs your psychiatrist or family physician may prescribe, and so the situation rapidly becomes quite complex. In my mind, the question becomes: Is this tremor so severe and disabling that it justifies adding a potent cardiac drug? Is there another way to deal with this side effect without adding more drugs? Would a reduction in dose be indicated? Sometimes the beta-blocker may be justified; sometimes it may not be necessary.

The same kind of reasoning applies to antidepressants. Sometimes a second drug is necessary to combat a side effect, but often it is not the best choice. Let's assume that you are being treated with fluoxetine (Prozac) for depression. Three common side effects of Prozac include insomnia, anxiety, and sexual problems. Let's examine how your doctor might handle each of these.

- If you are overly stimulated from Prozac and you are having trouble sleeping, your doctor may add a small dose of a second, more sedative antidepressant at night. For example, 50 to 100 mg of trazodone (Desyrel) is often used. This is a pretty good approach, because the trazodone differs from most sleeping pills in that it is not addictive. However, you may also be able to combat the excessive stimulation by taking a smaller dose of Prozac and by taking it earlier in the day. Then you might not need to add a second drug. Keep in mind, too, that the excessive stimulation from Prozac tends to occur when you first start taking it and may also disappear after a week or two.
- Prozac can cause anxiety or agitation, especially when you first start taking it. Your doctor may want to add a benzodiazepine (minor tranquilizer) such as clonazepan (Klonopin) or alprazolam (Xanax) to combat the nervousness. But the benzodiazepines can be addictive when taken daily for more than three weeks, and anxiety can usually be managed without adding one of these agents. A reduction in the dose of the Prozac will often help. The effectiveness of the SSRI antidepressants such as Prozac does not seem to depend on the dose, so there is little justification for taking a dose that creates excessive discomfort. The passage of time will often help as well, since the anxiety from Prozac seems to diminish or disappear after the first few weeks.

  Some patients develop a second wave of nervousness and restlessness after they have been on Prozac for a number of weeks or months. Sometimes this pattern of agitation is called "akathisia"—a syndrome in which your arms and legs become so extremely restless that you simply cannot sit still. This intensely uncomfortable side effect is quite common with the neuroleptic drugs used to treat schizophrenia but occurs much less often with most antidepressants. Prozac leaves your blood very slowly, however, so the levels increase more and more during the first five weeks that you are taking it. Even though a particular dose of Prozac, such as 20 mg or 40 mg per day, may have been fine at first, after a month or so that same dose may become much too high for you. A dramatic reduction in dose might greatly reduce the side effects without reducing the antidepressant effects at all. However, many patients with akathisia have to be taken off the Prozac and switched to another medication because the akathisia has become so severe and uncomfortable. Your doctor may add another drug temporarily to combat akathisia, but it seems prudent to reduce the dose of Prozac or to go off the drug entirely if akathisia develops.

- As noted above, as many as 40 percent of men and women on Prozac (as well as the other SSRI antidepressants) develop sexual problems, including a loss of interest in sex as well as difficulties having an orgasm. Your doctor might want to add one of several drugs (bupropion, buspirone, yohimbine, or amantadine) currently being used to try to combat these sexual side effects. Once again, the potential benefit should be weighed against the hazards of these agents, and alternative strategies can be considered. I have rarely, if ever, kept a patient on an SSRI indefinitely, so most patients have elected just to put up with this side effect, knowing it would not be a long-term problem. If the SSRI is causing a dramatic improvement in mood and there are no other side effects, the loss of interest in sex for several months may be an acceptable trade-off. But of course these are subjective issues, and you will have to make your own decision about this after discussing your options with your physician.

In the next chapter, you will see that I recommend against combination drug therapies for most patients taking antidepressants. If you take more than one drug at a time, you increase the chances for dangerous drug interactions. In addition, the second medication may create new and different side effects. In most cases, if you and your doctor work together and use a little common sense, it will not be necessary to treat antidepressant drug side effects by adding additional drugs.

How Can I Prevent Potentially Dangerous Interactions between Antidepressants and Other Drugs, Including Nonprescription Drugs?

In recent years doctors have become increasingly aware that certain types of drugs may interact with each other in ways that can be dangerous. Two drugs may be quite safe and have few or no side effects if you take either one separately; but if you take the two drugs at the same time, there could be serious consequences because of how the two drugs interact with each other.

This problem of drug interactions has become increasingly important in recent years for two reasons. First, there is an increasing trend among psychiatrists to prescribe more than one psychiatric drug at a time to many of their patients. This is not an approach with which I am entirely comfortable, but it is nevertheless very common. Each new drug raises the possibility of drug interactions, since different psychiatric drugs can interact with each other in potentially dangerous ways. And, as noted in the last chapter, more and more patients are being put on antidepressant drugs (as well as other types of psychiatric drugs) for prolonged periods of time, sometimes indefinitely. This is also not an approach with which I am comfortable, and I have found that long-term drug treatment for depression is not necessary for most patients. But many psychiatrists do prescribe drugs for prolonged times—the practice is in vogue. And if you do take a psychiatric drug for a long time, eventually you will probably receive one or more prescriptions from other doctors for other medical problems. For example, your doctor might prescribe a medication for an allergy, high blood pressure, pain, or an infection. In addition, you might take an over-the-counter medication for a cold, a cough, a headache, or an upset stomach. Now the possibility of drug interactions has to be considered, because these drugs may interact with the psychiatric drug you have been taking.

Of course, it goes without saying that psychiatric drugs can also interact with tobacco and alcohol as well as street drugs such as cocaine or amphetamines. In some cases these interactions can also be quite dangerous and even fatal. Some antidepressants interact in extremely dangerous ways with commonly used drugs, including over-the-counter medications. I am not trying to be overly alarmist here. With a little education and good teamwork with your physician, you can take an antidepressant safely.

In this section I will explain why and how drug interactions happen. In addition, in Chapter 20, I will describe a number of important drug interactions for each drug or category of drug you might be taking. Remember that knowledge about these drug interactions is rapidly evolving. New information comes out almost on a daily basis. Make certain each doctor you see has a complete and accurate list of every drug you are taking, including any over-the-counter (nonprescription) drugs you take. Ask your doctor if there are any drug interactions that could be important. Ask your pharmacist the same thing. If they are not sure, ask them to check it out for you. It is virtually impossible to keep all potential drug interactions in your mind, because so much new information is constantly emerging. References and computer programs that list dangerous drug interactions are readily available to help with this task. If you are appropriately assertive and have a little education about the topic, you will be in a better position to have an intelligent discussion with your doctor about interactions among the drugs you are taking.

You will see in Chapter 20 that I have prepared detailed charts listing drug interactions for specific antidepressants or mood stabilizers you may be taking. So, for example, if you are taking Prozac, you can review the table that lists its drug interactions. This should take only a minute or two.

You may think that you shouldn't have to study these charts, because your doctor should know all about any dangerous drug interactions and ensure that nothing bad happens to you. There are several problems with this line of reasoning. First, though your doctor may be extremely knowledgeable, she or he is also human and cannot keep up with all the new information that is emerging, no matter how smart she or he may be. Second, even if your doctor told you about every conceivable drug interaction, there is no way you could remember all of them! And third, in this era of managed care, doctors are having to manage more and more patients, and you may get only a few minutes with your prescribing physician at infrequent intervals to review your symptoms and the dose of the medication. There may simply not be enough time to discuss all the possible drug interactions you need to know about.

How and Why Do These Drug Interactions Occur?

There are four basic ways that two drugs can interact. First, one drug can cause the level of a second drug in your blood to increase—sometimes to an alarming degree, even though you are taking only a "normal" dose of both drugs. What are the consequences of a sudden increase in the level of a drug in your blood? First, you may experience more side effects, since they are usually related to the dose. Second, many psychiatric drugs lose their effectiveness when the dose is too high or too low. And third, there can be toxic and even fatal reactions when the blood level of any drug becomes too high.

A second type of drug interaction is just the opposite. One drug can cause the level of another drug in your blood to decrease. This can cause the second drug to become ineffective, even though you are taking a normal dose. You and your doctor may wrongly conclude that the drug does not work for you when the real problem is that your blood level is too low.

A third type of interaction is when two drugs each have similar effects or side effects that intensify each other. Suppose, for example, that you are being treated for high blood pressure and then you begin to take a psychiatric drug that also lowers blood pressure as a side effect. The result could be that you might experience a sudden drop in blood pressure and possibly even faint when you suddenly stand up.

A fourth and more ominous type of drug interaction is not related to changes in blood levels but simply to toxic effects of certain drug combinations. In other words, two drugs that are safe when taken separately may lead to extremely dangerous interactions when you take them together.

Now let's examine the first two types of drug interactions in more detail. Why does one drug sometimes cause the level of a second drug to increase or fall dramatically? Well, a simple way to think about it would be to imagine that you are trying to fill a bathtub with water. If the plug is out, the water will have a tendency to go out as fast as it comes in. As a result, the water level in the tub will not go up high enough to take a bath, no matter how long you leave the faucet on. In contrast, if the plug is in the tub and you don't turn the water off, the tub will overflow.

Now compare your body to the bathtub. (I do not mean to imply that you have a bad figure!) The medicine you take each day is like the water coming into the tub. Certain enzyme systems in your liver can be compared to the hole in the bottom of the tub. These enzymes in your liver chemically change drugs into other substances (called "metabolites'') that your kidneys can get rid of more easily. This process is called ''metabolism.'' Metabolites of the drugs you take usually end up in your urine.

When you add a second drug, your liver may metabolize the first drug more slowly. This would be comparable to plugging up the hole at the bottom of the tub. And so, as you keep taking the first medicine, your blood level gets too high, in just the same way that the water in the tub gets too high and eventually spills over the side. Or the second drug you take could have the opposite effect of making the hole in the bottom of the tub much bigger. In this case, your liver's metabolism speeds up and rids your body of the first drug much faster. In this case, you may keep taking the same dose of the first drug each day but your blood level remains too low to have the desired antidepressant effect. In this case, the water goes out of the tub just as fast as it comes in.

That's pretty much the basic principle. The drugs that are likely to interact with each other are those that are metabolized by the ''cytochrome P450'' enzyme systems in the liver. There are many of these enzyme systems, and different kinds of drugs are metabolized by different enzyme systems. Only certain drugs or combinations of drugs will stimulate or inhibit any of these enzyme systems. Psychiatric drugs can interact with other psychiatric and nonpsychiatric drugs, such as antibiotics, antihistamines, or painkillers. In other words, psychiatric drugs can affect other drugs your doctor may prescribe (such as a pill for high blood pressure), in exactly the same way that those other drugs can have an impact on any psychiatric drugs you may be taking. The bottom line is that the level of any drug you are taking might become too high or too low if you are also taking another drug at the same time.

Let me now give you some specific examples of these drug interactions. Suppose you are taking one of the new selective serotonin reuptake inhibitors called paroxetine (trade name Paxil). This drug is very similar to Prozac. Now suppose that the paroxetine is not working very well, which sometimes happens, and you are still feeling depressed. Your doctor might decide to add a second antidepressant. If your doctor chooses desipramine (trade name Norpramin), the paroxetine you are taking will have the effect of "plugging up the tub." Now your body will not be able to metabolize the new drug (desipramine) very well. As a result, your blood level of desipramine may increase to three to four times higher than expected. Most psychiatrists are aware of this drug interaction and will be careful to prescribe desipramine in a tiny dose if a patient is taking an SSRI like paroxetine. But if your psychiatrist was not aware of this particular drug interaction and decided to give you a "normal" dose of desipramine, you could develop a toxic level of desipramine in your blood.

Is this serious? Well, there are three potential problems. First, desipramine is not effective at excessively high blood levels. Second, there will be many more side effects at high levels. And third, in rare instances, excessive blood levels of desipramine can trigger abnormal heart rhythms and occasionally even cause death.

Is this type of drug interaction rare? No. The levels of antidepressants can sometimes increase or decrease quite dramatically when combined with common prescription or over-the-counter drugs you might take without thinking twice. The tables in Chapter 20 will delineate the interactions most important to any antidepressant you might be taking.

Finally, some toxic and dangerous drug interactions do not necessarily depend on doses or blood levels. For example, many of the newer antidepressants such as Prozac have powerful effects on the serotonin systems in the brain. The monoamine oxidase inhibitors (MAOIs) also affect the serotonin systems in the brain, but through a different mechanism. The antidepressant tranylcypromine (trade name Parnate) is an example of one of these MAOI drugs. If you take Prozac and Parnate at the same time, the combination could trigger an extremely dangerous reaction known as the "serotonin syndrome." The symptoms can include fever, muscle rigidity, and rapid changes in blood pressure, along with agitation, delirium, seizures, coma, and death. Obviously, this combination of drugs should not be given!

You will see in Chapter 20 that many medications can be dangerous if you are taking an MAOI. The list of forbidden drugs includes many antidepressants, some decongestants (especially if they contain dextromethorphan, a common ingredient of cold preparations), antihistamines, local anesthetics, some anticonvulsants, some painkillers such as meperidine (Demerol), antispasmodics including cyclobenzaprine (Flexeril) and weight-loss preparations. Some of these drugs will cause the serotonin syndrome described above, and some of them will cause another dangerous reaction known as a "hypertensive crisis." In extreme cases, the symptoms of a hypertensive crisis include brain hemorrhage, paralysis, coma, and death. Certain common foods such as cheese are also on the "forbidden" list if you are taking one of the MAOIs, because they can cause a hypertensive crisis as well.

Many doctors do not prescribe the MAOIs because of concerns about these toxic interactions. You may also think: "Well, I will just take a safer drug so I won't have to worry." This makes good sense, since many safer medications are available. However, many commonly prescribed antidepressants can cause dangerous interactions. For example, two common antidepressants, nefazodone (trade name Serzone) and fluvoxamine (trade name Luvox) should not be combined with several commonly prescribed drugs because these particular combinations can trigger an abnormal heart rhythm that may result in sudden death. The drugs include terfenadine (trade name Seldane and used for allergies), astemizole (trade name Hismanal and used for allergies), or cisapride (trade name Propulsid, a stimulant for the gastrointestinal tract).

I do not mean to give the impression that it is dangerous to take antidepressant drugs. To the contrary, they are usually quite safe and effective, and the catastrophic drug interactions I have described are fortunately rare. In addition, most psychiatrists go to great lengths to educate themselves about recent developments and try to keep up with new information about side effects and drug interactions. But in the real world we live in, no doctor is perfect and no doctor can have comprehensive knowledge about all possible drug interactions. For example, your primary care physician may not be familiar with some new antidepressant your psychiatrist has prescribed. And so a little research on your part will be helpful. As an enlightened consumer, you can read about any antidepressant medicine you are taking in Chapter 20 and in other readily available references such as the *Physician's Desk Reference* (*PDR*). You can find these books at any library, bookstore, or pharmacy. You can also find the *PDR* at your doctor's office. You can also review the drug insert that comes with the medication. It doesn't take more than five or ten minutes to review this information. Then you can ask informed questions and bring out the best in your physician. The teamwork can give you a safer and better experience with your antidepressant. This is definitely one case where an ounce of prevention can be worth more than a pound of cure.

Chapter 20

The Complete Consumer's Guide to Antidepressant Drug Therapy*

(Notes and References appear on pages 682–687.)

In this chapter I will give you practical information about the costs, doses, side effects, and drug interactions for all the currently available antidepressant and mood-stabilizing drugs. I would recommend you use this chapter as a reference source rather than trying to read it all at once—there is just too much detailed information to digest at one sitting. If you want to learn about a particular drug that you or a family member may be taking, the Table of Antidepressants on pages 514–515 will help you locate the information you need in this chapter. Let's assume, for example, that you are taking fluoxetine (Prozac). You can read the section on the SSRI antidepressants starting on page 547. In addition,

*I would like to thank Joe Bellenoff, M.D., a psychopharmacology fellow at Stanford University Medical School, and Greg Tarasoff, M.D., a senior psychiatric resident at Stanford, for helpful suggestions during the revision of this chapter. In addition, much useful information was obtained form the excellent *Manual of Clinical Psychopharmacology*, Third Edition, by Alan F. Schatzberg, M.D., Jonathan Cole, M.D., and Charles DeBattista, D.M.H., M.D. (Washington: American Psychiatric Press, 1997). This scholarly but highly readable book is an invaluable reference. I highly recommend it for individuals who would like more information on the medications currently used in the treatment of emotional problems.

Table of Antidepressants

Antidepressant Drug Class	*Chemical Name (and Trade Name)*[a]	*Page #*
Tricyclic Antidepressants		524
	amitriptyline (Elavil, Endep)	
	clomipramine (Anafranil)	
	desipramine (Norpramin, Pertofrane)	
	doxepin (Adapin, Sinequan)	
	imipramine (Tofranil)	
	nortriptyline (Aventyl)	
	protriptyline (Vivactil)	
	trimipramine (Surmontil)	
Tetracyclic Antidepressants		524
	amoxapine (Asendin)	
	maprotiline (Ludiomil)	
SSRI Antidepressants		547
	citalopram (Celexa)	
	fluoxetine (Prozac)	
	fluvoxamine (Luvox)	
	paroxetine (Paxil)	
	sertraline (Zoloft)	
MAO Inhibitors		564
	isocarboxazid (Marplan)	
	phenelzine (Nardil)	
	selegiline (Eldepryl)	
	tranylcypromine (Parnate)	
Serotonin Antagonists		599
	nefazodone (Serzone)	
	trazodone (Desyrel)	
Other Antidepressants		605
	bupropion (Wellbutrin)	605
	venlafaxine (Effexor)	611
	mirtazapine (Remeron)	615

Antidepressant Drug Class	*Chemical Name (and Trade Name)*[a]	*Page #*
Mood Stabilizers		617
	carbamazepine (Tegretol)	640
	gabapentin (Neurontin)	651
	lamotrigine (Lamictal)	652
	lithium (Eskalith)	617
	valproic acid (Depakene) and divalproex sodium (Depakote)	634

[a]Many of the antidepressants are now available as generic brands (see Table 20–1). Only the trade names of the original brands are listed in this table.

the section on drug costs starting on this page, as well as the information starting on page 659, should be of general interest to all readers.

Costs of Antidepressant Medications

We often think that more expensive means better, but this is not always the case with antidepressants. As it turns out, there are some very dramatic differences in the costs of the different antidepressants that do not reflect differences in effectiveness. In other words, sometimes a drug that is much cheaper will be just as effective, or even more effective, than another drug that costs more than forty times more. Therefore, if the cost of the medication is a concern for you, then a little education may save you a great deal of money.

The costs and doses of the most commonly prescribed antidepressants and mood stabilizing agents are listed in Table 20–1 on pages 518–523. Note that I am quoting *the cheapest wholesale price* for each antidepressant drug in Table 20–1. The retail price you pay for the same medication at the drug store will probably be higher. If you choose a different brand of the same medication, it may be higher yet. Please keep this in mind in all of the following discussions of drug costs.

If you compare the costs of the different types of drugs

and the different doses, it will provide you with some interesting information. You will see, for example, that many of the older tricyclic and tetracyclic drugs are now available generically. When a drug is first manufactured, the drug company gets a seventeen-year patent so it can market the drug exclusively. The relatively high cost of the newer drugs that are still protected by patents helps to cover the costs of the research, development, and testing. After the patent expires, other companies can compete and manufacture the drug, and so the price goes down drastically.

You will see in Table 20–1 that these so-called "generic" medications are much less costly than the newer drugs that are still under patent. Let's assume that your doctor prescribes a dose of 150 mg per day of imipramine for your depression. The cost of the three 50-mg pills you will take will be less than 10 cents per day, or roughly $3 per month. This is because imipramine is now available generically. In contrast, if your doctor prescribes two 20-mg Prozac pills per day, your cost will be nearly $4.50 per day or $135 per month—over forty times more than the imipramine. And if she or he prescribes four Prozac pills—the maximum dose—your cost will be $270 per month. This is a steep price for many people. Don't forget these are *wholesale* prices—you may pay even more.

Is Prozac forty to a hundred times more effective than imipramine? Definitely not. As you will learn below, most of the antidepressants tend to be comparably effective. Research studies have not confirmed that Prozac is any more effective than imipramine—in fact it may be slightly less effective for severe depressions. However, the big advantage of Prozac is that it has fewer side effects (such as dry mouth or sleepiness) than imipramine. This may be quite important to some people and may make the price difference worthwhile. On the other hand, you will learn that Prozac has some side effects of its own, such as problems with sexual functioning (difficulty achieving orgasm) in as many as 30 percent to 40 percent of patients, and possibly more. People who don't like this particular side effect might actually prefer the cheaper medication.

You will also see in Table 20–1 that pills which contain a larger quantity of a particular drug are not necessarily more expensive than pills which contain a smaller quantity. This is especially true if you are taking one of the newer drugs that is still under patent, so you may be able to save money by buying pills containing a larger dose. For example, you will see in Table 20–1 that the cost of a hundred nefazodone (Serzone) tablets is $83.14 for the 100-mg size. The price for a hundred tablets of the larger sizes (150 mg to 250 mg) is exactly the same. So if you are taking a large dose, say 500 mg per day, you could either take five of the 100-mg pills (cost of $4.16 per day) or two of the 250-mg pills (cost of $1.66 per day).

In addition, you can often save money by buying a larger size of a medication and breaking a pill in half. So to continue with the same example, if you are taking 250-mg pills, it will cost you approximately half as much if you purchase 500-mg pills and break them in half.

For the generic drugs, things are different. On the average, the costs are low overall and depend on the dose, and the savings at higher doses are not so drastic. In addition, because so many different companies manufacture these drugs, the prices for the different doses are not always entirely consistent—sometimes a smaller dose will actually cost more than a larger dose. For example, look at the pricing structure for the tricyclic antidepressant, desipramine (trade name Norpramin) on page 518. You will see that a hundred of the 10-mg pills cost $15.75, while a hundred of the 25-mg pills costs only $7.14. So the larger pill is actually cheaper. This is because different companies manufacture the two sizes.

To make things even more confusing, there are other cases where a larger dose costs substantially more and you can save money by taking a smaller size. For example, take another look at the costs of desipramine on page 518. You will see that a hundred 75-mg desipramine pills cost $12.42, and that a hundred 150-mg desipramine pills cost $109.95 (again, because of different manufacturers). So you

Table 20–1. Names, Doses, and Costs of Antidepressant Medications

Chemical Name[a]	*Trade (Brand) Name*[b]	*Available Sizes (mg) & Cheapest Wholesale Cost per 100 Pills*[c]		*Daily Dose Range*[d]	*Are Generics Available*[e]
		Tricyclic Antidepressants			
amitriptyline	Elavil	10 mg	$1.73	75–300 mg	Yes
		25 mg	$1.85		
		50 mg	$2.78		
		75 mg	$3.53		
		100 mg	$4.28		
		150 mg	$2.09		
clomipramine	Anafranil	25 mg	$78.29	150–250 mg	No
		50 mg	$105.57		
		75 mg	$138.97		
desipramine	Norpramin	10 mg	$15.75	150–300 mg	Yes
		25 mg	$7.14		
		50 mg	$10.91		
		75 mg	$12.42		
		100 mg	$40.89		
		150 mg	$109.95		
doxepin	Sinequan	10 mg	$3.98	150–300 mg	Yes
		25 mg	$4.43		
		50 mg	$6.60		
		75 mg	$8.93		
		100 mg	$11.25		
		150 mg	$14.96		

imipramine hydrochloride	Tofranil	10 mg	$1.88	150–300 mg	Yes
		25 mg	$2.33		
		50 mg	$3.08		
imipramine pamoate	Tofranil-PM (sustained release)	75 mg	$103.67	150–300 mg	No
		100 mg	$136.29		
		125 mg	$169.95		
		150 mg	$193.73		
nortriptyline	Aventyl	10 mg	$11.55	50–150 mg	Yes
		25 mg	$15.90		
		50 mg	$19.43		
		75 mg	$24.83		
protriptyline	Vivactil	5 mg	$46.46	15–60 mg	No
		10 mg	$67.36		
trimipramine	Surmontil	25 mg	$64.08	150–300 mg	No
		50 mg	$108.14		
		100 mg	$157.20		
Tetracyclic Antidepressants					
amoxapine	Asendin	25 mg	$32.87	150–450 mg	Yes
		50 mg	$53.44		
		100 mg	$89.16		
		150 mg	$43.87		
maprotiline	Ludiomil	25 mg	$19.43	150–225 mg[f]	Yes
		50 mg	$29.10		
		75 mg	$40.88		

Table 20-1. Cont.

Chemical Name[a]	*Trade (Brand) Name*[b]	*Available Sizes (mg) & Cheapest Wholesale Cost per 100 Pills*[c]		*Daily Dose Range*[d]	*Are Generics Available*[e]
SSRI Antidepressants					
citalopram	Celexa	20 mg	$161.00	20–60 mg	No
		40 mg	$168.00		
fluoxetine	Prozac	10 mg	$218.67	10–80 mg	No
		20 mg	$224.54		
fluvoxamine	Luvox	50 mg	$198.67	50–300 mg	No
		100 mg	$204.37		
paroxetine	Paxil	10 mg	$189.33	10–50 mg	No
		20 mg	$189.20		
		30 mg	$214.80		
sertraline	Zoloft	50 mg	$176.23	25–200 mg	No
		100 mg	$181.33		
MAO Inhibitors					
phenelzine	Nardil	15 mg	$40.24	15–90 mg	No
selegiline	Eldepryl	5 mg	$215.90	20–50 mg	No
tranylcypromine	Parnate	10 mg	$45.80	10–50 mg	No
isocaboxazid	Marplan	10 mg	unavailable	10–50 mg	unavailable

Serotonin Antagonists					
nefazodone	Serzone	100 mg	$83.14	300–500 mg	No
		150 mg	$83.14		
		200 mg	$83.14		
		250 mg	$83.14		
trazodone	Desyrel	50 mg	$5.03	150–300 mg	Yes
		100 mg	$11.70		
		150 mg	$58.43		
Other Antidepressants					
bupropion	Wellbutrin	75 mg	$62.17	200–450 mg	No
		100 mg	$82.96		
venlafaxine	Effexor	25 mg	$105.53	75–375 mg	No
		37.5 mg	$108.68		
		50 mg	$111.93		
		75 mg	$118.66		
		100 mg	$125.78		
	Effexor XR (Extended Release Capsules)	37.5 mg	$193.88	75–375 mg	No
		75 mg	$217.14		
		150 mg	$236.53		
mirtazapine	Remeron	15 mg	$198.00	15–45 mg	No

Table 20-1. Cont.

Chemical Name[a]	*Trade (Brand) Name*[b]	*Available Sizes (mg) & Cheapest Wholesale Cost per 100 Pills*[c]		*Daily Dose Range*[d]	*Are Generics Available*[e]
Mood Stabilizers[g]					
lithium	Eskalith	150 mg	$7.63	900–1500 mg[h]	Yes
		300 mg	$5.25		
		600 mg	$13.23		
	Lithobid, Eskalith CR (sustained release)	300 mg	$15.53		
		450 mg	$35.80		
carbamazepine	Tegretol	100 mg	$14.67	800–1200 mg	Yes
		200 mg	$10.08		
valproic acid	Depakene	250 mg	$12.98	750–3000 mg	Yes
divalproex sodium	Depakote[i]	125 mg	$30.95	750–3000 mg	No
		250 mg	$60.76		
		500 mg	$112.08		
lamotrigine	Lamictal	25 mg[j]	—	50–150 mg[k]	No
		100 mg	$175.54		
		150 mg	$184.43		
		200 mg	$193.33		
gabapentin	Neurontin	100 mg	$37.80	900–2000 mg	No
		300 mg	$94.50		
		400 mg	$113.40		

[a]If your doctor prescribes the chemical or ''generic'' name on the prescription, your pharmacist can often substitute an inexpensive brand that can be much less costly than the trade-name drugs.

[b]Only the brand name of the original drug is listed. Generic versions of these drugs have their own brand names.

[c]Cost source: *Mosby's GenRx, 1998 (8th Edition): The Complete Reference Guide for Generic and Brand Drugs*. St. Louis: Mosby. The average wholesale price for 100 pills of the least expensive brand currently available is listed. This is the price your local retail pharmacist would have to pay for the product without any special discounts. Your cost will be more and will depend on the markup by your pharmacist.

[d]The doses would be used for the treatment of an episode of depression. Some patients may benefit from doses higher or lower than the normal range. If prolonged treatment is necessary following recovery, a smaller dose may be sufficient. Always consult with your doctor before changing the dose.

[e]These are drugs with generic brands available in 1998. More of the current antidepressant drugs will become available as generic brands when their original drug patents expire.

[f]Maprotiline should not exceed 175 mg per day if a patient is kept on the drug for an extended period. The manufacturer suggests that the dose should not exceed a maximum of 225 mg for periods of up to six weeks.

[g]The doses of several mood stabilizers must be monitored by blood tests and will therefore be highly individualized for each patient, depending on your age, gender, weight, diagnosis, and individual metabolism, as well as other medications you may be taking.

[h]Higher doses may be required during acute mania because the body appears to metabolize lithium more rapidly during manic episodes.

[i]This is also available as Depakote Sprinkle (125 mg), which can be sprinkled onto food.

[j]The price of the 25 mg Lamictal was not listed in *Mosby's GenRx* (1998 edition).

[k]This is the recommended dose range for epilepsy when given in conjunction with valproic acid. When given alone, the recommended dose range for epilepsy is 300 mg to 500 mg per day.

can save lots of money by taking two 75-mg pills instead of one 150-mg pill. Again, this is because different companies manufacture the 75-mg and 150-mg sizes. This may strike you as odd, but the pricing structure in some instances is completely out of whack.

If you or a family member is taking an antidepressant, make sure you study Table 20–1 and discuss these cost issues with your druggist. A little quick and easy research on your part may result in large savings.

Another important point, not illustrated in the table, is that the cost of the same generic drug and dose can vary greatly because the generics often have so many different manufacturers. In Table 20–1, I have always listed the least costly generic brand of each pill; other more costly versions of the same pill are not listed. For example a hundred 50-mg imipramine pills manufactured by the drug company HCFA FFP will cost only $3.08. Because this was the lowest-priced generic brand, I listed it in Table 20–1. In contrast, a hundred of the same size imipramine manufactured by Novartis, another drug company, will cost $74.12—more than twenty times more. Keep in mind that if your doctor prescribes the antidepressant by its chemical name (as listed in Table 20–1), your druggist will have the freedom to provide you with the least costly generic brand if one is available.

My goal is not to promote any one drug or class of drug. All antidepressants have merit, and they all have some drawbacks. The key point is this: more expensive does not always mean better. If you review the costs of these drugs, you can work with your doctor and pharmacist to choose the medication and brand that will make the most sense for you.

Specific Kinds of Antidepressants

Tricyclic and Tetracyclic Antidepressants

The first drugs listed in the Table of Antidepressants on page 514 are called "tricyclic" and "tetracyclic" antidepressants. The tricyclic and tetracyclic antidepressants differ slightly in their chemical structures. "Tri" means three and "tetra" means four. "Cyclic" refers to a circle or ring. The tricyclic compounds consist of three linked molecular rings, while the tetracyclics consist of four.

You will see that eight tricyclic and two tetracyclic antidepressants are listed in the Table of Antidepressants. The eight tricyclic drugs include amitriptyline (Elavil), clomipramine (Anafranil), desipramine (Norpramin), doxepin (Sinequan), imipramine (Tofranil), nortriptyline (Aventyl), protriptyline (Vivactil), and trimipramine (Surmontil). These eight tricyclic drugs used to be the most widely prescribed antidepressants. They are still among the most effective of all the antidepressants. Many of them are also the least expensive because generic brands have become available. However, tricyclics tend to have more side effects than the newer drugs, and so they are less popular than they used to be. By the same token, they have been prescribed for several decades and have a long track record of reasonably good effectiveness and safety.

The two tetracyclic antidepressant medications listed on the table are called amoxapine (Asendin) and maprotiline (Ludiomil). These two tetracyclics were synthesized and released after the tricyclics had been in use for some time. It was hoped that they would represent significant improvements in treatment, either because of increased effectiveness for certain types of depression, or because of fewer side effects.

Unfortunately, these expected improvements did not really materialize. For the most part, the effectiveness, mechanism of action, and side effects of the eight tricyclic and the two tetracyclic antidepressants are quite similar.

Doses for the Tricyclic and Tetracyclic Antidepressants. Table 20–1 on pages 518–523 lists the costs and dose ranges of the eight tricyclic and the two tetracyclic antidepressant medications. As noted above, many of them are inexpensive because they are no longer on patent, and so generic brands are readily available. Don't be fooled into thinking the cheaper antidepressants are less effective, however. A number of studies suggest they may be slightly more effective than many of the newer antidepressants such as Prozac.

The most common error your doctor is likely to make is to prescribe a dose of a tricyclic antidepressant that is too low. This statement may run against your grain if you feel you should take the lowest dose possible. In the case of tricyclics, if the prescribed dose is too low, the medication will not be effective. If you insist on taking a dose that is too low, you may be wasting your time. It simply will not help you. On the other hand, dosages above those recommended in Table 20–1 can be dangerous and may lead to a worsening of your depression.

Having said that, let me also say that there are cases in which people do respond to doses that are smaller than those listed (especially the elderly), and there are also times when people may need larger doses. One reason for this is that there can be considerable differences in how rapidly people metabolize antidepressant drugs. These differences are partially genetic, and are due to levels of certain enzymes in your liver, as described previously. If you are a "fast metabolizer" you will need a larger dose to maintain an effective blood level, and if you are a "slow metabolizer" you will need a smaller dose. In addition, you will learn below about other drugs that can make tricyclic blood levels fall and lose their effectiveness or increase and become more toxic.

If you suspect you may be taking an inappropriately large or small dose, review the dose ranges in Table 20–1 and discuss your concerns with your physician. Blood-level testing for most of the tricyclic antidepressants is readily available, so your doctor may order a blood test to make sure that the dose you are taking is neither too high nor too low for you.

The best way to begin taking a tricyclic medicine is to start out with a small dose and to increase the amount each day until a dose within the normal therapeutic range is achieved. This buildup can usually be completed within one or two weeks. For example, a typical daily dose schedule for imipramine, one of the most commonly prescribed tricyclic antidepressant drugs listed in Table 20–1, might be the following:

Day one—50 mg at bedtime;
Day two—75 mg at bedtime;
Day three—100 mg at bedtime;
Day four—125 mg at bedtime;
Day five—150 mg at bedtime.

You and your doctor may prefer to build up the dose a bit more gradually. Doses of up to 150 mg per day can be conveniently taken once a day at night. The antidepressant effect will last all day long, and the most bothersome side effects will occur at night, when they will be least noticed. If doses larger than 150 mg per day are required, the additional medicine should be given in divided doses during the daytime.

For the more sedating tricyclic antidepressants, up to half the maximum indicated dose may be taken on a once-per-day basis before bedtime. This dosage promotes sleep. Several of the tricyclic antidepressants, including desipramine, nortriptyline, and protriptyline, can be stimulating. They can be taken in divided doses in the morning and at noon. If taken too late in the day, they may interfere with sleep.

If you reduce the dose of a tricyclic antidepressant or if you decide to stop taking the medicine, it is best to reduce the dose gradually and never abruptly. Sudden discontinuation of any antidepressant may result in side effects. These include upset stomach, sweating, headache, anxiety, or insomnia. Usually, you can go off a tricyclic antidepressant safely and comfortably by tapering the dose gradually over a one- or two-week period.

Side Effects of the Tricyclic Antidepressants. The most frequent side effects of the tricyclic antidepressants are listed in Table 20–2 on pages 530–532. You will see in this table that all the tricyclic antidepressants have quite a number of side effects, and this is their greatest drawback.

The most common side effects include sleepiness, dry mouth, a mild hand tremor, temporary light-headedness when you suddenly stand up, weight gain, and constipation. They can also cause excessive sweating, difficulties with sex, twitches or jerking when you fall asleep at night, and a number of other effects listed in Table 20–2. Most of these side effects are not dangerous, but they can be annoying.

You learned earlier that the side effects of antidepressant drugs can be predicted if you know how strongly they block histamine receptors, alpha-adrenergic receptors, and muscarinic receptors (also called cholinergic receptors) in the brain. You can see from Table 20–2 that each antidepressant has a different profile of side effects depending on its action on these three receptor systems in the brain.

Blockade of the brain's histamine receptors makes you hungry and sleepy. Table 20–2 indicates that four of the tricyclic antidepressants (amitriptyline, clomipramine, doxepin, and trimipramine) have rather strong effects on the histamine receptors. Consequently, these four antidepressants are more likely to make you feel sleepy and hungry. If you are having trouble sleeping, this side effect could be a benefit, but if you are already feeling sluggish and unmotivated, these drugs may make things worse for you. If you have been losing weight due to depression, the appetite boost could be beneficial. However, if you are overweight, you might have to pay more attention to your diet and exercise more in order to avoid weight gain, which can be demoralizing. Since there are now many available antidepressants that do not cause weight gain, it might be better to switch to one of them. You can see in Table 20–2 that three of the tricyclics (desipramine, nortriptyline, and protriptyline) have only weak effects on the histamine receptors. These antidepressants will be less likely to cause sleepiness and weight gain. There are many antidepressants in other categories as well that do not cause sleepiness and weight gain.

You may also recall that blockade of the brain's alpha-adrenergic receptors causes a drop in blood pressure. This can result in temporary light-headedness or dizziness when you suddenly stand up because your leg veins become more relaxed, and blood temporarily pools in your legs. As a result, your heart temporarily does not have enough blood to pump up to your brain, and so your vision may get black and you may feel dizzy or woozy for a few seconds. Antidepressants with relatively strong effects on the brain's alpha-adrenergic receptors will be more likely to cause dizziness when you suddenly stand up. You will see in Table 20–2 that many tricyclics have strong effects on alpha-adrenergic receptors, but that two of them (desipramine and nortriptyline) have only weak effects on them. Consequently, these two drugs are less likely to cause dizziness or a drop in blood pressure.

Finally, blockade of the brain's muscarinic receptors causes side effects such as dry mouth, constipation, blurred vision, difficulties getting your urine flow started, and a speeding up of the heart, even when you are resting. Because of these effects on the heart, the tricyclic medications in Table 20–2 with the strongest effects on these muscarinic receptors may not be advisable for patients with cardiac problems. Drugs with strong anticholinergic effects can also create problems with memory. Many patients have told me that they cannot remember a word they want to use, or they forget someone's name when they take these drugs. The memory effects are dose-related and should disappear when you stop taking the drug.

You can see that two of the tricyclic drugs in Table 20–2 (desipramine and nortriptyline) have relatively weak anticholinergic effects. These two drugs will be the least likely to cause side effects like dry mouth and forgetfulness. These two also tend to have weaker effects on the histaminic and alpha-adrenergic receptors. Because they have fewer side effects, they are among the most popular tricyclic antidepressants.

The effects of antidepressant drugs on these three brain receptor systems do not completely explain all their side effects. In the right-hand column, I have listed many of the more common or significant side effects for each drug. For example, you will see that some of them can cause skin rashes.

Table 20–2. Side Effects of Tricyclic Antidepressants[a]

Note: This list is not comprehensive. In general, side effects that occur in 5% or 10% or more of patients are listed, as well as rare but dangerous side effects.

Side Effect[b]	*Sedation and Weight Gain*[c]	*Light-Headedness and Dizziness*	*Blurred Vision, Constipation, Dry Mouth, Speeded Heart, Urinary Retention*	*Common or Significant Side Effects*
Brain Receptor	*histamine (H_1) receptors*	*alpha-adrenergic (α_1) receptors*	*muscarinic (M_1) receptors*	
amitriptyline (Elavil, Endep)	+++	+++	+++	dizziness; speeded heart; abnormal ECG; dry mouth; constipation; weight gain; trouble urinating; blurred vision; ringing in ears; sweating; weakness; headache; tremor; tiredness; insomnia; confusion
clomipramine (Anafranil)	++ to +++	+++	++ to +++	dizziness; speeded heart; abnormal ECG; dry mouth; upset stomach; loss of appetite; constipation; weight gain; trouble urinating; menstrual changes; disturbed sexual

				functioning; blurred vision; sweating; weakness; muscle cramps; tremor; tiredness; insomnia; anxiety; headache; rash; seizures
desipramine (Norpramin, Pertofrane)	+	+	+ to ++	dry mouth; rashes; agitation; anxiety; headache; insomnia; stimulation
doxepin (Adapin, Sinequan)	+++	+++	++ to +++	dizziness; speeded heart; dry mouth; constipation; weight gain, blurred vision; sweating; sleepiness
imipramine (Tofranil)	++	++ to +++	++ to +++	dizziness; speeded heart; abnormal ECG; dry mouth; constipation; weight gain; trouble urinating; blurred vision; sweating; weakness; headache; tiredness; insomnia; anxiety; stimulation; rash; seizures; sensitivity to light; stimulation
nortriptyline (Aventyl)	+ to ++	+	++	dry mouth; constipation; tremor; weakness; confusion; anxiety or stimulation

Table 20-2. Cont.

Side Effect[b]	*Sedation and Weight Gain*[c]	*Light-Headedness and Dizziness*	*Blurred Vision, Constipation, Dry Mouth, Speeded Heart, Urinary Retention*	*Common or Significant Side Effects*
Brain Receptor	*histamine (H_1) receptors*	*alpha-adrenergic (α_1) receptors*	*muscarinic (M_1) receptors*	
protriptyline (Vivactil)	0 to +	+ to ++	+++	dizziness; decreased or increased blood pressure; abnormal ECG; nausea; constipation; blurred vision; sweating; weakness; insomnia; stimulation; headache
trimipramine (Surmontil)	+++	++ to +++	++ to +++	dizziness; decreased or increased blood pressure; abnormal ECG; dry mouth; constipation; weight gain; blurred vision; sweating; weakness; headache; tremor; sleepiness; confusion; intolerance to heat or cold

[a]The + to +++ ratings in this table refer to the likelihood that a particular side effect will develop. The actual intensity of the side effect will vary among individuals and will also depend on how large the dose is. Reducing the dose can often reduce side effects without reducing effectiveness.

[b]Many side effects, if troublesome, can be minimized by a reduction in dosage. Side effects are usually greatest in the first few days and tend to disappear later.

[c]The drugs that are the most sedative may also have greater antianxiety effects. In other words, they may calm you and make you less nervous. When given at night, the sedative agents help reduce insomnia.

Some tricyclics, most notably clomipramine (Anafranil), can cause seizures, and so this drug would not be a good choice for individuals with epilepsy.

If you and your doctor are choosing one of the antidepressants listed in Table 20–2, you might want to consider the side-effect profile when making your choice. This is because all these medications are comparably effective, so their side effects may be the most important criteria for you in deciding among them. So if you are having trouble sleeping at night, one of the more sedative antidepressants may be useful. These sedative agents are also somewhat calming and so they might be helpful if you are experiencing anxiety.

Many of the side effects of the tricyclic antidepressants listed in Table 20–2 occur in the first few days. With the exception of dry mouth and weight gain, these side effects frequently diminish as you become accustomed to the drug. If you can simply put up with the side effects, many of them will disappear after a few days. If the effects are strong enough to make you uncomfortable, your doctor may decide to reduce the dose, which usually helps.

Some side effects suggest you are taking an excessive dose. These include difficulty in urination, blurred vision, confusion, severe tremor, substantial dizziness, or increased sweating. A dose reduction for such symptoms is definitely indicated. A stool softener or laxative can help if constipation develops. As noted above, light-headedness is most likely to occur when you stand up suddenly, because the blood flow to your brain is temporarily diminished. The dizzy feeling usually persists for only a few seconds. If you get up more carefully and slowly, or if you exercise your legs before standing (by tightening and then relaxing your leg muscles, as when you run in place), this should not be a problem. The movement of your legs causes your leg muscles to "pump" the blood back up to your brain. Support stockings can also help.

Some patients describe feeling "strange," "spaced out," or "unreal" for several days when they first start taking a tricyclic antidepressant. In my experience, one of the tricyclics called doxepin (Sinequan) seems more likely to cause this "spaced out" effect. When patients report feeling strange on the first day or two of taking an antidepressant, I usually advise them to stick with it. In nearly all cases the sensation disappears completely within a few days.

If you give patients sugar pills (placebos) they think are antidepressants, they will also report side effects that are similar to the side effects reported by patients taking antidepressants. For example, in one study, 25 percent of the patients taking clomipramine reported difficulties sleeping, so you might conclude that this drug causes insomnia in a quarter of those who take it. However, 15 percent of the patients in the same study who received only placebo also reported insomnia. So the likelihood of insomnia actually caused by clomipramine would be 25 percent minus 15 percent, or 10 percent. Clearly, this side effect is "real," but it is somewhat less common than you might at first expect.

Such studies indicate that many "side effects" may not actually be caused by the medication you are taking. Some side effects may result from fears about the medication, or from the depression itself, or from other stressful events in your life such as a conflict with your spouse, rather than from the drug itself.

Side Effects of the Tetracyclic Antidepressants. You can see in Table 20–3 on pages 536–537 that the side effects of the tetracyclic antidepressants are similar to those of the tricyclic antidepressants. However, they have some special side effects of their own that should be considered if you are taking one of these drugs. Maprotiline (Ludiomil) appears to be more likely than the eight tricyclic antidepressant drugs to cause seizures, a particularly troublesome side effect. Although the likelihood of seizures is low, patients with a history of seizures or head trauma should probably avoid this drug. Recent studies suggest that the likelihood of seizures with maprotiline is significantly greater when the dose is increased too rapidly, or when patients are kept on higher-than-recommended doses (225 to 400 mg per day) for more than six weeks.[16] Therefore the manufacturer has suggested that maprotiline should be started and increased very slowly, and that the dose should be maintained at no more than 175 mg per day if patients take this drug for more than six weeks.

Amoxapine (Asendin) has a distinct and troublesome type of side effect not shared with most other antidepressants. This is because one of its metabolites blocks dopamine receptors in the brain, much like antipsychotic drugs such as chlorpromazine (Thorazine) and many others which are used in the treatment of schizophrenia. Thus, patients who take amoxapine can in rare instances develop some of the same types of side effects that occur in patients who take antipsychotic drugs. For example, women may experience galactorrhea (the production of milk by the breast.) Any of several so-called ''extrapyramidal'' reactions can also develop. One of them, called akathisia, is a motor restlessness syndrome. This is an unusual kind of muscular ''itchiness''—your arms or legs feel intensely restless and so you cannot sit still. You feel the compulsion to keep moving or pacing about. Akathisia is uncomfortable but not dangerous.

In rare instances amoxapine can also cause symptoms that mimic Parkinson's disease. Symptoms include passive inactivity, a ''pill-rolling'' tremor of the thumb and fingers while at rest, decreased swinging of the arms when walking, stiffness, stooped posture, and others. If these symptoms develop, notify your doctor right away. She or he will probably want you to stop the drug and try an alternative medication. Although alarming, these symptoms are not dangerous and should disappear when you stop taking the amoxapine.

However, a more serious side effect of amoxapine (as well as many other antipsychotic drugs) is called ''tardive dyskinesia.'' Patients with tardive dyskinesia develop involuntary, repetitious movements of the face, especially the

Table 20–3. Side Effects of Tetracyclic Antidepressants[a]

Note: This list is not comprehensive. In general, side effects that occur in 5% or 10% or more of patients are listed, as well as rare but dangerous side effects.

Side Effect	*Sedation and Weight Gain*	*Light-Headedness and Dizziness*	*Blurred Vision, Constipation, Dry Mouth, Speeded Heart, Urinary Retention*	*Common or Significant Side Effects*
Brain Receptor	*histamine* (H_1) *receptors*	*alpha-adrenergic* (α_1) *receptors*	*muscarinic* (M_1) *receptors*	
amoxapine (Asendin)	++	++	+ to ++	dizziness; speeded heart; dry mouth; stomach upset; constipation; trouble urinating; blurred vision; rashes; tremor; tiredness; insomnia; EPS[b]; lactation; restlessness; excessive stimulation; tardive dyskinesia; galactorrhea; NMS[c]
maprotiline (Ludiomil)	++	+	+	dry mouth; constipation; weight gain; blurred vision; rashes; sleepiness; seizures; stimulation; sensitivity to light; edema (swelling of ankles)

[a]The + to +++ ratings in this table refer to the likelihood that a particular side effect will develop. The actual intensity of the side effect will vary among individuals and will also depend on how large the dose is. Reducing the dose can often reduce side effects without reducing effectiveness.

[b]EPS = extrapyramidal symptoms (described in text) including akathisia and dystonic reactions and tardive dyskinesia.

[c]NMS = neuroleptic malignant syndrome. This is a potentially fatal reaction that also occurs in reaction to antipsychotic drugs (also known as neuroleptics). The symptoms include increased fever, rigid muscles, altered mental status, irregular pulse or blood pressure, rapid heart, profuse sweating, and abnormal heart rhythms.

lips and tongue. The abnormal movements can also involve the arms and legs. Once it begins, tardive dyskinesia sometimes becomes irreversible or difficult to treat. The risk appears to be the highest among elderly women, but it can occur with any patient. The risk of tardive dyskinesia also increases the longer you have been on the drug, but it can develop after only a brief period of treatment at a low dose.

Finally, as if that weren't enough to frighten you, amoxapine can, in rare cases, cause a dreaded complication known as neuroleptic malignant syndrome, or NMS. NMS consists of high fever, delirium, and muscle rigidity, along with changes in blood pressure, heart rate and rhythm, and sometimes death. All these risks should obviously be carefully balanced against any potential benefits of amoxapine; it may sometimes be difficult to justify using this medication when so many equally effective and safer drugs are readily available.

Tricyclic and Tetracyclic Antidepressant (TCA) Drug Interactions. I described the problem of drug interactions in Chapter 19. Briefly, when you are taking more than one drug, there is a possibility that the drugs may interact in ways that will be detrimental to you. One drug may cause the level of the second drug to increase or decrease in your blood. As a result, the second drug may cause excessive side effects (if its blood level gets too high) or it may become ineffective (if its blood level falls). In addition, sometimes the interaction of two drugs can lead to toxic reactions that are quite dangerous.

A number of drug interactions for the tricyclic and tetracyclic antidepressants are listed in Table 20–4 on pages 540–547. This list is not comprehensive, but it does include many of the more common or important interactions. If you are taking any other medications along with a TCA, it would be wise to review this table. Note that both prescription and nonprescription drugs are listed, including many psychiatric and nonpsychiatric drugs as well. In addition, you should ask your physician and pharmacist if there are any drug interactions among the drugs you are taking.

You can see in Table 20–4 that smoking cigarettes and drinking alcohol can both cause the blood level of a TCA to fall, thus reducing the likelihood that the drug will be effective. Your doctor may need to do a blood test to find out if your blood level is adequate. In addition, alcohol can enhance the sedative effects of the tricyclic antidepressants, a combination that can be quite hazardous if you are driving or operating dangerous machinery.

Certain antidepressants can be particularly hazardous for individuals with specific medical conditions. In particular, the tricyclics can be dangerous to individuals with cardiovascular disease, including those with a previous heart attack, abnormalities in heart rhythm, or high blood pressure. Special precautions should also be taken for individuals with thyroid disease. Make sure you mention any medical problems you have to the doctor who is prescribing your antidepressant so that she or he can take the proper precautions.

As noted above, several of the tricyclic and tetracyclic antidepressants can cause seizures in rare instances. An incidence of seizures as high as 1 percent to 3 percent has been reported with clomipramine, imipramine, and maprotiline.[17] These estimates may be overly high. At any rate, the risk can be reduced by making sure the dose is not excessive and by raising the dose gradually. Nevertheless, these drugs should be used with caution, if at all, in individuals with a history of seizure disorders, head trauma, or other neurologic disorders associated with seizures. In addition, caution should be used if these drugs are combined with other drugs that can lower the seizure threshold, such as the major tranquilizers (neuroleptics) and others. Rapid withdrawal from sedative agents, such as alcohol, minor tranquilizers, and barbiturates can also trigger seizures, and so clomipramine, imipramine, and maprotiline should be used with great caution in combination with these agents.

Table 20–4. Drug Interaction Guide for Tricyclic and Tetracyclic Antidepressants (TCAs)[a]

Note: The drugs in the left-hand column can interact with TCAs. The comments describe the types of interactions. This list is not exhaustive; new information about drug interactions comes out frequently. If you are taking a TCA and any other medication, ask your doctor and pharmacist if there are any drug interactions you should know about.

Antidepresssants

Drug	*Comment*
tricyclic and tetracyclic antidepressants (TCAs can interact with other TCAs)	desipramine causes an ↑ in other TCAs—abnormal heart rhythms can result
SSRIs	TCA levels can ↑ (as much as 2- to 10-fold); abnormal heart rhythms can result; SSRI levels can also ↑
MAOIs	serotonin syndrome [b] [especially clomipramine (Anafranil)]; low blood pressure; hypertensive reactions
serotonin antagonists, including trazodone (Desyrel) and nefazodone (Serzone)	nefazodone may cause low blood pressure
bupropion (Wellbutrin)	↑ in risk of seizures; extreme caution required
venlafaxine (Effexor)	probably okay; in theory TCA could cause ↑ in venlafaxine blood levels
mirtazapine (Remeron)	information not yet available

Antibiotics

Drug	*Comment*
chloramphenicol (Chloromycetin)	TCA levels and toxicity may ↑
doxycycline (Vibramycin)	TCA levels and effectiveness may ↓
isoniazid (INH, Nydrazid)	TCA levels and toxicity may ↑

Antifungal Agents	
Drug	*Comment*
imidazoles such as fluconazole (Diflucan), itraconazole (Sporanox), ketoconazole (Nizoral) and miconazole (Monistat vaginal suppositories or cream)	TCA levels may ↑, especially nortriptyline
griseofulvin (Fulvicin)	TCA levels may ↓

Diabetes Medications	
Drug	*Comment*
insulin	greater-than-expected drop in blood sugar
oral hypoglycemic drugs	greater-than-expected drop in blood sugar

Medical Conditions	
Condition	*Comment*
glaucoma	highly anticholinergic TCA can trigger attacks of narrow-angle glaucoma; symptoms include eye pain, blurred vision, and halos around lights
heart disease	use TCA with extreme caution; may trigger abnormal heart rhythms
liver disease	use TCA with caution; the metabolism by the liver may be impaired, with excessively high blood levels and increased side effects and toxic effects
seizure disorder	use TCA with caution; TCA may cause ↑ in seizures (TCA lowers the seizure "threshold")
thyroid disease	use TCA with caution in patients with thyroid disease or those taking thyroid medication; may trigger abnormal heart rhythms

Medications for Abnormal Heart Rhythms	
Drug	*Comment*
disopyramide (Norpace)	abnormal heart rhythms
epinephrine	TCA may enhance the effects, leading to rapid heart, abnormal heart rhythms, and ↑ in BP
quinidine	blood levels of quinidine and TCA may ↑; abnormal heart rhythms and weakened heart muscle can lead to congestive heart failure

Medications for High Blood Pressure	
Drug	*Comment*
beta-blockers such as propranolol (Inderal)	beta-blockers may cause increased depression; TCA may cause greater-than-expected drop in BP
clonidine (Catapres)	TCA [e.g., desipramine (Norpramin)] may reduce effectiveness of clonidine because blood levels ↓
calcium channel blockers	BP drop may be greater than expected
guanethidine (Ismelin)	may lose antihypertensive effect when combined with TCA [e.g., desipramine (Norpramin)]
methyldopa (Aldomet)	BP drop may be greater than expected, especially with amitriptyline (Elavil); some TCAs [e.g., desipramine (Norpramin)] may reduce the antihypertensive effect
prazosin (Minipress)	BP may ↑ because levels of prazosin may ↓

Drug	*Comment*
reserpine (Serpasil)	may cause greater-than-expected drop in BP; may also cause excessive stimulation
thiazide diuretics such as hydrochlorothiazide (Dyazide)	blood-pressure drop may be greater than expected; effects of TCA may increase

Medications for Low Blood Pressure (for patients in shock)

Drug	*Comment*
epinephrine	TCA may enhance the effects, leading to rapid heart, abnormal heart rhythms, and ↑ in BP

Mood Stabilizers and Anticonvulsants

Drug	*Comment*
carbamazepine (Tegretol)	blood levels of TCA and carbamazepine may ↓; TCA can make seizures more likely
lithium (Eskalith)	may enhance antidepressant effects
phenytoin (Dilantin)	blood levels of TCA may ↑ or ↓; TCA can make seizures more likely
valproic acid (Depakene)	↑ in blood levels of amitriptyline (Elavil) and valproic acid

Pain Medications and Anesthetics

Drug	*Comment*
acetaminophen (Tylenol)	TCA levels may ↑; acetaminophen levels may ↓
aspirin	TCA levels may ↑
halothane (Fluothane)	TCA levels may ↑; TCA with strong anticholinergic effects may cause abnormal heart rhythms

Pain Medications and Anesthetics cont.	
Drug	*Comment*
cyclobenzaprine (Flexeril) (a muscle relaxant used to treat muscle spasm)	may cause abnormal heart rhythms
methadone (Dolophine)	may have greater-than-expected narcotic effect; for example, desipramine (Norpramin) may double the blood level of methadone
meperidine (Demerol)	greater-than-expected narcotic effect; lower doses of meperidine or another painkiller may be needed
morphine (MS Contin)	greater-than-expected narcotic effect and sedation; TCA levels may ↓
pancuronium (Pavulon)	abnormal heart rhythms, especially TCA with strong anticholinergic effects

Sedatives and Tranquilizers	
Drug	*Comment*
alcohol	May have enhanced sedative effects. This could be hazardous when driving or operating dangerous machinery. May cause TCA levels to ↓
barbiturates (such as phenobarbital)	enhanced sedative effects; may cause TCA levels to ↓
buspirone (BuSpar)	enhanced sedative effects as described above
chloral hydrate (Noctec)	TCA levels may ↓
ethchlorvynol (Placidyl)	Temporary mental confusion has been reported when combined with amitriptyline (Elavil), but could conceivably occur with other TCAs as well

Drug	*Comment*
major tranquilizers (neuroleptics)	levels of TCA and phenothiazine neuroleptics [such as chlorpromazine (Thorazine)] may ↑, leading to more side effects and greater potency; abnormal heart rhythms have been observed with thioridazine (Mellaril), clozapine (Clozaril), and pimozide (Orap)
minor tranquilizers (neuroleptics)	enhanced sedative effects

Stimulants (Pep Pills) and Street Drugs

Drug	*Comment*
amphetamines (''speed'' or ''crank'') **cocaine** **benzedrine** **benzphetamine (Didrex)** **dextroamphetamine (Dexedrine)** **methamphetamine (Desoxyn)** **methylphenidate (Ritalin)**	These drugs may boost the blood levels and effects of some TCA [(e.g., imipramine (Tofranil), clomipramine (Anafranil), desipramine (Norpramin)] and vice versa; abnormal heart rhythms and increased blood pressure have been observed with cocaine, but seem possible when any stimulants are combined with TCA

Weight-Loss and Appetite-Suppression Medications

Drug	*Comment*
fenfluramine (Pondimin)	Possible serotonin syndrome when combined with clomipramine; increased TCA levels

Other Medications	
Drug	*Comment*
antihistamines	increased drowsiness; it is safer to use antihistamines that are not sedative
acetazolamide (Diamox)	TCA blood levels may ↑; blood pressure may fall
birth control pills and other medications containing estrogen	TCA blood levels may ↑, with greater side effects; higher doses of estrogen may reduce the effects of TCA
caffeine (in coffee, tea, soda, chocolate)	TCA blood levels may ↑
charcoal tablets	TCA blood levels may ↓ due to poor absorption from the stomach and intestinal tract
cholestyramine (Questran)	TCA blood levels may ↓
cimetidine (Tagamet)	TCA blood levels may ↑ (greater side effects)
disulfiram (Antabuse)	TCA blood levels may ↑ (greater side effects); in two reported cases, disulfiram plus amitriptyline (Elavil) caused a severe brain reaction (organic brain syndrome) with mental confusion and disorientation
ephedrine (can be found in Bronkaid, Marax, Primatene, Quadrinal, Vicks Vatronol nose drops, and several other asthma and cold medications)	TCA may block the ↑ in BP ordinarily caused by ephedrine; ephedrine levels and effects may ↓
high fiber diet	TCA blood levels may ↓ due to poor absorption from the stomach and intestinal tract
liothyronine (T3, Cytomel)	can enhance the effects of TCA; abnormal heart rhythms can result; TCA blood levels may ↑

Drug	*Comment*
prochlorperazine (Compazine)	TCA blood levels may ↑ with increased side effects and toxic effects
psyllium (Metamucil)	TCA blood levels may ↓ due to poor absorption from the stomach and intestinal tract
scopolamine (Transderm)	may cause ↑ in TCA blood levels
L-dopa (Sinemet)	absorption of TCA from the stomach and intestinal tract into the blood may ↓; effects of both TCA and L-dopa may ↓
theophylline (Bronkaid)	TCA blood levels may ↑
tobacco (smoking)	TCA blood levels may ↓

[a]Information in this table was obtained from several sources including the *Manual of Clinical Psychopharmacology*[1] and *Psychotropic Drugs Fast Facts.*[17] These excellent references are highly recommended.

[b]This is a dangerous and potentially fatal syndrome which includes rapid changes in vital signs (fever, oscillations in blood pressure), sweating, nausea, vomiting, rigid muscles, myoclonus, agitation, delirium, seizures, and coma.

Selective Serotonin Reuptake Inhibitors (SSRIs)

Currently, the most popular antidepressant drugs are the selective serotonin reuptake inhibitors, or SSRIs. At this time, five SSRIs are prescribed in the United States. These include citalopram (Celexa), the newest SSRI which was released in the U.S. in 1998, fluoxetine (Prozac), the first SSRI which was released in 1988, and fluvoxamine (Luvox), paroxetine (Paxil), and sertraline (Zoloft). The effects of these SSRIs on the brain are much more specific and selective than the older tricyclic and tetracyclic drugs discussed above. Instead of interacting with many different systems in the brain, these drugs have selective effects on nerves that use serotonin as a transmitter substance.

When it first appeared on the market, there was a great deal of excitement about Prozac because it was chemically quite different from the older antidepressants. Unlike the tricyclic and tetracyclic drugs, it has specific effects on the serotonergic nerves in the brain. Since a serotonin deficiency was hypothesized to be the cause of depression, it was hoped that Prozac would be dramatically more effective than the tricyclic and tetracyclic drugs which seemed to affect so many different systems in the brain in a less specific manner. It was also expected that Prozac (and the other SSRIs) would have fewer side effects than the tricyclic and tetracyclic drugs. This is because Prozac does not have such strong effects on the histaminic, alpha-adrenergic, and muscarinic receptors.

Only one of these two hopes was fulfilled. Prozac and the other four SSRIs do cause significantly fewer side effects than the tricyclic and tetracyclic antidepressants and are more pleasant to take. For example, they are less likely to cause sleepiness, weight gain, dry mouth, dizziness, and so on. They are also much safer since they are less likely to have adverse effects on the heart, and they are much less likely to result in death if a patient intentionally or unintentionally takes an overdose. The biochemists who created these new drugs deserve credit in this regard.

Unfortunately, the SSRIs are not more effective than the older drugs. As many as 60 percent to 70 percent of depressed patients will improve when treated with SSRIs, and these percentages are no better than the older drugs. Among chronically depressed patients, the probability of responding appears to be lower. The SSRIs also appear to be slightly less effective than the older tricyclic antidepressants for more severely depressed patients. In addition, the amount of improvement is often only partial—the patient may become less depressed, but may not return to full self-esteem and joyous daily living. This is a problem for all the antidepressants, and not just the SSRIs. Although they are no more effective, the SSRIs are dramatically more expensive than the older drugs. In addition, the SSRIs have some new and different side effects described below that were not publicized when they were first released.

Because of their favorable safety record and diminished side effects, though, the SSRIs have truly captured the antidepressant market. More money was spent on Prozac in 1995 ($2.5 billion) than was spent on all other antidepressants in 1991 ($2.0 billion). One reason for the upsurge in popularity is that primary care physicians now feel comfortable prescribing antidepressants because the SSRIs are so safe. As a result, many depressed patients who would not think of going to a psychiatrist or psychologist receive SSRIs from their family physicians.

Because the SSRIs are used so widely and because they have received so much media attention, many people believe they are incredibly powerful and almost magically effective. But this is simply not the case, as noted above. For some depressed people, the SSRIs can be very effective. For many others, they are only somewhat effective. And often they do not seem to have any antidepressant effects at all. It is the same story with all of our currently available antidepressants—they are valuable tools to fight depression but they are often not the entire answer and they are certainly not a panacea for what ails you.

The fact that the SSRIs are not more effective than the older drugs has caused scientists to reconsider the validity of the "serotonin" theory of depression which I described in Chapter 17. You will recall that according to this theory, a deficiency of serotonin in the brain causes depression, and an increase in serotonin should reverse it. If this theory were valid, the SSRIs should cause depressed patients to become undepressed almost immediately—but Prozac can take as many as five to eight weeks to become effective. Regardless of what causes depression or why antidepressants work, the SSRIs have been helpful to many depressed individuals.

Doses of SSRIs. The doses of the five SSRIs are listed in Table 20–1 on page 520. Unlike the older antidepressants, which are often prescribed in doses that are too low, the SSRIs are often prescribed in doses that are unnecessarily high. Because they have so few side effects, doctors feel comfortable prescribing high doses and may prescribe more than is really needed. For example, although 20 mg to 80 mg per day was the dose range initially recommended for Prozac, a single dose of 10 mg per day will be sufficient for many patients. Once they are feeling better, many patients need only 5 mg per day, or even less. These smaller doses are much less expensive and will produce fewer side effects.

These low doses are effective because Prozac stays in the body for a much longer period of time than most other drugs—as long as several weeks. When you take Prozac, your blood level continues to increase each day because the Prozac leaves your body so slowly. After a while your blood level becomes quite high. This is why you may need only a tiny dose if you have been taking Prozac for several weeks or more.

To understand this better, let's go back to the bathtub analogy I introduced in Chapter 19 to explain drug interactions. Let's imagine that the Prozac you are taking is like the water going into the bathtub, but the hole in the bottom of the tub is tiny. Over time, the water level increases, because more water goes into the tub than goes out. The water level can be compared to the level of the Prozac in your blood. After four to five weeks, the water level finally gets up to the correct therapeutic range. Now you can turn the faucet down quite a bit so that the level in the tub does not continue to increase and overflow. This would be analogous to reducing the dose of Prozac after you have been on it for several weeks. Paradoxically, you are now taking much smaller doses than when you first started taking the Prozac, but your blood level is far higher.

Technically, we say that "steady state" has been reached. Steady state means that the blood level remains more or less constant, because the amount you take each day is similar to the amount that your body eliminates each day. The other four SSRIs do not have this property, because they leave the body much faster than Prozac. You generally cannot reduce the doses after several weeks.

The effectiveness of very low doses of Prozac is now

well known among the psychiatric profession, but I first learned this from my patients soon after Prozac was released onto the market. Many patients reported that after they had been on Prozac for a month or two, they seemed to need only tiny doses, often as little as one tenth of a pill per day, and sometimes even less. At first I thought these patients had overly lively imaginations, but soon many patients were reporting the same thing. I advised them to take one Prozac pill, grind it up, and dissolve it in water or apple juice to store in the refrigerator. Then they adjusted their dose of Prozac by drinking a certain amount of the fluid each day. So, for example, if you have dissolved one 20 mg pill in some apple juice and you drink one tenth of the juice each day, this would correspond to a dose of 2 mg per day. But if you try this, make sure you label the juice clearly so that no one drinks your Prozac for breakfast! Also, make sure you talk it over with your doctor and that she or he approves of what you are doing.

It is also important for you to know that after you stop taking Prozac, it will stay in your body for a long time because it leaves your body so slowly. This would be like a bathtub that takes an extraordinarily long time to empty out after you pull the plug because the drain is plugged up. After you are no longer taking the Prozac, significant levels will remain in your blood for as many as five weeks or more before the drug is entirely cleared out of your system. Many medications can be dangerous to mix with Prozac. You must not take these specific medications until you have been off the Prozac entirely for at least five weeks. For example, tranylcypromine (Parnate) is an antidepressant known as an MAO inhibitor that will be discussed below. Tranylcypromine (as well as other MAO inhibitors) can cause dangerous and potentially fatal reactions if mixed with Prozac. After you stop taking Prozac, a delay of at least five to eight weeks will be necessary before you can safely start taking tranylcypromine.

The other SSRIs, such as citalopram (Celexa), fluvoxamine (Luvox), sertraline (Zoloft) and paroxetine (Paxil), leave the body more rapidly than Prozac but they are still metabolized rather slowly. For example, if you stop taking one of these drugs, it will take your body approximately one day to get rid of one half of the amount in your body. It will take approximately four to seven days for most or all of the drug to leave your body. This is much faster than Prozac. Therefore, these other SSRIs drugs do not build up to such high levels in your blood after you have been taking them for more than a few weeks. Because they go in and out of your blood more rapidly, they are usually taken several times per day, whereas Prozac can be taken once a day.

Age can also influence your dose requirements if you are taking an SSRI. For example, levels of citalopram (Celexa), fluoxetine (Prozac), and paroxetine (Paxil) are approximately twice as high in older individuals (over 65 years of age) than in younger individuals. If you are taking one of these drugs and you are over 65, you will need a lower dose. Blood levels of sertraline (Zoloft) are also higher in older individuals, although the differences are not as pronounced. In contrast, fluvoxamine (Luvox) blood levels do not seem to be affected by age.

Sometimes gender can play a role as well. For example, the blood levels of fluoxetine (Prozac) are 40 percent to 50 percent lower in males than in females. Similarly, young men develop blood levels of sertraline (Zoloft) that are 30 percent to 40 percent lower, on the average, than young women. Men may need relatively higher doses of these drugs, whereas women may need relatively lower doses.

Health problems can also influence your dose requirements. Individuals with liver, kidney, or heart disease may not get rid of SSRIs as rapidly, and so smaller doses may be needed. Make sure you ask your doctor about this if you are being treated for a liver, kidney or heart ailment.

Side Effects of SSRIs. The most frequent side effects of the five SSRIs are listed in Table 20–5 on pages 553–554. As noted above, the side effects of the SSRIs are milder than the older drugs, and this is the reason for their enor-

Table 20–5. Side Effects of SSRI Antidepressants

Note: This table was adapted with permission from Preskorn[23] and from the prescribing information for citalopram. Only the more common side effects of each drug are listed. The numbers in the table represent the percent of patients receiving the drug who reported each side effect minus the percent of patients on placebo who reported the same side effect. For example, if 20% of patients on Prozac reported nervousness as a side effect, while 10% of patients on placebo reported this same side effect, the number 10% would appear in this chart. This would be an estimate of the "true" nervousness actually caused by Prozac. For each side effect, the drug or drugs with the highest percentages are indicated in boldface.

	fluoxetine (Prozac)	*fluvoxamine (Luvox)*	*paroxetine (Paxil)*	*sertraline (Zoloft)*	*citalopram (Celexa)*
# of patients on drug	1730	222	421	861	1063
# of patients on placebo	799	192	421	853	466
General Symptoms					
headache	**5%**	**3%**	0%	1%	—[a]
dizziness	4%	1%	**8%**	5%	—
nervousness	**10%**	**8%**	5%	4%	1%
tiredness	6%	**17%**	**14%**	8%	8%
difficulty sleeping	**7%**	4%	**7%**	**8%**	1%
weak or fatigued muscles	6%	6%	**10%**	3%	—
tremor	**6%**	**6%**	**6%**	**8%**	2%
Mouth, Stomach, and Intestinal Tract					
dry mouth	4%	2%	**6%**	**7%**	**6%**
loss of appetite	**7%**	**9%**	5%	1%	2%
nausea or upset stomach	**11%**	**26%**	**16%**	**14%**	7%
diarrhea	5%	0%	4%	**8%**	3%
constipation	1%	**11%**	5%	2%	—

	fluoxetine (Prozac)	*fluvoxamine (Luvox)*	*paroxetine (Paxil)*	*sertraline (Zoloft)*	*citalopram (Celexa)*
Other					
excessive sweating	5%	0%	9%	6%	2%
Sexual					
loss of interest in sex delayed or no orgasm	Specific comparative data on the sexual side effects of the SSRIs were not available. However, it appears that 30% to 40% of patients receiving SSRIs do experience some sexual side effects.[b]				

[a]A dash means that the incidence of this side effect was not greater than placebo.
[b]During the initial drug testing studies patients were not explicitly asked about sexual side effects. Consequently, the estimates of sexual side effects in the *PDR* are too low.

mous popularity. They are less likely than the tricyclic antidepressants to cause dry mouth, constipation, or dizziness. They do not stimulate the appetite when you first start taking them; if anything, some patients taking SSRIs lose weight in the beginning. Unfortunately, when the SSRIs are taken for a prolonged period of time, their side effects sometimes increase. For example, some patients taking these agents report increases in appetite and weight gain after a while, even though they lost weight at first.

Some of the most common and troublesome side effects of the SSRIs include nausea, diarrhea, cramping, heartburn, and other signs of stomach upset. Approximately 20 percent to 30 percent of patients reported these symptoms in the earliest studies with the SSRIs.[18] You will see in Table 20–5 that fluvoxamine (Luvox) is the most likely to cause constipation, whereas sertraline (Zoloft) is more likely to cause diarrhea. Patients taking paroxetine (Paxil) and sertraline (Zoloft) are more likely to complain of a dry mouth because of the anticholinergic effects of these drugs. In some studies, as many as 20 percent of the patients taking paroxetine (Paxil) reported dry mouth. (However, the percentages in the table are much lower because the placebo effects have been subtracted out.)

Most of these effects on the stomach and intestinal tract tend to occur in the first week or two and then disappear as the body adjusts to the medicine. In addition, if you start the SSRI at a low dose and then increase the dose gradually, these side effects are less likely to occur. It can also help if you take the medication with meals. (The tricyclic and tetracyclic drugs discussed in the previous section can also be taken with meals to minimize any adverse effects on the stomach and gastrointestinal tract.)

The SSRI drugs can occasionally cause headaches when you first start to take them. In Table 20–5 the rates for headache seem to be the highest for fluoxetine (Prozac) and fluvoxamine (Luvox); in contrast, the rates for citalopram (Celexa), paroxetine (Paxil), and sertraline (Zoloft) appear to be no greater than the rates of headaches reported by patients who take placebos. Excessive sweating has also been reported, especially with paroxetine (Paxil), but this is not usually severe. Patients taking high doses of the SSRIs may also complain of tremor, and this side effect seems to be equally common among all of the SSRI drugs.

Although initially reported as a "rare" side effect, it is now clear that delayed time to orgasm is quite common for men and women taking SSRIs. Some patients also complain of a loss of interest in sex or an inability to achieve an erection. These side effects were reported in fewer than 5 percent of patients during the premarketing research trials. However, now that the drugs are widely used, it has become clear that these side effects are far more common than reported in clinical trials and can occur in 30 percent or more of patients. The sexual side effects may be a reasonable trade-off if the drug helps you overcome your depression. Keep in mind that a loss of interest in sex can also be a symptom of depression itself. In addition, you will probably not need to stay on the drug indefinitely. Once you are feeling better and you stop taking the SSRI, your sexual functioning should return to normal.

You might wonder why these side effects were not noted in the premarketing research studies. At the 1998 Stanford

Psychopharmacology Conference, one of the speakers jokingly mentioned that the drug companies seem to have a "don't ask, don't tell" policy about certain kinds of adverse effects, including sexual side effects. I guess the idea is that what you don't know won't hurt you. I think this policy is unfortunate, because the FDA (and potential consumers) may be given an overly rosy picture about the effectiveness, side effect profile, and safety of a new drug. After the drug has been in widespread use for several years, a different picture often emerges.

The effects on sex are so predictable that one of these drugs, paroxetine (Paxil) is now recognized as an effective treatment for men who experience premature ejaculation (having orgasms too rapidly during sex). Some people do not experience a delayed orgasm on SSRIs. Others experience it but are not bothered, and some actually view it as a benefit. What is important is that if this feels like a problem for you, you should discuss it with your doctor before discontinuing the medication on your own. It might be possible to reduce the dose without a loss of the antidepressant effects.

Several drugs can be combined with an SSRI in an attempt to combat the sexual difficulties. Four which show promise include bupropion (Wellbutrin, in doses of up to 225 mg to 300 mg per day), buspirone (BuSpar; 15 to 30 mg per day), yohimbine (5 mg three times daily), or amantadine (100 mg three times daily).

Citalopram (Celexa), one of the newest SSRIs on the American market, may have fewer sexual side effects than the other SSRIs. You can see in Table 20–5 that it does appear to have fewer side effects in general than the other four SSRIs. In addition, there is the hope that it will be more effective for severe depressions than the SSRIs. It will be interesting to see if citalopram (Celexa) is more effective and does actually have fewer side effects after the drug has been in widespread use for a period of time. Sometimes marketing claims when drugs are first released turn out not to be supported by clinical experience or by subsequent research by independent investigators.

Among the SSRIs, fluoxetine (Prozac) appears to be the most activating (stimulating), although fluvoxamine (Luvox) seems almost as likely to cause this side effect. Because fluoxetine (Prozac) is stimulating, it is sometimes given in the morning and at noon, rather than at bedtime. The stimulation can often be a benefit to depressed patients who feel tired, sluggish, and unmotivated. On the other hand, fluoxetine (Prozac) and fluvoxamine (Luvox) can also cause anxiety or jitteriness in as many as 10 percent to 20 percent of patients. These side effects can sometimes create additional difficulties for depressed patients who already have these kinds of symptoms.

The stimulating effects of fluoxetine (Prozac) are not necessarily bad, even for anxious patients. Anxiety and depression nearly always go hand in hand to a certain extent, and many patients need treatment for both kinds of problems. Patients with significant anxiety, such as chronic worrying, panic attacks, or agoraphobia, are often the ones who complain that fluoxetine (Prozac) makes them feel more nervous initially. I often tell these patients that the nervousness they feel is a good thing, because it shows the drug is working in the brain. I encourage them to stick with it, because in a few weeks or less they may notice a significant improvement in their depression as well as their anxiety. Most anxious patients have been able to stick with the fluoxetine (Prozac), and the predicted improvement often does occur. This illustrates how a positive attitude can sometimes help patients overcome drug side effects.

Although any of the SSRIs can cause trouble with sleeping, not all of them are as stimulating as fluoxetine (Prozac). In fact, paroxetine (Paxil) and fluvoxamine (Luvox) can be quite sedating for some patients. In other words, these drugs will tend to relax or tire you, instead of stimulating you the way fluoxetine (Prozac) does. In fact, paroxetine (Paxil) is sometimes given two hours before bedtime so that the maximum sleepiness will occur at the time that you ordinarily go to sleep. Paroxetine (Paxil) or fluvoxamine (Luvox) might be good choices if insomnia is a major aspect of your depression. Note, however, that patients taking paroxetine (Paxil) are also somewhat more likely to complain of weak or fatigued muscles. Citalopram (Celexa) and sertraline (Zoloft) appear to be halfway in-between—they do not typically cause excessive stimulation or sedation, but are more neutral in this respect.

In the section below on serotonin antagonists, I will describe an antidepressant called trazodone (trade name Desyrel) which has calming, sedative properties. Trazodone can be given in small doses (50 to 100 mg at bedtime) to patients who are taking SSRIs. This has three potential benefits: (1) the calming effects of trazodone will reduce the nervousness caused by the SSRIs; (2) the trazodone can be given at bedtime to improve sleep; (3) trazodone may sometimes boost the antidepressant effects of the SSRI and increase the likelihood of recovery.

In spite of these advantages, I usually try to treat patients with one drug at a time. This avoids any extra side effects and minimizes the possibility of adverse drug interactions. In my experience, treatment with one drug at a time is usually successful. If you reduce the dose of any SSRI, you can often minimize the side effects without having to add additional drugs. I will address the problem of using more than one drug at the same time toward the end of this chapter.

For example, if you are starting fluoxetine (Prozac) and you are bothered by nervousness, insomnia, or upset stomach, you can take a lower dose and increase the dose more gradually. In addition, if you have been on fluoxetine (Prozac) for several weeks or more, there is an excellent chance you can reduce the dose, often quite dramatically. This will often minimize the side effects without interfering with the antidepressant effects of this drug. As noted above, this is because levels of fluoxetine (Prozac) build up after a period of time, so the same dose may produce far more side effects because your blood level has become so much higher. There is really no need for large doses or excessively high blood levels of any of the SSRIs, because low doses have been shown to be just as effective as high doses.

SSRI Drug Interactions. A number of common drug interactions for the SSRIs are listed in Table 20–6 on pages 560–563. You will see in Table 20–6 that many other psychiatric drugs can interact with the SSRIs, including antidepressants, major and minor tranquilizers, and mood stabilizers. Important interactions with nonpsychiatric drugs are also listed. If you are taking an SSRI and one or more additional drugs at the same time, it would be wise to review this table. Make sure you also ask your physician and pharmacist if there are any drug interactions you should be aware of. This includes prescription drugs as well as nonprescription drugs that are sold over the counter.

As you can see, SSRIs have a tendency to cause the blood levels of other antidepressants to increase. This is because the SSRIs slow down the metabolism of these other drugs in the liver, as discussed in Chapter 19. In some cases, this could be dangerous. For example, the combination of an SSRI with a tricyclic antidepressant can potentially cause abnormal heart rhythms. Although this complication is rare, the effects on the heart can be serious. The combination of an SSRI with bupropion (Wellbutrin) can increase the risk of seizures—an uncommon but serious side effect of bupropion. However, as noted above, bupropion is often added to an SSRI in low doses to try to minimize the sexual side effects of the SSRIs. This can usually be done safely. Make sure you inform your physician if you have any history of head trauma or seizures, because this particular drug combination may not be advisable for you.

As mentioned in Chapter 19, the interaction of an SSRI with an MAOI antidepressant is extremely dangerous regardless of the dose of either drug. This combination should be avoided because it can result in the potentially lethal "serotonin syndrome" described in Chapter 19. In addi-

Table 20–6. Drug Interaction Guide for SSRI Antidepressants.[a]

Antidepressants	
Drug	*Comment*
tricyclic and tetracyclic antidepressants	SSRIs can cause TCA levels to ↑; abnormal heart rhythms can result
SSRI antidepressants	not usually combined; ↑ in SSRI blood levels can result
monoamine oxidase inhibitors (MAOIs)	serotonin syndrome[b]
serotonin antagonists [trazodone (Desyrel) and nefazodone (Serzone)]	blood levels of nefazodone or trazodone and their metabolite (mCPP) may ↑ and cause anxiety
bupropion (Wellbutrin)	↑ risk of seizures; caution required
venlafaxine (Effexor)	may cause ↑ in levels of venlafaxine
mirtazapine (Remeron)	no information available as yet
Antihistamines	
Drug	*Comment*
terfenadine (Seldane) and astemizole (Hismanal)	fluvoxamine (Luvox) may ↑ levels of terfenadine and astemizole; fatal heart rhythms can occur
cyproheptadine (Periactin)	may reverse the effects of SSRIs
Diabetes Medications	
Drug	*Comment*
tolbutamide (Orinase)	fluvoxamine (Luvox) may ↑ levels of tolbutamide; low blood sugar may result
insulin	fluvoxamine (Luvox) may cause ↓ in blood sugar; insulin levels may need to be adjusted

Heart and Blood Pressure Medications

Drug	*Comment*
digoxin (Lanoxin) and digitoxin (Crystodigin)	↑ in blood levels of digitoxin and potential toxic effects including mental confusion
medications for high blood pressure	levels of beta-blockers including metoprolol (Lopressor) and propranolol (Inderal) also used for angina may ↑, leading to excessive heart slowing and ECG abnormalities; calcium channel blockers including nifedipine (Procardia) and verapamil (Calan) may also ↑, leading to more potent effects on blood pressure
medications for abnormal heart rhythms	SSRI may ↑ risk of abnormal heart rhythms when combined with drugs to control heart rhythms, such as flecainide (Tambocor), encainide, mexiletine (Mexitil), and propafenone (Rythmol)

Other Psychiatric Drugs

Drug	*Comment*
benzodiazepines (minor tranquilizers) including alprazolam (Xanax), diazepam (Valium) and others	levels of benzodiazepines may ↑; excessive drowsiness or confusion; lower doses of benzodiazepines may be needed, fluvoxamine (Luvox) has strongest effect, but problems have also been reported with fluoxetine (Prozac); clonazepam (Klonopin) and temazepam (Restoril) may be safer than alprazolam (Xanax) and diazepam (Valium)

Other Psychiatric Drugs cont.	
Drug	*Comment*
buspirone (BuSpar)	may enhance the effects of SSRIs; however, fluoxetine (Prozac) may reduce the effectiveness of BuSpar, some patients with obsessive compulsive disorder who received this combination experienced a worsening of symptoms
lithium	↑ or ↓ levels may result; may lead to lithium toxicity at normal lithium levels
L-tryptophan	can cause agitation, restlessness, and upset stomach as well as the serotonin syndrome
major tranquilizers (neuroleptics) such as haloperidol (Haldol), perphenazine (Trilafon) and thioridazine (Mellaril)	blood levels of major tranquilizer may ↑ leading to increased side effects; fluvoxamine (Luvox) may be the safest SSRI to combine with neuroleptics; risperidone (Risperdal) and clozapine (Clozaril) may block the antidepressant effects of the SSRIs
methadone (Dolophine)	fluvoxamine (Luvox) leads to ↑ in blood levels
mood stabilizers and anticonvulsants	SSRIs, especially fluvoxamine (Luvox) and fluoxetine (Prozac), can cause ↑ in levels of carbamazepine (Tegretol) and phenytoin (Dilantin). The combination of either SSRI with phenytoin can cause phenytoin toxicity

Other Medications	
Drug	*Comment*
alcohol	increased drowsiness

Drug	*Comment*
caffeine (in coffee, tea, soda, chocolate)	fluvoxamine (Luvox) causes levels to ↑; excess nervousness may result
cisapride (Propulsid)	fluvoxamine (Luvox) may ↑ levels of cisapride; fatal heart rhythms can occur
cyclosporine (Sandimmune; Neoral) (an immunosuppressive drug used in organ transplants)	levels of cyclosporine may ↑
dextromethorphan (a cough suppressant in many over-the-counter medications)	hallucinations reported with fluoxetine (Prozac), possible with any SSRI
tacrine (Cognex)	fluvoxamine (Luvox) leads to ↑ in blood levels
tobacco (smoking)	levels of fluvoxamine (Luvox) may ↓
theophylline (Bronkaid)	fluvoxamine (Luvox) leads to ↑ in blood levels and can produce toxic effects, including excess nervousness
warfarin (Coumadin) (a blood-thinner)	fluvoxamine (Luvox) may ↑ levels of warfarin (Coumadin); increased bleeding may result. The increased bleeding can also result without any changes in the prothrombin test (this bleeding test is used to monitor the dose of warfarin). This is because the SSRIs can also impair clotting through their effects on blood platelets, whereas warfarin affects the clotting proteins

[a]Information in this table was obtained from several sources including the *Manual of Clinical Psychopharmacology*[1] and *Psychotropic Drugs Fast Facts.*[17] These excellent references are highly recommended.

[b]This is a dangerous and potentially fatal syndrome which includes rapid changes in vital signs (fever, oscillations in blood pressure), sweating, nausea, vomiting, rigid muscles, myoclonus, agitation, delirium, seizures, and coma.

tion, remember that both the SSRIs and the MAOIs can require a considerable period of time to clear out of your body after you have stopped taking them. If you stopped taking Prozac and then started an MAOI several weeks later, it could trigger the serotonin syndrome because Prozac would still be present in your bloodstream. Similarly, if you were to start Prozac within two weeks of stopping an MAOI, this might also trigger the serotonin syndrome. The effects of the MAOIs last only one to two weeks, so you will not have to wait as long when you switch from an MAOI to an SSRI as when you switch in the opposite direction.

A number of other important interactions which are listed in the table involve common drugs that many people might take for a cold or flu, diabetes, high blood pressure, allergies, and so on. For example, dextromethorphan is a cough suppressant in many over-the-counter cold preparations. When combined with an SSRI, dextromethorphan can cause visual hallucinations. This has been reported with fluoxetine (Prozac) but could theoretically occur with any SSRI. You will also see that two common antihistamines, terfenadine (Seldane) and astemizole (Hismanal), can produce abnormal and potentially fatal heart rhythm abnormalities when combined with certain SSRIs, and a third antihistamine called cyproheptadine (Periactin) can block the antidepressant effects of an SSRI.

Make sure you review this table if you are taking an SSRI. If you have any questions, discuss them with your doctor and pharmacist. The SSRIs are safe for the overwhelming majority of individuals who take them. With a little good teamwork between you and your doctor, your experience with an SSRI can be positive.

MAO Inhibitors

The Table of Antidepressants on pages 514–515 lists four drugs known as monoamine oxidase inhibitors (MAOIs). They include isocarboxazid (Marplan), phenelzine (Nardil), selegiline (Eldepryl), and tranylcypromine (Parnate). You may recall from Chapter 17 that the MAOIs fell into relative disuse when the newer and safer compounds were developed. They are probably vastly underutilized because they can be quite dangerous if mixed with a number of common foods (such as cheese) and medicines (including many common over-the-counter cold, cough, and hay fever drugs) and because they require fairly sophisticated medical skills on the part of the prescribing doctor.

In recent years the MAOIs have experienced a much-deserved resurgence of popularity because they are often remarkably effective for patients who do not respond to other kinds of antidepressants. Many of these patients have experienced so many years of chronic depression that their illness has become an unwelcome lifestyle. The beneficial effects of the MAOIs can sometimes be quite impressive.

The MAOIs can also be particularly effective in an "atypical depression" that is characterized by the following types of symptoms:

- overeating (as opposed to a loss of appetite in classic depression);
- fatigue and sleeping too much (rather than trouble with sleeping);
- irritability or hostility (in addition to the depression);
- extreme sensitivity to rejection.

Patients with this form of depression sometimes also emphasize chronic feelings of fatigue as well as a "leaden paralysis." It is not clear whether this really represents a subtype of depression or simply a particular group of symptoms that any depressed individual might experience.

Nevertheless, studies conducted at Columbia University suggest that the MAOIs may actually be better than the cyclic antidepressants for patients with these kinds of symptoms. The MAOIs can also be remarkably effective when high levels of anxiety accompany the depression, including phobias (such as social phobia), panic attacks, or hypochondriacal complaints. Patients with recurrent obsessive thoughts and compulsive, ritualistic, nonsensical habits (such as recurrent hand-washing or repetitive checking of door locks) may also experience relief when treated with MAOIs.

The MAOIs can also be helpful when chronic anger or impulsive self-destructive behavior accompanies the depression. Patients with these features are sometimes diagnosed as having "borderline personality disorder." Although these individuals can sometimes be quite difficult to treat, I have seen many who were dramatically helped by the MAOIs. Of course, all patients who take MAOIs must agree to follow the dietary restrictions and medication guidelines religiously. If a patient is unreliable or will not agree to this, other types of medications should be used instead.

The mechanism of action of the MAOIs is different from that of the other antidepressant drugs. You learned in Chapter 17 that most antidepressants act by blocking the pumps for neurotransmitters at the nerve endings. As a result, the levels of the chemical transmitters such as serotonin, norepinephrine, or dopamine build up in the synaptic regions. In contrast, the MAOIs seem to work by preventing the breakdown of chemical messengers within the nerves. As a result, levels of serotonin, norepinephrine, and dopamine build up inside the nerve terminals and these messengers are released into the synapses in much higher concentrations when the nerves fire. This results in a greater stimulation of the nerves at the other side of the synaptic junctions.

The MAOIs require careful medical management and close teamwork with your doctor. They are well worth the effort because they can sometimes lead to profound mood transformations, even when other drugs have been ineffective. Because they may cause increases in blood pressure, they are not usually recommended for individuals over sixty years of age or individuals with heart problems. In addition, they are not usually prescribed for individuals with significant cerebrovascular disorders, such as strokes or aneurysms, or individuals with brain tumors. Paradoxically, though, they can sometimes be used with individuals with high blood pressure because they usually cause the blood pressure to fall.[19] Consultation with a cardiologist would be necessary to make sure there are no dangerous interactions with your other blood pressure medications.

Like other antidepressants, the MAOIs usually require at least two or three weeks to become effective. Your doctor will probably want to obtain a medical evaluation before starting you on this type of drug. This evaluation may include a physical examination, a chest X ray, an electrocardiogram, a blood count, blood chemistry tests, and a urinalysis.

Doses of MAOIs. The doses of the MAOIs are listed in Table 20–1 on page 520. The two most commonly prescribed drugs for depression and anxiety are tranylcypromine (Parnate) and phenelzine (Nardil). One of the MAOIs, isocarboxazid (Marplan), is no longer available in the United States but is available in some other countries including Canada. In addition, selegiline (Eldepryl) is rarely used for depression but is often used in small doses (5 mg to 10 mg per day) in the treatment of Parkinson's disease. It is just starting to be used for depression and some other psychiatric disorders, although in higher doses than for Parkinson's disease, as indicated in Table 20–1. Although the Food and Drug Administration (FDA) has not yet approved selegiline for use in psychiatric disorders, recent studies indicate that it can also be effective for patients with atypical depression as well as those with chronic, severe depression.

A common prescribing error with the MAOIs is to give too big a dose too soon. For example, you will see in Table 20–1 on page 520 that the usual dose range for tranylcypromine (Parnate) is 10 mg to 50 mg per day. Some doctors prescribe larger doses, but I have seen many patients respond to just one or two pills per day. Because the MAOIs can have some toxic side effects, I think it is prudent to start them at low doses, to increase very slowly, and not to push the dose too high. I usually start the patient on just one pill per day of an MAOI for the first week, and then increase to two pills per day. If the patient does not respond to a reasonable dose, say three or four pills per day of tranylcypromine or phenelzine, I usually do not increase the dose further but instead try an alternative medication along with a different psychotherapeutic strategy.

How long should you stay on an MAOI if it does not seem to be working? It seems obvious to me that if you have not had a fairly dramatic response after three or four weeks, as confirmed by your weekly scores on the mood test in Chapter 2, then you have probably given the drug a fair trial. You might respond better to another type of drug or to the cognitive therapy techniques described in this book.

How long should you stay on an MAOI if you do respond favorably? As with any antidepressant, you will have to discuss this with your physician, and many different approaches are currently in vogue. Some physicians believe that patients need antidepressants indefinitely to correct a "chemical imbalance," but I have not usually found it necessary to keep patients on MAOIs or other antidepressants indefinitely. I have found that patients nearly always do well when they discontinue their MAOIs after a reasonable period of feeling good. Sometimes this may be as short as three months, sometimes as long as six to twelve months.

As with most antidepressants, you should taper off an MAOI gradually so there will be no withdrawal effects. Tapering too rapidly has caused some patients to experience sudden manic reactions. Suddenly going off selegiline can cause nausea, dizziness, and hallucinations, so one has to be especially careful to taper slowly.

What if you go off the MAOI and then get depressed again in the future? If you have responded to an MAOI in the past, you may respond more rapidly if you take the same MAOI again in the future. In my practice I have had many patients who experienced a positive response to an MAOI (usually Parnate) and continued to feel undepressed for many years after they stopped taking the drug. Eventually, a few of them became depressed again and called for a "tune-up" appointment. I always gave them the first available appointments. If they sounded quite depressed, I told them to start the medication again. I also told them to start doing their psychotherapy homework again, especially the exercise of writing down and challenging their negative thoughts. When I saw them a few days later, many of them were already feeling better. Some of them told me that they began to improve in as little as one day or less when they took the MAOI for the second time. I believe that the medication as well as the cognitive therapy contributed to the rapid improvement.

I have not seen this rapid response with other types of antidepressants and do not know why it sometimes happens with MAOIs. Several patients explained that their bodies seemed to "recognize" the effects of the MAOI right away, especially the pleasurable stimulation that tranylcypromine (Parnate) causes. This helped them "remember" what it was like not to feel depressed. In a few cases, the improvement in mood came within an hour or two of the first pills they took. In the majority of cases, one or two cognitive therapy sessions seemed to reverse the relapse of depression.

Side Effects of MAOIs. The most frequent side effects are listed in Table 20–7 on pages 572–573. As noted above, tranylcypromine (Parnate) tends to be stimulating. The stimulating effects of tranylcypromine (Parnate) can be especially helpful to depressed individuals who feel tired, lethargic, and unmotivated. Tranylcypromine (Parnate) may provide them with some much-needed "go power." Because tranylcypromine (Parnate) tends to be stimulating, it can also cause insomnia. In order to minimize the insomnia, the entire dose can be taken once a day in the morning or in divided doses in the morning and at noon. The latest recommended time to take tranylcypromine (Parnate) is 6:00 P.M. Phenelzine (Nardil) is less stimulating than tranylcypromine (Parnate) and may be an attractive option for patients who feel too stimulated by tranylcypromine (Parnate).

The other side effects of the MAOIs are similar to those of the tricyclic and tetracyclic drugs described previously, but they are usually mild, especially when the MAOIs are taken in low doses. As you can see in Table 20–7, the MAOIs do not have strong effects on the muscarinic receptors (you will recall that these are also called cholinergic receptors). Consequently, they are not likely to cause dry mouth, blurred vision, constipation, or a delay in starting the urine flow. Weight gain also does not seem to be so much of a problem with these drugs, although some patients experience an increased appetite. Weight gain appears to be less of a problem with tranylcypromine (Parnate) than phenelzine (Nardil). Because tranylcypromine is stimulating, it may actually reduce your appetite, as do some of the SSRIs including fluoxetine (Prozac).

Some patients experience light-headedness when standing suddenly because these drugs have relatively strong effects on the alpha-adrenergic receptors. If dizziness does develop, the interventions described previously can help. These include: (1) ask your doctor if you can lower the dose—often you can still maintain the antidepressant effect; (2) get up more slowly and exercise your legs by walking in place immediately when you stand; (3) wear support stockings; (4) drink adequate fluids and make sure you eat enough foods with salt to maintain your body's electrolytes.

Like most antidepressants, the MAOIs can sometimes cause a rash, although I do not recall ever seeing this. A loosening of the stool or constipation might also occur. Some patients report an upset stomach. Taking the medication with meals can alleviate this. Some patients report muscle twitches, but this is usually not dangerous. If you experience muscle pains, cramps, or tingling fingers—side effects I have never observed—a daily dose of 50 to 100 mg of vitamin B_6 (pyridoxine) may help. This is because MAOI drugs may interfere with pyridoxine metabolism, so taking extra pyridoxine may compensate for this effect. Some doctors recommend taking vitamin B_6 routinely if you are on an MAOI.

The MAOIs can sometimes interfere with sexual functioning, especially in higher doses. Some patients experience a decreased interest in sex and difficulties maintaining an erection or achieving orgasm. In this regard, the MAOIs are a lot like the SSRIs described previously. The sexual side effects may result from their effects on the serotonin receptors in the brain, but this is not known for sure. Although the sexual side effects can be disconcerting, these difficulties may be a worthwhile trade-off if the medication is having a beneficial effect on your mood. You should be reassured that the sexual side effects are dose-related and usually disappear entirely when you are no longer taking the MAOI.

One young man I treated actually found the sexual side effects to be helpful. He reported that he had always had a problem with premature ejaculation. Once he started taking tranylcypromine (Parnate), the problem disappeared. In fact, he reported he could make love for prolonged periods without any danger at all of having a premature orgasm. He said his girlfriend thought this was a great miracle, and he advised me to buy stock in the company that manufactured the drug!

One pleasurable side effect of an MAOI is an excessively positive reaction to the drug. In other words, quite a number of patients not only overcome their depressions but begin to feel euphoric or high. This is not necessarily bad, but in some cases may become so extreme that the patient experiences the symptoms of mild mania. In the rare patient with a history of bipolar manic-depressive illness (patients with previous extreme highs and lows that were not caused by drugs or alcohol), there is the possibility that an MAOI might trigger a full-blown manic episode. This is actually true of most antidepressants, and not just the MAOIs.

Table 20–7. Side Effects of Monoamine Oxidase Inhibitors[a]

Note: This list is not comprehensive. In general, side effects that occur in 5% or 10% or more of patients are listed, as well as rare but dangerous side effects.

Side Effect	*Sedation and Weight Gain*	*Light-Headedness and Dizziness*	*Blurred Vision, Constipation, Dry Mouth, Speeded Heart, Urinary Retention*	*Common or Significant Side Effects*[b]
Brain Receptor	*histamine (H_1) receptors*	*alpha-adrenergic (α_1) receptors*	*muscarinic (M_1) receptors*	
isocarboxazid (Marplan)	+	+++	0 to +	headache; changes in heart rhythm and rate; overactivity or mania; tremor; jittery; confusion; memory problems; insomnia; edema; weakness; sweating; upset stomach; delayed orgasm
phenelzine (Nardil)	+	+++	0 to +	dizziness; headache; fatigue; trouble sleeping; weakness; tremor; twitching; dry mouth; upset stomach; constipation; weight gain; delayed orgasm; jittery; euphoria; trouble starting urine; swelling; sweating; rash

selegiline (Eldepryl)	0	+	+	(limited information available);[c] nausea; weight loss; delayed orgasm; confusion; dry mouth; dizziness; possibly other side effects
tranylcypromine (Parnate)	0 to +	+++	0 to +	overstimulation; euphoric or manic feelings; restlessness; anxiety; trouble sleeping; tiredness or weakness; twitching; tremor; muscle spasms; upset stomach; loss of appetite; constipation; diarrhea; headache; delayed orgasm; numbness or tingling; swelling; racing heart; blurred vision

[a]The + to +++ ratings in this table refer to the likelihood that a particular side effect will develop. The actual intensity of the side effect will vary among individuals and will also depend on how large the dose is. Reducing the dose can often reduce side effects without reducing effectiveness.

[b]Many of the side effects of the MAOIs can often be reduced or eliminated by reducing the dose. They usually have very few side effects, and can often be quite effective, at small doses.

[c]This is because this drug is usually prescribed for patients with Parkinsonism who take many other drugs, and also have many symptoms due to their illness. Therefore, it is difficult to determine how frequently selegiline would cause side effects in depressed individuals. At higher doses, the side effects of selegiline are probably very similar to the other MAOIs.

If you do start to feel unusually happy, it would be wise to keep in touch with your prescribing doctor to make sure these feelings do not get out of hand. In my experience, this is not usually a serious problem—the euphoric feelings provide a welcome relief from the depression and tend to diminish in a week or so. The euphoric feelings also respond to a reduction in dose.

Dr. Alan Schatzberg and his colleagues[1] have pointed out that some patients may seem drunk or intoxicated when taking MAOIs. Patients may also feel confused and have trouble with coordination. These adverse reactions are more likely when the doses are pushed to very high levels. Obviously, the dose should be reduced immediately if these toxic effects develop. I have personally never seen these effects because I have never pushed the doses of MAOIs to unusually high levels.

Two of the MAOI drugs, phenelzine (Nardil) and isocarboxazid (Marplan), can have negative effects on the liver. Therefore, your doctor may want to do a blood test to monitor levels of certain enzymes that reflect liver function before you start these drugs, and then again every few months while you are taking them. Patients with liver disease or abnormal liver function tests are usually advised not to take any of the MAOIs, including tranylcypromine (Parnate).

Dr. Alan F. Schatzberg and his colleagues[1] have pointed out that selegiline (Eldepryl) may have fewer side effects than the other MAOI drugs, at least at low doses. At low doses, selegiline seems less likely to cause dizziness when standing, sexual problems, or difficulties sleeping. However, selegiline is much more expensive than the other MAOIs, and in most cases the other MAOIs will do the job just as effectively. In addition, the side effects of all the MAOI antidepressants tend to be minimal at lower doses. In my experience, many patients have responded favorably to low doses of the MAOIs, so selegiline may not really have any significant advantages over the two older and cheaper drugs.

As you will learn next, all the MAOIs can cause dangerous blood pressure elevations when patients ingest the forbidden foods. Selegiline is less likely to have this effect, but only if the selegiline is taken in small doses (10 mg per day or less). Larger doses of selegiline are often needed for psychiatric problems. At these higher doses it is necessary to observe the same dietary precautions that you would observe with any of the MAOIs. This is unfortunate because it was initially hoped that depressed patients would be able to take selegiline and not have to restrict their diets so religiously.

Hypertensive and Hyperpyretic Crises. In rare cases, the MAOIs can produce two types of serious toxic reactions if they are not used properly. This is why so many doctors avoid using them. With good education and preventive medications, the MAOIs can be administered safely, but you will need to study this section quite carefully if you are taking an MAOI.

One of the dangerous reactions is called a "hypertensive crisis." "Hyper" means high and "tensive" refers to blood pressure, so a hypertensive crisis is a sudden increase in your blood pressure. Increases in blood pressure are not usually dangerous and can occur in many normal situations even if you are not taking medications. For example, when you are lifting weights, your blood pressure can easily go into the range of 180/100 or higher at the moment you are straining and exerting maximum effort to raise the barbell. Our bodies are used to these temporary elevations in blood pressure. However, if you were on an MAOI and you ate one of the forbidden foods, your blood pressure might increase to dangerous levels and remain elevated for an hour or more. If you continued to eat the forbidden foods that interact with the MAOIs, sooner or later a vessel in your brain could rupture because of the mechanical stress. This would cause a stroke, certainly an excessive price to pay for taking an antidepressant.

The initial symptoms of a ruptured or leaking vessel in

your brain can include an excruciating headache, a stiff neck, nausea, vomiting, and sweating. As the bleeding continues, paralysis, coma, and death can occur. Because of the danger of hypertensive reactions, your doctor will check your blood pressure at each session. The risk of a stroke is higher in individuals over sixty because our arteries become less resilient with age and are more likely to tear or rupture when subjected to the stress of a sudden increase in blood pressure. Regardless of your age, you will need to monitor your blood pressure and watch your diet carefully when taking an MAOI.

These hypertensive crises are sometimes also called "noradrenergic crises" because they are thought to be due to an excessive release of norepinephrine. Norepinephrine is a transmitter substance used by nerves in your brain and in your body. Hypertensive crises usually occur if you eat certain forbidden foods containing a substance called tyramine or if you take one of the forbidden drugs that I will describe in detail below. If you are careful, the risk of a serious hypertensive crisis is very small.

The other dangerous reaction to an MAOI is called a "hyperpyretic crisis." "Pyretic" refers to fire, or fever. The patient with a hyperpyretic crisis may develop a high fever along with a number of alarming symptoms that can include sensitivity to light, rapid changes in blood pressure, rapid breathing, sweating, nausea, vomiting, rigid muscles, jerking and twitching, confusion, agitation, delirium, seizures, shock, coma, and death. A hyperpyretic crisis is sometimes also called a "serotonin syndrome" because it is due to an abnormal and dangerous increase in levels of serotonin in the brain. A hyperpyretic crisis occurs when the patient takes certain forbidden medications that must not be combined with the MAOIs. These drugs cause an increase in levels of serotonin in the brain. Obviously, a hyperpyretic crisis requires immediate discontinuation of the MAOI along with emergency medical treatment. The treatment may include intravenous fluids and treatment with the serotonin antagonist, cyproheptadine (Periactin), at a dose of 4 mg to 12 mg.

Several decades ago when MAOIs were first available, doctors were not as aware of the blood pressure elevations that resulted from eating foods containing tyramine or from taking the kinds of drugs described below, and so these hypertensive reactions were more common and severe. Now doctors and patients are much more aware of the problem and the risk is much smaller. In fact, extreme hypertensive and hyperpyretic reactions are quite rare. I am personally aware of only one patient, treated by a colleague in Boston, who developed a stroke due to a hypertensive crisis (noradrenergic syndrome) while taking an MAOI. I have had about half a dozen patients over the years who paged me because they suddenly developed elevated blood pressure. I told each of them to go to a local hospital emergency room for observation. In every case, the blood pressure quickly subsided without any treatment aside from observation. None of these patients experienced any adverse effects. I have never seen a patient who developed a hyperpyretic crisis (serotonin syndrome) while on an MAOI.

This is because we know a great deal about what causes these two kinds of reactions and how they can be avoided. If you are taking an MAOI, you will need to educate yourself by studying the following sections carefully. You will have to avoid taking certain types of drugs and exercise a little self-discipline in your diet in order to be safe. You will find it is well worth the extra effort required to protect yourself.

How to Avoid a Hypertensive or Hyperpyretic Crisis. There are two important keys to preventing a hypertensive or hyperpyretic crisis if you are taking an MAOI. First, you must obtain a blood-pressure cuff and monitor your own blood pressure carefully. Second, you must carefully avoid certain foods or medications (including some street drugs) that will predictably trigger these reactions. I will describe these forbidden foods and medicines in detail below. You will see that the substances that can trigger a hypertensive crisis are somewhat different from the substances that can trigger a hyperpyretic crisis.

You can obtain a blood-pressure cuff at your local pharmacy so you can monitor your own blood pressure whenever you want. Practice using the cuff. Although it may seem a little awkward or confusing at first, you will find that it gets pretty easy to take your blood pressure after you have practiced a few times. In my practice I have required every patient taking an MAOI to do this. In the rare situation where a patient did not want to go to the trouble of obtaining a cuff and learning how to use it, I have refused to prescribe an MAOI.

Initially you can monitor your blood pressure once a day or even twice a day if you are so inclined. After you have been taking the MAOI for a couple weeks, you will not need to monitor your blood pressure so frequently. Once a week will usually be sufficient. You can check your blood pressure if you forget and eat one of the forbidden foods. You can also check it if you feel woozy or nauseous or if you get an excruciating or severe headache. We all get headaches from time to time, and they rarely ever indicate a stroke. However, if you have a blood-pressure cuff, you can check your blood pressure and make sure it is not dangerously elevated.

If your blood pressure goes up to a dangerous level, you should call your doctor or go to an emergency room. How much elevation is dangerous? The blood pressure consists of two numbers. The higher number is called the "systolic" blood pressure and the lower number is called the "diastolic" blood pressure. A value of 120/80, for example, would be considered normal for most people. Most emergency room doctors would not be particularly concerned until these numbers reach the range of 190 to 200 over 105 to 110. At that level, they might observe you carefully and monitor your blood pressure every few minutes. Most of the time, the elevated blood pressure will subside without treatment. If your blood pressure continues to rise, the ER doctor could give you an antidote (such as phentolamine or prazosin) to lower your blood pressure back into a safe range.

The best time to take your blood pressure is about one to one and a half hours after you have taken the medication. About 25 percent of my patients have noted modest blood pressure elevations at this time even if they have not eaten any of the forbidden foods in Table 20–8 on pages 580–581 or taken the medicines in Table 20–9 on pages 584–590. These increases were not usually extreme or dangerous—a 20- or 30-point elevation in the systolic blood pressure was typical. Nevertheless, in those cases, I have recommended stopping the medication because these patients seemed overly sensitive to the effects of the MAOI on their blood pressure. It just did not seem worth the worry and risk, especially since a different antidepressant might be just as effective.

Foods to Avoid. Hypertensive crises may occur if you eat foods (see Table 20–8) that contain a substance known as tyramine. If you are taking an MAOI, too much tyramine can interfere with your brain's ability to regulate your blood pressure. Tyramine causes nerves to release more norepinephrine into the synaptic regions that separate them from the postsynaptic nerves. These postsynaptic nerves may become overly stimulated when too much norepinephrine is released. Because these nerves help to regulate blood pressure, all the extra norepinephrine that is released can cause a dangerous and sudden increase in blood pressure.

You will recall from Chapter 17 that an enzyme called monoamine oxidase (MAO) is located inside the presynaptic nerves. This enzyme usually destroys any excess norepinephrine that builds up inside these nerves and prevents these nerves from releasing too much norepinephrine when they fire. But the MAOI drugs block this enzyme, and so the norepinephrine levels inside these nerves increase sub-

Table 20–8. Foods and Beverages to Avoid If You Are Taking a Monoamine Oxidase Inhibitor (MAOI)[a]

Foods to Avoid Completely

Cheese, particularly strong or aged cheese (cottage cheese and cream cheese are allowed)

Beer and ale: particularly tap beers, beers from microbreweries and strong ales

Red wine: especially Chianti wine

Brewer's yeast tablets or yeast extracts (breads and cooked forms of yeast are safe. The yeast extracts from health food stores are dangerous. Yeast extracts may be found in certain soups. Some powdered protein diet supplements contain yeast extracts.)

Pods of fava beans, also called Italian green beans (regular green beans are safe)

Meat or fish that is smoked, dried, fermented, unrefrigerated, or spoiled, including:

- fermented or air-dried sausages, such as salami and mortadella (some experts state that bologna, pepperoni, summer sausage, corned beef, and liverwurst are safe)[17]
- pickled or salted herring
- liver (beef or chicken), especially old chicken liver (fresh chicken liver is safe)

Overripe bananas or avocados (most fruits are completely safe)

Sauerkraut

Some soups, including those made from beef bouillon or Asian soup stocks (e.g., miso soup). (Tinned and packet soups are felt to be safe, unless made from bouillon or meat extracts)

Foods or Beverages that May Cause Problems in Large Amounts

White wine or clear alcohol, such as vodka or gin

Sour cream

Yogurt: must be pasteurized and less than 5 days old to be safe

Soy sauce

NutraSweet (the artificial sweetener)

Chocolate

Caffeine in beverages (coffee, tea, and soda) and chocolate

Foods or Beverages Once Thought to Cause Problems which Are Probably Safe in Small Amounts
Figs (avoid overripe figs)
Meat tenderizers
Caviar, snails, tinned fish, pate
Raisins

[a]Modified from B. McCabe and M. T. Tsuang, "Dietary Considerations in MAO Inhibitor Regimens," *Journal of Clinical Psychiatry* 43 (1982): 178–181.

stantially. When you eat foods containing tyramine, all that extra norepinephrine suddenly spills into the synaptic region, causing a massive stimulation of the nerves that regulate your blood pressure.

If you watch your diet carefully, the likelihood is good that you will experience no adverse blood-pressure elevation. The most common trigger is cheese, especially strong cheese. You will have to give up pizza and grilled cheese sandwiches for a while if you are taking an MAOI.

Most of the forbidden foods contain the breakdown products of protein—including tyramine. So, for example, freshly cooked chicken is perfectly safe, but cooked leftover chicken that has been sitting for a couple days can be dangerous because tyramine forms when the meat decomposes. One of my patients on tranylcypromine (Parnate) ate some leftover chicken that had been in the refrigerator for several days. Soon after eating it, he experienced a significant elevation in blood pressure. This was because the chicken had partially decomposed due to effects of bacteria. Fortunately, he was not harmed, but this experience served as a useful warning to be careful. The fermented or partially decomposed meats on the list in Table 20–8, such as strong sausage or smoked fish, as well as strong cheese, may contain large amounts of tyramine and can be especially dangerous. Some experts also advise against eating Chinese food while taking MAOIs. This may be due to the soy sauce, the monosodium glutamate, or other ingredients.

How much tyramine is necessary to cause a hypertensive reaction? This varies quite a bit from person to person. On average, foods containing at least 10 mg of tyramine will be sufficient to cause a hypertensive crisis if you are taking phenelzine (Nardil). As little as 5 mg of tyramine may be sufficient if you are taking tranylcypromine (Parnate). What foods contain this amount of tyramine? Well, most beers contain less than 1.5 mg of tyramine, and many contain less than 1 mg, so you would have to drink several beers to run a significant risk. However, some ales contain 3 mg of tyramine per serving, and some tap beers can also be particularly risky. For example, one serving of Kronenbourg, Rotterdam's Lager, Rotterdam's Pilsner, or Upper Canadian Lager contains between 9 and 38 mg of tyramine[17]. So even one glass of these beers could be dangerous.

Cheeses can also vary greatly. Processed American cheese contains only about 1 mg of tyramine per serving, but Liederkranz, New York State cheddar, English Stilton, blue cheese, Swiss cheese, aged white cheese and Camembert all contain more than 10 mg per single serving.[17]

Suppose you eat one of the forbidden foods by accident, and then you check your blood pressure and discover that it does not go up. What does this mean? There is a lot of individual variation in the sensitivity to the effects of the forbidden foods. You may be one of those individuals who is significantly less likely to react with an elevation in blood pressure. However, you should not become complacent, because these hypertensive reactions are unpredictable. If you cheat and eat the forbidden foods from time to time, it is a lot like playing Russian roulette. You may get away with it for a while and then discover that you have experimented once too often. For example, you may eat a piece of pizza on nine separate occasions without any increase in blood pressure, and conclude that it is safe to eat pizza. But this can be very misleading, because the tenth time you eat a piece of pizza you may experience a sudden and severe increase in blood pressure. It is not known why this happens, but it does underscore the importance of consistent self-discipline if you are taking an MAOI.

Medications and Drugs to Avoid. A number of prescription drugs, nonprescription drugs, and street drugs that can cause a hypertensive or hyperpyretic crisis when combined with MAOIs are listed in Table 20–9 on pages 584–590. These reactions are especially dangerous and so you must carefully avoid these drugs. Some of the medications that interact with MAOIs do not cause such severe reactions. For example, caffeine may cause you to become more jumpy and jittery than usual. Moderate amounts of caffeine are reasonably safe, however. (You may think of caffeine as more of a food than a drug, but it is a mild stimulant.)

The list of drugs that interact with MAOIs includes:

- most antidepressants—virtually any of them can be dangerous;
- many antiasthma drugs;
- many common cold, cough allergy, sinus, decongestant, and hay fever medications that contain sympathomimetic agents (discussed in detail below) or dextromethorphan, the cough suppressant. You will have to check labels carefully, because many over-the-counter drugs contain these substances;
- drugs used in the treatment of diabetes—they may become more potent than usual if you are taking an MAOI, and can cause your blood sugar to fall more than expected;
- some drugs used in the treatment of low or high blood pressure—both types of drugs can in some cases cause blood pressure elevations when combined with MAOIs;
- mood stabilizers and anticonvulsants;
- some painkillers, including some local and general anesthetics;
- sedatives (including alcohol) and tranquilizers—they may have more pronounced effects than usual when you are

(continues on page 591)

Table 20–9. Prescription Drugs and Over-the-Counter Medications to Avoid If You Are Taking a Monoamine Oxidase Inhibitor (MAOI)[a]

Note: This list is not exhaustive; new information about drug interactions comes out frequently. If you are taking an MAOI and any other medication, ask your doctor and pharmacist if there are any drug interactions.

Antidepressants

Drug	*Comment*
tricyclic antidepressants,[b] especially desipramine (Norpramin, Pertofrane) and clomipramine (Anafranil)	Some (e.g., clomipramine) may cause a hyperpyretic crisis or seizures; others (e.g. desipramine) may cause a hypertensive crisis
tetracyclic antidepressants, especially bupropion (Wellbutrin)	hypertensive crisis (noradrenergic syndrome)
SSRIs (all are extremely dangerous)	hyperpyretic crisis (serotonin syndrome)
other MAOIs	hyperpyretic crisis (serotonin syndrome); hypertensive crisis (nonadrenergic syndrome)
serotonin antagonists, including trazodone (Desyrel) and nefazodone (Serzone)	hyperpyretic crisis (serotonin syndrome)
mirtazapine (Remeron)	hypertensive crisis (noradrenergic syndrome)
venlafaxine (Effexor)	hypertensive crisis (noradrenergic syndrome)

Asthma Medicines

Drug	*Comment*
ephedrine, a bronchodilator contained in Marax, Quadrinal, and other asthma drugs	hypertensive crisis
inhalants which contain albuterol (Proventil, Ventolin), metaproterenol (Alupent, Metaprel), or other beta-adrenergic bronchodilators	blood pressure elevations and a rapid heart; beclomethasone and other nonsystemic steroid inhalers are generally safer
theophylline (Theo-Dur), a common ingredient in asthma drugs	rapid heart and anxiety

Cold, Cough, Allergy, Sinus, Decongestant, and Hay Fever Medications (including tablets, drops, or sprays)

Drug	*Comment*
antihistamines: terfenadine (Seldane-D)	can cause an increase in MAOI blood levels
dextromethorphan can be found in many cold and cough medications, especially any drug with DM or Tuss in its name. These include Bromarest-DM or -DX, Dimetane-DX cough syrup, Dristan Cold & Flu, Phenergan with Dextromethorphan, Robitussin-DM, several Tylenol coid, cough, and flu preparations, and many others	hyperpyretic crisis (serotonin syndrome); may also cause brief episodes of psychosis or bizarre behavior
ephedrine can be found in Bronkaid, Primatene, Vicks Vatronol nose drops and several other asthma and cold medications.	hypertensive crisis (noradrenergic syndrome)
oxymetazoline (Afrin) nose drops or sprays used to treat nasal decongestion	hypertensive crisis (noradrenergic syndrome)
phenylephrine can be found in Dimetane, Dristan decongestant, Neo-Synephrine nasal spray and nose drops, and many other similar preparations, including some eye drop medications	hypertensive crisis (noradrenergic syndrome)

Cold, Cough, Allergy, Sinus, Decongestant, and Hay Fever Medications (including tablets, drops, or sprays) cont.

Drug	*Comment*
phenylpropanolamine is contained in Alka-Seltzer Plus Cold and Night-Time Cold medicine, Allerest, Contac decongestants, Coricidin D decongestants, Dexatrim appetite pills, Dimetane-DC Cough syrup, Ornade Spansules, Robitussin-CF, Sinarest, St. Joseph Cold Tablets, Tylenol Cold medicine, and many others	hypertensive crisis (noradrenergic syndrome)
pseudoephedrine can be found in Actifed, Allerest No Drowsiness formula, Benadryl combinations, CoAdvil, Dimetane-DX Cough syrup, Dristan Cold Maximum Strength, Robitussin-DAC syrup, Robitussin-PE, Seldane-D tablets, Sinarest No Drowsiness, Sinutab, Sudafed, Triaminic Nite Light, and numerous Tylenol allergy, sinus, flu, and cold preparations, as well as several Vicks products including NyQuil, to mention just a few	hypertensive crisis (noradrenergic syndrome)

Diabetes Medications

Drug	*Comment*
insulin	may cause a greater drop in blood sugar
oral hypoglycemic agents	as above

Medications for Low Blood Pressure (for patients in shock)

Drug	*Comment*
sympathomimetic amines including: • **dopamine (Intropin)** • **epinephrine (Adrenalin)** • **isoproterenol (Isuprel)** • **metaraminol (Aramine)** • **methyldopa (Aldomet)** • **norepinephrine (Levophed)**	hypertensive crisis (noradrenergic syndrome) because these drugs cause blood vessels to constrict

Medications for High Blood Pressure

Drug	*Comment*
guanadrel (Hylorel) **guanethidine (Ismelin)** **hydralazine (Apresoline)** **methyldopa (Aldomet)** **reserpine (Serpasil)**	These blood-pressure medications may cause a paradoxical increase in blood pressure when combined with MAOIs.
beta-blockers	may be more potent when combined with MAOIs, leading to a greater-than-expected drop in blood pressure and dizziness when standing
calcium channel blockers	appear to be reasonably safe when combined with MAOIs. Check with your doctor and monitor blood pressure closely. Watch for a greater-than-expected drop in blood pressure
diuretics	watch for a greater-than-expected drop in blood pressure. May increase blood level of MAOI

Mood Stabilizers	
Drug	*Comment*
carbamazepine (Tegretol)	hyperpyretic crisis (serotonin syndrome); MAOI may cause carbamazepine levels to fall, so epileptics may experience seizures
lithium (Eskalith)	can cause hyperpyretic crisis (serotonin syndrome) in animal studies

Painkillers and Anesthetics	
Drug	*Comment*
anesthetics: general	Tell your anesthesiologist you are on an MAOI. If possible, discontinue the MAOI two weeks before elective surgery Muscle relaxants such as succinylcholine and tubocurarine may have a more pronounced or prolonged effect. General anesthetics such as halothane may lead to excitement, excessive depression of the brain, or hyperpyretic reactions
anesthetics: local	Some contain epinephrine or other sympathomimetics—make sure you tell your dentist you are taking an MAOI
cyclobenzaprine (Flexeril) (a muscle relaxant used to treat muscle spasm)	hyperpyretic crisis (serotonin syndrome) or severe seizures
meperidine (Demerol)	A single injection can cause seizures, coma, and death (serotonin syndrome). Most other narcotics, including morphine and codeine, have been used safely with MAOIs

Sedatives and Tranquilizers

Drug	*Comment*
alcohol	May have enhanced sedative effects, especially when combined with phenelzine (Nardil). This could be hazardous when driving or operating dangerous machinery
barbiturates (such as phenobarbital)	enhanced sedative effects as described above
buspirone (BuSpar)	enhanced sedative effects as described above
major tranquilizers (neuroleptics)	enhanced sedative effects as described above; some neuroleptics may cause a drop in blood pressure when combined with MAOIs
minor tranquilizers (benzodiazepines) such as alprazolam (Xanax), diazepam (Valium) and others	enhanced sedative effects as described above
sleeping pills	enhanced sedative effects as described above
L-tryptophan	hyperpyretic crisis (serotonin syndrome); blood pressure elevations; disorientation, memory impairment, and other neurologic changes

Stimulants (Pep Pills) and Street Drugs

Drug	*Comment*
amphetamines (speed or crank) **cocaine** **benzedrine** **benzphetamine (Didrex)** **dextroamphetamine (Dexedrine)** **methamphetamine (Desoxyn)** **methylphenidate (Ritalin)**	the hypertensive crisis (noradrenergic syndrome) is possible; methylphenidate is considered somewhat less risky than the amphetamines

Weight-Loss and Appetite-Suppression Medications

Drug	*Comment*
pemoline (Cylert)	drug interactions have not been studied in humans; great caution should be used; some experts report that pemoline has been combined with MAOIs in some cases[1]
fenfluramine (Pondimin)	hyperpyretic crisis (serotonin syndrome)
phendimetrazine (Plegine)	hypertensive crisis (noradrenergic syndrome)
phentermine and some over-the-counter meds	hypertensive crisis (noradrenergic syndrome)
phenylpropanolamine (Acutrim)	hypertensive crisis (noradrenergic syndrome)
stimulants (listed above)	hypertensive crisis (noradrenergic syndrome)

Other MAOI Drug Interactions

Drug	*Comment*
caffeine (in coffee, tea, soda, chocolate)	Probably safe in moderate amounts; avoid large amounts; may cause blood pressure elevations, a racing heart, and anxiety
disulfiram (Antabuse) (used to treat alcoholism)	Severe reactions when mixed with an MAOI
L-dopa (Sinemet) (used to treat Parkinson's disease)	hypertensive crisis (noradrenergic syndrome)

[a]Information in this table was obtained from several sources including the *Manual of Clinical Psychopharmacology*[1] and *Psychotropic Drugs Fast Facts*.[17] These excellent references are highly recommended.

[b]Many patients have been successfully treated with a combination of an MAOI and a tricyclic antidepressant under close observation, but such drug combinations are dangerous and require a high level of expert supervision.

taking an MAOI. The increased sleepiness could be hazardous if you are driving;

- L-tryptophan—the natural amino acid;
- stimulants (pep pills) and street drugs;
- many weight-loss (appetite suppressing) medications;
- caffeine, which is present in coffee, tea, many sodas, hot cocoa, and chocolate. Caffeine is also present in a number of prescription and nonprescription medications such as Cafergot suppositories and tablets, Darvon Compound-65, NōDōz, Fiorinal, Excedrin, and many other cold or pain preparations;
- Disulfiram (Antabuse), used to treat alcoholism;
- Levo-dopa, used in the treatment of Parkinson's disease.

Drugs that are categorized as sympathomimetics are particularly dangerous because they are contained in many over-the-counter drugs for common ailments such as colds. They are called sympathomimetics because they tend to mimic the effects of the sympathetic nervous system, which is involved in the control of blood pressure.

Several sympathomimetic drugs are found in large numbers of prescription and over-the-counter cold preparations, cough medicines, decongestants, and hay fever medications. These include ephedrine, phenylephrine, phenylpropanolamine and pseudoephedrine. For example, **ephedrine** can be found in Bronkaid, Primatene, Vicks' Vatronol nose drops, and several other cold and asthma medications. **Phenylephrine** can be found in Dimetane, Dristan decongestants, Neo-Synephrine nasal spray and nose drops, and many other similar preparations. **Phenylpropanolamine** is contained in Alka-Seltzer Plus Cold Medicines, Contac decongestants, Coricidin D decongestants, Dexatrim appetite suppressant pills, Dimetane-DC Cough syrup, Ornade Spansules, Robitussin-CF, Sinarest, St. Joseph Cold Tablets, and many other cold medicines. **Pseudoephedrine** can be found in Actifed, Advil Cold & Sinus, Allerest No-Drowsiness formula, Benadryl combinations, Dimetane-DX Cough syrup, Dristan Cold Maximum Strength, Robitussin-DAC syrup, Robitussin-PE, Seldane-D tablets, Sinarest No Drowsiness, Sinutab, Sudafed, Triaminic Nite Light, and numerous Tylenol allergy, sinus, flu, and cold preparations, as well as several Vicks products including NyQuil, to mention just a few.

Some cold and cough preparations contain **dextromethorphan**. This is not a sympathomimetic drug, but a cough suppressant. Dextromethorphan is on the list of forbidden medications because it can cause a hyperpyretic crisis. Dextromethorphan can be found in any drug with "DM" or "Tuss" in its name, as well as many preparations without these suffixes. A few examples are Bromarest-DM or -DX, Dimetane-DX Cough syrup, Dristan Cold & Flu, Phenergan with Dextromethorphan, Robitussin-DM, several Tylenol cold, cough, and flu preparations, and many other medications as well.

Because so many common over-the-counter medications contain sympathomimetics or dextromethorphan, it is nearly impossible to keep up with all of them. You can best protect yourself by reading the warning labels that come with these medications and by checking with your doctor or pharmacist before you combine any with an MAOI.

Diabetics taking MAOIs need to know that the MAOIs may also cause blood levels of insulin as well as some oral hypoglycemic agents to increase. As a result, your blood sugar may fall more than expected. This can cause a hypoglycemic reaction, with dizziness, faintness, sweating, and so forth, because your brain does not get enough sugar from your blood. Your doctor may have to adjust the doses of your diabetic medications if you are on an MAOI.

Any of the MAOIs can lower your blood pressure, and so they can intensify the effects of other blood-pressure medications your doctor has prescribed, including diuretics and beta-blockers. The MAOIs can also cause the blood levels of a number of blood-pressure medications to increase. This also tends to intensify their effects. As noted above, some blood-pressure medications can have the paradoxical effect of causing an increase in blood pressure if you are taking an MAOI. Make sure you let your doctor know about the MAOI. Many major tranquilizers (neuroleptics) can also cause blood pressure to fall, and MAOIs can increase this effect as well.

Some painkillers must be avoided if you are taking an MAOI. For example, a single injection of the painkiller meperidine (Demerol) has been known to cause seizures, coma, and death in patients taking MAOIs. Other opiates, including morphine, are thought to be safer. Most mild nonprescription painkillers, such as aspirin or Tylenol, are also thought to be safe as long as they contain no caffeine. However, cyclobenzaprine (Flexeril), which is commonly used to treat local muscle spasm, can cause fever, seizures, and death. This drug should be avoided entirely.

Many local and general anesthetics can also interact with the MAOIs. Some local anesthetics contain epinephrine or other sympathomimetic drugs that can create hypertensive reactions. Inform your dentist that you are taking an MAOI so she or he can choose a local anesthetic that will be safe for you. If you require elective surgery while on an MAOI, it would be best to discontinue the MAOI for one or two weeks prior to the surgery. Some general anesthetics, such as halothane, can cause excitement or excessive sedation as well as hyperpyretic reactions when combined with an MAOI. The muscle relaxants used by anesthesiologists, such as succinylcholine or tubocurarine, may also have more potent effects. Make sure you inform your anesthesiologist if you are taking an MAOI.

Sedative drugs, including alcohol, major tranquilizers (neuroleptics) and minor tranquilizers, barbiturates and sleeping pills, can interact with MAOIs. This is especially true for phenelzine (Nardil). Because phenelzine also tends to be sedating, it can enhance the effects of any other sedative agent. You should try to avoid combining MAOIs with sedative drugs because the sleepiness you experience could be hazardous, especially if you are driving or operating dangerous machinery.

L-tryptophan is another sedative agent that should not be combined with MAOIs because it can cause a hyperpyretic crisis (serotonin syndrome). L-tryptophan is an essential amino acid that is present in certain foods such as meats and dairy products. It used to be available in health food stores and has been actively promoted as a natural sedative agent to help people with insomnia. It has also been used as a treatment for depression, but the evidence for its antidepressant effects is meager at best. Following ingestion, L-tryptophan rapidly accumulates in the brain, where it is converted into serotonin. If the dose of L-tryptophan is large enough, you will begin to feel sleepy. If you are taking an MAOI, the increase in brain serotonin may be massive. This is because your brain cannot metabolize the excess serotonin when you are on an MAOI, so the levels of serotonin can escalate to dangerous levels, triggering the serotonin syndrome.

However, some researchers have purposely treated depressed patients with an MAOI plus 2 to 6 grams per day of L-tryptophan in an attempt to make the MAOI treatment more effective. The purpose of these augmentation strategies is to convert a drug nonresponder into a drug responder. Some studies have indicated that this combination can be more potent than treatment with an MAOI alone. Such a treatment is somewhat dangerous, and should probably be administered by experts and reserved for patients with very difficult, resistant depressions.[20] Dr. Jonathan Cole and his colleagues have given doses of 3 to 6 grams of L-tryptophan to patients who had been taking an MAOI for several weeks or more.[1] They observed some early signs of the serotonin syndrome in these patients, suggesting the potential benefits of this drug combination may not be worth the risk.

In animal studies, the combination of lithium with an MAOI can also cause the serotonin syndrome. This is because lithium causes L-tryptophan to enter the brain more rapidly. L-tryptophan is present in the foods we eat, and a large meal can contain as much as 1 gram of L-tryptophan. If you combine lithium with an MAOI, you may get a large increase in serotonin in your brain following meals. However, some doctors have added lithium to an MAOI if the MAOI has not been effective, in just the same way they might add L-tryptophan to try to augment the antidepressant effect of the MAOI. If you receive lithium plus an MAOI, you must be monitored closely to make sure you do not develop any symptoms of the serotonin syndrome, such as fever, tremor, jerking of the muscles, or confusion.

MAOIs are often combined with lithium for another reason. Bipolar patients with abnormal episodic mood elevations as well as depression are often maintained indefinitely on lithium or another mood stabilizer, as described below. During the depressed phase of the cycle, many bipolar patients will need an antidepressant as well as lithium to reverse the depression. The MAOIs, as well as many other kinds of antidepressants, have been used safely and successfully in this way. However, patients need to be monitored closely for signs of hyperpyretic crises as well as episodes of mania, which can occur on rare occasions when bipolar patients receive antidepressants.

Stimulants, pep pills, and weight-loss pills are especially dangerous when combined with MAOIs. Some of these drugs are categorized as sympathomimetics, and they can cause hypertensive crises. For example, methylphenidate (Ritalin), which is widely used for the treatment of attention deficit disorder in children and adults, is a sympathomimetic that could have this effect. Several commonly abused street or prescription drugs are also sympathomimetics. These include the amphetamines such as Benzedrine, Dexedrine, and Methedrine (also known as "speed" or "crank") and cocaine. Amphetamines used to be prescribed for weight loss, but their abuse potential is so high that most doctors no longer prescribe them for this purpose. However, a number of the newer popular weight-loss drugs can also be quite dangerous when mixed with MAOIs. For example, phentermine (Adipex; Fastin) can cause hypertensive reactions and fenfluramine (Pondimin), the controversial weight-loss drug that was recently in vogue, can lead to hyperpyretic crises.

As you know, caffeine is also a mild stimulant. It can cause racing of the heart, an irregular heartbeat, or increased blood pressure if you are taking an MAOI. Although coffee, tea, soda, and chocolate all contain caffeine, they are not strictly forbidden, especially in moderate amounts, because their effects are usually mild. Nevertheless, you should avoid caffeine in large quantities because it could precipitate a hypertensive crisis. Some experts recommend a daily maximum of two cups of coffee or tea, or two sodas. In addition, if you monitor your blood pressure with your own blood-pressure cuff, as described above, you can see whether that cup or two of coffee you love in the morning is actually causing a rise in blood pressure. If so, then you should cut down or give up caffeine completely while you are on the MAOI.

You can see in Table 20–9 that L-dopa (levodopa), which is used in the treatment of Parkinson's disease, can also cause increases in blood pressure when combined with an MAOI. However, patients with Parkinson's disease are sometimes treated with the MAOI selegiline, as well as other medicines. If these patients receive an MAOI along with L-dopa, the L-dopa should be started at a very small dose and increased slowly while checking the blood pressure.

As noted above, most of the forbidden drugs have warning labels to indicate they can be dangerous when combined with some antidepressant medications. If you are taking an MAOI, check the warning labels carefully before you take any new drug, and always check with your druggist or doctor as well. For a detailed list of drugs that cause hypertensive reactions for patients on MAOIs, see pages 157–160 of *Psychotropic Drugs Fast Facts* by Drs. Jerrold S. Maxmen and Nicholas G. Ward.[17] The *Physician's Desk Reference (PDR)*[21] also lists dangerous drug interactions for any prescription medication you may be taking. It is available in any library, drugstore, or medical clinic.

The lists of forbidden foods and medications may seem somewhat confusing or overwhelming. If your doctor prescribes an MAOI, she or he can give you a card to carry in your wallet that lists the foods and drugs to avoid. When in doubt, you can check the card. Some experts advise patients on MAOIs to carry Med-Alert cards so that any emergency room doctors will know that they are taking an MAOI in case they are in an accident or found unconscious and in need of emergency treatment. Then the doctors can take appropriate precautions when administering anesthesia or prescribing other drugs for you.

Remember that the chemical effects of an MAOI remain in your body for as much as one to two weeks after you stop taking it. This is why you must continue to observe the drug and dietary precautions for at least two weeks after you have taken your last MAOI. I would suggest that you actually wait a bit longer. Then you can begin to eat the forbidden foods, such as cheese, in small amounts at first, followed by blood-pressure checks. If your blood pressure is not affected, you can gradually increase the amount you eat until your diet is back to normal. Similarly, if you are switching from an MAOI to another antidepressant, you will have to be completely drug-free for two weeks after you take your last MAOI before starting the new antidepressant.

The same is true if you are starting an MAOI after you have taken another medication—you will have to wait for a period of time, depending on which medication you took. You will recall that you have to wait at least five weeks before starting an MAOI after going off Prozac because this drug remains in your blood for a prolonged time. Most of the other SSRIs are cleared out of your body more rapidly than Prozac, and so a two-week waiting period is usually sufficient. Some antidepressant drugs, such as nefazodone (Serzone) and trazodone (Desyrel), leave your body even faster, and you may have to wait only one week after taking them before starting an MAOI. Always check with your physician before making any changes in your medications.

Well, by now you may be asking whether it is worth it to take a drug like an MAOI which may seem so complicated and dangerous. This question is especially relevant these days, when so many newer and safer drugs are available. Usually, I would try at least two other drugs first. The SSRI drugs, in particular, often help the same types of patients who used to benefit from the MAOIs. I would like to emphasize, however, that in my experience, the MAOIs can usually be administered safely. I have prescribed them for many patients over the years. When doses are kept at a modest level, the side effects tend to be minimal. And when the MAOIs do work, their effects can be quite phenomenal.

In fact, some of my most impressive successes with medications have been with these MAOI drugs, especially tranylcypromine (Parnate). In addition, I have used these drugs with difficult patients who had experienced many unsuccessful treatments with drugs as well as psychotherapy. When these individuals did improve, the degree of improvement was sometimes extreme. These positive experiences with MAOIs have made a strong impression on me. I believe the enthusiasm of the physicians who use the MAOIs is quite justified. If your physician suggests a medication of this type, it might prove to be well worth the necessary extra effort (taking your blood pressure daily), sacrifice (no pizza!), and self-discipline (avoiding certain foods and medicines).

One last note is that a newer and safer MAOI drug, moclobemide, is being marketed in other parts of the world, including Canada, Europe, and South America. Unlike the MAOIs described above, the effects of moclobemide do not persist after you stop taking it. In addition, it does not seem to interact with tyramine in the diet to nearly the same degree. Dr. Alan Schatzberg and his colleagues[1] have pointed out that moclobemide appears to have very few side effects and that the risk of serious drug interactions is relatively low. Psychiatrists hope that moclobemide or another new MAOI called brofaromine will eventually be marketed in the United States.

Serotonin Antagonists

Two antidepressant drugs in the table on pages 514–515 are classified as "serotonin antagonists." They are trazodone (Desyrel) and nefazodone (Serzone). Their mechanism of action appears to be somewhat different from most other antidepressants. Trazodone and nefazodone can boost serotonin by blocking its reuptake at nerve synapses, much like the SSRIs described above. However, these drugs have less potent effects on the serotonin pump than the SSRIs, or even the older tricyclic antidepressants, and this is probably not how these drugs work.

As described in Chapter 17, trazodone and nefazodone appear to block some of the serotonin receptor sites on postsynaptic nerve membranes. At least fifteen different kinds of serotonin receptors have been discovered in the brain. The two receptors that are blocked by trazodone and nefazodone are called 5-HT_{2A} and 5-HT_{2C} receptors. 5-HT is simply shorthand for serotonin; the number and letter after the 5-HT identify the specific type of receptor. Trazodone and nefazodone indirectly stimulate another type of serotonin receptor called the 5-HT_{1A} receptor. This receptor is thought to be important in depression, anxiety, and violence. According to one theory, the stimulation of these 5-HT_{1A} receptor sites might explain the antidepressant effects of trazodone and nefazodone. In addition, trazodone and nefazodone are effective anti-anxiety drugs. If you tend to be nervous and worried, like many depressed individuals, these medications may be especially helpful for you.

Doses of Trazodone and Nefazodone. The starting dose for trazodone is 50 to 100 mg per day. Most patients will do well on 150 mg to 300 mg per day. The starting dose for nefazodone is 50 mg twice per day. The doses of both drugs can be increased very slowly over several weeks to a maximum of 600 mg per day.

Nefazodone and trazodone have short half-lives. The

half-life is the time it takes your body to get rid of half of the drug that is in your system. A drug with a short half-life leaves the blood fairly rapidly and must be taken two or three times per day. In contrast, a drug like Prozac, with an extremely long half-life, leaves your body slowly and needs to be taken only once per day.

As with any antidepressant, you should monitor your mood with a test like the one in Chapter 2 while taking trazodone and nefazodone. This will show whether the drugs are working, and to what extent. If you have not improved substantially after three or four weeks, it may be wise to switch to another drug. Although withdrawal symptoms are quite rare for these medications, it is wise to taper off nefazodone and trazodone slowly, rather than stopping them suddenly. This is good advice with any antidepressant.

Side Effects of Trazodone and Nefazodone. The most common side effects of these two drugs are listed in Table 20–10 on page 601. One common side effect is stomach upset (such as nausea). This side effect is also common with the SSRIs and other drugs that stimulate the serotonin systems in the brain. The upset stomach is more likely when nefazodone and trazodone are taken on an empty stomach, and so it can be helpful to take them with food, just like the SSRIs.

Trazodone and nefazodone can also cause dry mouth in some patients. Both drugs can also cause a temporary drop in blood pressure when you stand up, resulting in dizziness or light-headedness. Trazodone is much more likely to cause these problems than nefazodone. Elderly people are more prone to dizziness and fainting, and so nefazodone may be a better choice for them. As discussed above, several things can alleviate this problem: get up more slowly; walk in place when you get up so as to ''pump'' blood back to your heart from your legs; use support stockings; and take adequate amounts of fluid and salt to prevent any dehydration. Talk to your doctor if you have problems with

Table 20–10. Side Effects of Serotonin Antagonists[a]

Note: This list is not comprehensive. In general, side effects that occur in 5% or 10% or more of patients are listed, as well as rare but dangerous side effects.

Side Effect	*Sedation and Weight Gain*	*Light-Headedness and Dizziness*	*Blurred Vision, Constipation, Dry Mouth, Speeded Heart, Urinary Retention*	*Common or Significant Side Effects*
Brain Receptor	*histamine (H_1) receptors*	*alpha-adrenergic (α_1) receptors*	*muscarinic (M_1) receptors*	
nefazodone (Serzone)	+ to ++	++	+	dry mouth and throat; headache; tiredness; insomnia; nausea; constipation; weakness; dizziness; blurred vision; abnormal vision; confusion
trazodone (Desyrel)	+++	++ to +++	0	dizziness; dry mouth and throat; upset stomach; constipation; blurred vision; headache; fatigue; sleepiness; confusion; anxiety; priapism (rare; see text)

[a]The + to +++ ratings in this table refer to the likelihood that a particular side effect will develop. The actual intensity of the side effect will vary among individuals and will also depend on how large the dose is. Reducing the dose can often reduce side effects without reducing effectiveness.

dizziness or other side effects; she or he may be able to lower the dose.

Another major side effect of trazodone is that it makes you sleepy. This is why it is best taken at night. If you are taking another antidepressant, your doctor may also prescribe a small dose of trazodone at bedtime in order to promote sleep. This is because some antidepressants, such as Prozac and the MAOIs, tend to be stimulating and may interfere with sleep. Trazodone is not addictive and it will not cause dependency or addiction the way some sleeping pills do. The calming, sedative effects of trazodone also help to reduce anxiety. If you tend to be worried and high-strung, this may be a good drug for you. Nefazodone is much less sedating than trazodone, and is not a useful medication for insomnia. In fact, it can occasionally have the opposite effect of causing restlessness, in much the same way that the SSRIs do.

Another adverse side effect of trazodone is called "priapism." Priapism is an involuntary erection of the penis. Fortunately, this side effect is quite rare, occurring in approximately one male patient out of 6,000. It has been reported in only a few hundred cases so far. Personally, I have never seen a case of priapism, but men who take trazodone should be aware that it is remotely possible. If the priapism is not treated right away, it can lead to damage to the penis and permanent impotence (the inability to get an erection). Some patients require surgery to correct the priapism. Injecting a drug like epinephrine directly into the penis can sometimes counteract the priapism if you catch it quickly enough. If this unusual side effect does occur, or if you are beginning to notice an erection that will not go away, contact your doctor or go to an emergency room right away. Nefazodone, on the other hand, does not cause priapism.

Priapism sounds frightening, but I do not mean to discourage men from taking this medication. If you read the

Physician's Desk Reference carefully, you will see that there is a remote chance of a dangerous side effect from nearly any drug you might take, including aspirin. Priapism is a very unlikely side effect of trazodone and can be treated at any emergency room if you act rapidly when the symptom first develops.

Some patients taking these drugs report visual "trails" or afterimages when they are looking at objects that are moving. This side effect is also quite unusual and similar in some respects to the visual images reported by individuals who take LSD, but not dangerous. These visual trails are more common with nefazodone than with trazodone and occur in slightly more than 10 percent of patients taking this drug. They often improve over time.

Drug Interactions for Trazodone and Nefazodone. As noted earlier, some drug combinations can be dangerous because one drug causes the level of the other drug in your blood to become excessively high. Nefazodone has the effect of raising the blood level of a number of drugs. These include commonly prescribed drugs for anxiety, including many of the minor tranquilizers such as alprazolam (Xanax), triazolam (Halcion), buspirone (BuSpar) and others. As a result, you should be very cautious when combining these drugs with nefazodone, because you could become excessively sleepy.

Trazodone will also enhance the sedative effects of other sedative drugs because trazodone itself will make you sleepy. Consequently, trazodone or nefazodone can enhance the sedative effects of any drug that makes you sleepy, such as alcohol, barbiturates, sleeping pills, painkillers, some major tranquilizers (neuroleptics), and some antidepressants. Be very cautious if you combine any sedative agents with nefazodone or trazodone, especially if you are driving or operating dangerous machinery.

Nefazodone can increase the levels of several tricyclic

antidepressants in your blood, especially amitriptyline (Elavil), clomipramine (Anafranil), and imipramine (Tofranil), so the doses of these drugs may need to be lower than usual.

If nefazodone is combined with one of the SSRIs, there is the possibility that a metabolite of nefazodone called mCPP (m-chlorophenylpiperazine) could build up in your blood. This substance may lead to agitation or feelings of panic or unhappiness. If you are switching from an SSRI to nefazodone, mCPP could also build up because the effects of the SSRIs can persist in your body for several weeks after you stop taking them. Neither trazodone nor nefazodone should be combined with an MAOI antidepressant because this combination could trigger the serotonin syndrome (hyperpyretic crisis) described previously.

If you are taking nefazodone, make sure you inform your psychiatrist about any blood-pressure medication you are taking, and inform your general medical doctor as well. Your blood pressure may drop more than expected if you combine trazodone with a blood-pressure medication. If your blood pressure does drop too much, you may notice dizziness when you suddenly stand up. Many psychiatric medications can also lower the blood pressure, including many of the tricyclic antidepressants as well as a number of the major tranquilizers (neuroleptics). If these drugs are combined with trazodone or nefazodone, the drop in blood pressure may be pronounced.

Trazodone can also cause increased blood levels of the anticonvulsant, phenytoin (Dilantin) as well as the heart medication, digoxin (Lanoxin). These combinations can lead to toxic blood levels of phenytoin or digoxin. Make sure your doctor monitors your blood levels of phenytoin or digoxin carefully if you take trazodone, as excessively high levels can be dangerous.

The effects of trazodone on the blood thinner, warfarin (Coumadin) are unpredictable. The levels of warfarin may increase or decrease. If the warfarin levels increase, you may have a greater tendency to bleed, and if the warfarin decreases, your blood may have a greater tendency to clot. Your doctor can monitor any changes with blood tests and adjust the dose of warfarin if necessary.

Even more dangerous are the previously described interactions between nefazodone and two commonly prescribed antihistamines that are given for allergies (terfenadine, trade name Seldane) and astemizole (trade name Hismanal). Nefazodone causes the levels of these two antihistamines to increase, which can result in potentially fatal changes in heart rhythms. Nefazodone should not be combined with cisapride (trade name Propulsid, a stimulant for the gastrointestinal tract) for the same reason—sudden fatal heart failure can result.

Bupropion (Wellbutrin)

Three other types of antidepressant drugs are listed in the Table of Antidepressants on pages 514–515. These include bupropion (Wellbutrin), venlafaxine (Effexor), and mirtazapine (Remeron). They are somewhat different from each other and from the antidepressants already discussed.

Bupropion was supposed to be introduced in the United States in 1986, but its release was delayed until 1989 because a number of patients with bulimia (binge-eating followed by vomiting) who were treated with this drug had seizures. Further studies indicated that the danger of seizures was related to the dose of bupropion and that the risk was much lower in patients who did not have eating disorders, so the drug was released again. Because of the increased seizure risk with bupropion, the manufacturer recommends that this drug not be prescribed to anyone with a history of epilepsy, a major head injury, a brain tumor, bulimia, or anorexia nervosa.

Bupropion does not affect the serotonin system in the

brain. Instead, it seems to work by potentiating the norepinephrine system, much like the tricyclic antidepressant called desipramine (Norpramin). There is also some evidence that it may stimulate the dopamine system in the brain, but these effects are much weaker, and it is not clear whether they contribute to the antidepressant effects of bupropion. Nevertheless, bupropion is sometimes classified as a "combined noradrenergic-dopaminergic antidepressant," because of its effects on the norepinephrine and dopamine systems.

Bupropion is used to treat depressed outpatients and inpatients over the entire range of depression severity. Preliminary studies suggest that it may also be useful for a number of other problems, including smoking cessation, social phobia, and attention deficit disorder. These widespread effects of bupropion do not mean this drug is special, however. Nearly all antidepressants have been reported to be at least partially effective for a wide array of problems including depression, all of the anxiety disorders, eating disorders, anger and violence, chronic pain, and many other problems as well. One possible interpretation for these findings is that these drugs may not really be specific antidepressants. Instead, they may have widespread effects throughout the brain.

A new use for bupropion is to enhance the effects of the SSRI antidepressants. Suppose, for example, that you are taking a drug like Prozac but you have not responded to it adequately. Instead of switching you to a new drug, your doctor may add a low dose of bupropion in an attempt to enhance the effect of the Prozac. Bupropion, in doses of up to 225 mg to 300 mg per day, has been added to SSRI antidepressants in an attempt to combat the sexual side effects of SSRIs, such as loss of libido and difficulties having orgasms.

In my clinical experience, the effects of these drug combinations have often been disappointing. I would usually prefer to try another medication rather than combining drugs when a medication does not work. I am personally concerned that in some instances patients may be in danger of being overmedicated by physicians who are a bit too enthusiastic about adding more and more drugs in larger and larger doses. Also, because I rely so heavily on psychotherapeutic interventions in my own clinical work, I do not feel so much pressure to find a solution from drugs alone. Therefore I do not feel quite so much concern when one or more medications fails to work. I simply switch to another medication and continue to try a variety of new psychotherapeutic strategies, a combination that I find most successful.

Doses of Bupropion. You can see in Table 20–1 on pages 518–523 that the usual dose range for bupropion is 200 to 450 mg per day. At doses below 450 mg per day, the risk of seizures appears to be about four patients per 1000. However, the risk is ten times higher at doses above 450 mg per day—four patients per 100 will experience seizures. Whenever possible, it is good to keep the dose in the lower range to minimize the chance of seizures. In addition, no single dose should ever be greater than 150 mg.

Side Effects of Bupropion. The most common side effects of bupropion are listed in Table 20–11 on pages 608-609. Unlike the tricyclics, bupropion does not cause dry mouth, constipation, dizziness, or tiredness. It also does not stimulate the appetite. This is a big bonus for patients who have been bothered by weight gain. However, some patients have reported upset stomach (nausea).

Bupropion is also somewhat activating and can cause insomnia. Therefore, it may be relatively more effective for depressed patients who tend to feel tired, lethargic, and unmotivated—the stimulating effect may help get you moving. In this regard, it is similar to some of the tricyclic antidepressants (for example, desipramine), the SSRIs (for example, Prozac) and the MAOIs (for example, tranylcypromine).

Table 20–11. Side Effects of Other Antidepressants[a]

Note: This list is not comprehensive. In general, side effects that occur in 5% or 10% or more of patients are listed, as well as rare but dangerous side effects.

Side Effect	*Sedation and Weight Gain*	*Light-Headedness and Dizziness*	*Blurred Vision, Constipation, Dry Mouth, Speeded Heart, Urinary Retention*	*Common or Significant Side Effects*
Brain Receptor	*histamine (H_1) receptors*	*alpha-adrenergic (α_1) receptors*	*muscarinic (M_1) receptors*	
bupropion (Wellbutrin)	0 to +	0 to +	0 to +	dry mouth; sore throat; upset stomach; loss of appetite; stomach pain; sweating; headache; insomnia; restlessness; tremor; anxiety; sweating; dizziness; rash; ringing in ears; seizures
venlafaxine (Effexor)	0	0	0	dizziness; dry mouth and throat; upset stomach; loss of appetite; constipation; sweating; headache; drowsiness; insomnia; anxiety;

				weakness; tremor; blurred vision; problems with orgasm; loss of interest in sex; abnormal dreams; increased blood pressure
mirtazapine (Remeron)	+++	++	+ to ++	dry mouth; increased appetite and weight gain; constipation; sleepiness; dizziness. Warning: Consult your M.D. if you develop signs of an infection (such as fever). This could indicate a rare fall in the white cell count, a rare but dangerous side effect. Can also cause increased blood levels of cholesterol and triglycerides

[a]The + to +++ ratings in this table refer to the likelihood that a particular side effect will develop. The actual intensity of the side effect will vary among individuals and will also depend on how large the dose is. Reducing the dose can often reduce side effects without reducing effectiveness.

Drug Interactions for Bupropion. Because bupropion can substantially increase the risk of seizures, it should not be combined with other drugs that can also make a person more vulnerable to seizures. This includes many psychiatric drugs such as the tricyclic and tetracyclic antidepressants, the SSRIs, the two serotonin antagonists (trazodone and nefazodone), and many of the major tranquilizers (neuroleptics). In addition, there is a greatly increased risk of seizures when alcoholics suddenly stop drinking or when individuals abruptly stop taking minor tranquilizers (benzodiazepines such as Xanax or Valium), barbiturates, or sleeping pills. Bupropion is therefore especially risky for alcoholics and for individuals taking sedatives or tranquilizers regularly.

Many nonpsychiatric drugs (for instance, corticosteroids) can also increase the risk of seizures. Therefore, great caution must be exercised if bupropion is combined with any of these drugs, especially if the dose of bupropion is high. Make sure you check with your pharmacist or druggist about drug interactions if you are taking any other medication along with bupropion.

There are several other kinds of drug interactions you need to consider if you are taking bupropion:

- Barbiturates can cause the level of bupropion in the blood to fall. This could make the bupropion ineffective.
- Phenytoin (Dilantin) can also cause bupropion levels to fall, thus making the bupropion less effective. However, phenytoin is most often prescribed for epilepsy, and so patients taking phenytoin are not likely to receive bupropion.
- Cimetidine (Tagamet) may increase bupropion levels in the blood. This can increase the likelihood of side effects or toxic effects, including seizures.
- Bupropion must not be combined with the MAOIs because of the risk of a hypertensive crisis.

- L-dopa increases the side effects of bupropion; caution is required when these drugs are combined.

Venlafaxine (Effexor)

This is a relatively new antidepressant that is in a distinct class from other antidepressant medications. Released in 1994, it is called a "dual uptake inhibitor" or "mixed uptake inhibitor." This has a very simple meaning. It leads to increases in two types of chemical messengers (also called neurotransmitters) in the brain—serotonin and norepinephrine—by blocking the pumps that transport them back into the presynaptic nerves after they are released into the synapses.

As you will recall from Chapter 17, this capacity to increase levels of two different types of chemical messengers is not new. Many of the older and cheaper tricyclic antidepressants, such as Elavil (amitriptyline) also do this. The more important difference with venlafaxine is that it has fewer side effects because it does not stimulate the histaminic, alpha-adrenergic, and muscarinic brain receptors that cause tiredness, dizziness, dry mouth, and so forth. However, as you will see below, venlafaxine has quite a number of side effects of its own. Some of these, such as nausea, insomnia, and sexual difficulties, are similar to the SSRI antidepressants, and some (such as tiredness) are similar to the tricyclic antidepressants.

It has been claimed that the onset of action may be faster with venlafaxine because of its dual effects on serotonin and norepinephrine receptors. This does not seem likely, because the older tricyclic antidepressants also have dual effects on serotonin and norepinephrine receptors in the brain, but do not have rapid antidepressant effects. Research is now in progress to try to determine whether venlafaxine really does work any more rapidly.

Although a faster-acting antidepressant would represent an important breakthrough, we should probably not become too optimistic about this. Claims about the superior properties of new antidepressants have often not been substantiated by careful, systematic, independent research after the drugs have been available on the market for a period of time. In addition, you will see below that venlafaxine must be started at low doses and increased very slowly to prevent side effects from developing. For most patients, this will prevent the drug from having any rapid antidepressant effects.

Studies are in progress to examine the larger question of whether drugs with dual action have stronger antidepressant effects than SSRIs for certain types of patients, especially severely depressed patients who are hospitalized. This is important because the SSRIs (such as Prozac) which are now so popular have not been particularly effective for these patients. In one study, venlafaxine was more effective than Prozac in the treatment of inpatients with "melancholic" depression. "Melancholic" depression refers to a more severe depression with many organic features, such as waking up too early and a loss of appetite and sexual drive. Individuals with melancholic depressions may also have anhedonia along with feelings of guilt that can become extreme or even delusional. Anhedonia refers to a severe loss of the capacity to experience pleasure or satisfaction.

Like all antidepressants, venlafaxine is beginning to be used for a number of other disorders as well. These include chronic pain disorder as well as adult attention deficit disorder (ADD). Remember that all, or nearly all, antidepressants have been used for a great variety of disorders, so it is not likely that the effects of venlafaxine are superior for chronic pain or for ADD.

Doses of Venlafaxine. Some experts recommend starting venlafaxine at 18.75 mg twice per day, which is only half the starting dose recommended by the manufacturer, in order to minimize the likelihood that nausea will develop.[1] Then the daily dose can be slowly increased by

37.5 mg every third day until a total dose of 150 mg per day or above is reached. Most patients respond to a total dose of 75 mg to 225 mg per day. Higher doses tend to be more effective, but they are associated with more side effects.

Earlier when discussing SSRIs, we talked about the half-life of drugs—this is the time required by the body to eliminate one half of the drug in your body. Venlafaxine has a short half-life—meaning that it leaves your body in a matter of hours. Therefore, you must take the medication two or three times per day to maintain an adequate level in your bloodstream.

The manufacturer has recently marketed an extended (slow) release version of venlafaxine (called Effexor XR) that you need to take only once per day, which is more convenient. As you can see in Table 20–1 on page 521, the extended release capsules appear to be more costly, but this is really an illusion. For example, you can see in the table that the average wholesale price of a hundred of the 75 mg capsule of Effexor is $118.66, whereas the price of a hundred of the 75 mg extended-release capsules is $217.14, or almost twice as much. When I first saw these figures, I naturally concluded that the extended-release capsules were twice as expensive as the regular pills.

But let's see what happens in a real-life situation. Suppose your dose is 75 mg per day. You could take either one of the regular 37.5 mg pills in the morning and a second 37.5 mg pill in the evening, for a total cost of $2.17 per day, or one of the 75 mg extended-release pills once per day. As noted above, the cost of the 75 mg extended-release pills will also be $2.17 per day. Either way, Effexor is very expensive, since the daily dose may be as high as 375 mg per day. The high price is especially striking when you compare the cost of Effexor with the cost of many of the generic tricyclic antidepressants that are just as effective and available for less than ten cents per day.

As with any antidepressant, it is best to taper off venlafaxine slowly. At least two weeks are recommended, and some patients may require as much as four weeks.

Side Effects of Venlafaxine. The side effects of venlafaxine are listed in Table 20–11 on pages 608–609. As you can see, they are similar to the SSRI compounds described above. The most common side effects of venlafaxine are nausea, headache, sleepiness, insomnia, abnormal dreams, sweating, nervousness, and tremor. Venlafaxine can also cause the same types of sexual difficulties as the SSRIs, including a loss of interest in sex and difficulties achieving orgasm. These sexual side effects tend to be quite common, just as with the SSRIs. In spite of the claim that venlafaxine has fewer side effects than the older tricyclic antidepressants, this drug can nevertheless cause dry mouth and dizziness in some patients. The dizziness is particularly likely if you go off the drug too quickly.

One distinct type of side effect seen with venlafaxine is an increase in blood pressure. However, the blood-pressure increases are typically seen only at higher doses (225 mg per day or above). Nevertheless, if you have problems with your blood pressure, you and your doctor should monitor your blood pressure carefully, and this drug may not be a good choice for you. At doses less than 200 mg per day, the likelihood of an increase in blood pressure is only about 5 percent. The probability increases to 10 percent or 15 percent at doses greater than 300 mg per day. Blood-pressure increases of 20 to 30 mm of mercury have been observed, for example.

Drug Interactions for Venlafaxine. Because venlafaxine is relatively new, information about its interactions with other drugs is still relatively limited. Venlafaxine appears to be less likely to interact in adverse ways with other medications you are taking. Several drugs may cause blood levels of venlafaxine to increase, and so lower doses of venlafaxine may be needed. These include:

- some tricyclic antidepressants;
- the SSRI antidepressants;
- cimetidine (Tagamet).

Venlafaxine may cause the blood levels of several of the major tranquilizers to increase. These include trifluoperazine (Stelazine), haloperidol (Haldol), and risperidone (Risperdal), and so lower doses of these drugs may be needed. In theory, these drugs could also cause blood levels of Venlafaxine to increase.

Venlafaxine must not be combined with MAOI antidepressants because of the danger of the serotonin syndrome (hyperpyretic crisis) described on page 576. Remember that it takes up to two weeks for the effects of an MAOI to clear out of your body, so a two-week drug-free period will be required if you stop taking an MAOI and then start taking venlafaxine. In contrast, if you go off venlafaxine and then start taking an MAOI, a one-week drug-free period should be sufficient, because venlafaxine leaves the body fairly rapidly.

Mirtazapine (Remeron)

Mirtazapine (Remeron) was released in the United States in 1996. It also enhances both norepinephrine and serotonin activity, but through a different mechanism from venlafaxine. Premarketing studies suggest that mirtazapine may be effective for mildly depressed outpatients and for more severely depressed inpatients as well. It may also be particularly helpful for depressed patients who are very anxious or nervous.

Doses of Mirtazapine. The dose range for mirtazapine is 15 to 45 mg per day. Most physicians will prescribe a smaller amount at first (7.5 mg per day) and then slowly increase the dose. Because mirtazapine causes sleepiness in more than 50 percent of the people who take it, it can be given once a day at bedtime, usually in doses of 15 to 45 mg per day. Some physicians report that mirtazapine is less likely to cause less sleepiness when the dose is increased. This is the opposite of what you might expect intuitively. It is because the drug may have some stimulating effects at the higher doses. We will have to wait until there is more clinical experience with this drug to see if this is really true.

Side Effects of Mirtazapine. The side effects of mirtazapine are listed in Table 20–11 on pages 608–609. You can see that it blocks the histaminic, alpha-adrenergic, and muscarinic receptors in much the same way that the older tricyclic antidepressants do. Therefore, the side effect profile of mirtazapine is very similar to the tricyclics, especially amitriptyline, clomipramine, doxepin, imipramine and trimipramine (see Table 20–2). The more common side effects include tiredness (54 percent of patients) noted above, increased appetite (17 percent), weight gain (12 percent), dry mouth (25 percent), constipation (13 percent), and dizziness (7 percent). Keep in mind that these figures are somewhat inflated because they do not take into account the placebo effect. For example, 2 percent of patients on placebo also report weight gain, and so the true incidence of weight gain that can be attributed to the mirtazepine would be 12 percent minus 2 percent, or 10 percent. Mirtazepine is not likely to cause the stomach upset, insomnia, nervousness, and sexual problems commonly seen with the SSRIs such as Prozac.

Mirtazapine has some unique adverse effects not shared with other antidepressants. It can, in rare cases, cause your white blood cell count to fall. Because these cells are involved in fighting off infections, this could make you more vulnerable to a variety of infections. If you develop a fever while taking this drug, make sure you contact your physician immediately so that he or she can obtain a complete blood count. Mirtazapine can sometimes cause an increase in levels of blood fats such as cholesterol and triglycerides.

This could be a problem if you are overweight or have a heart condition or if your cholesterol and triglycerides levels are already elevated.

Drug Interactions for Mirtazapine. Because mirtazapine is relatively new, very little information about its drug interactions is available. It must not be combined with the MAOI antidepressants because of the risk of the serotonin syndrome (hyperpyretic crisis). Because it can be quite sedating, it will enhance the effects of other sedative drugs. These include alcohol, major and minor tranquilizers, sleeping pills, some antihistamines, barbiturates, many other antidepressants, and the antianxiety drug buspirone (BuSpar). The increased sleepiness you experience when these substances are combined with mirtazapine could lead to difficulties with coordination and concentration. This might be hazardous when driving or operating dangerous machinery.

Mood Stabilizers

Lithium

In 1949, an Australian psychiatrist named John Cade observed that lithium, a common salt, caused sedation in guinea pigs. He gave lithium to a patient with manic symptoms and observed dramatic calming effects. Tests of the effects of lithium in other manic patients yielded similar results. Since that time, lithium has slowly gained popularity throughout the world. It has been used successfully in the treatment of a number of conditions, including:

- Acute manic states. Although lithium is used to treat patients with severe mania, they will usually be treated with more potent, faster-acting drugs at the same time until the severe symptoms of mania have subsided. These other drugs include the antipsychotics (also known as major tranquilizers or neuroleptics) such as chlorpromazine (Thorazine), as well as benzodiazepines (also called "minor tranquilizers") such as clonazepam (Klonopin) or lorazepam (Ativan). These additional drugs are used until the mania has been brought under control. Once the severe manic symptoms subside, the other drugs are discontinued and the patient continues taking the lithium to prevent future mood swings.

- Recurrent manic and depressive mood swings in individuals with bipolar manic-depressive illness. Lithium has significant preventative effects, so that the likelihood of future manic episodes is reduced.
- Single episodes of depression. Lithium is sometimes added in smaller doses to an antidepressant drug that is not working in order to try to improve its effectiveness. I will describe this and other augmentation strategies later in the chapter.
- Recurrent episodes of depression in patients without manic mood swings. Lithium maintenance may help to prevent recurrences of depression following recovery. Some studies indicate that the preventative effects of long-term lithium treatment may be similar to the effects of long-term treatment with an antidepressant such as imipramine. However, this preventative effect on depression may not work for all patients. Lithium is probably more likely to prevent depressions in patients with a strong family history of bipolar (manic-depressive) illness.
- Individuals with episodic anger and irritability or outbursts of violent rage.
- Individuals with schizophrenia. Lithium can be combined with an antipsychotic medication, and the combination may be more effective than the antipsychotic medication alone. The improvement seems to occur in schizophrenic patients who also experience mania or

depression and in schizophrenic patients without any symptoms of mania or depression.

You should keep in mind that in all of these conditions, lithium is sometimes helpful but rarely ever curative. Like most medications, it is a valuable tool but not a panacea.

As noted above, manic-depressive illness is sometimes also called bipolar illness. "Bipolar" simply means "two poles." Patients with bipolar illness experience uncontrollable euphoric mood swings that often alternate with severe depressions. The manic phase is characterized by an extremely ecstatic, euphoric mood, inappropriate degrees of self-confidence and grandiosity, constant talking, nonstop hyperactivity, increased sexual activity, a decreased need for sleep, heightened irritability and aggressiveness, and self-destructive impulsive behavior such as reckless spending binges. This extraordinary disease usually develops into a chronic pattern of uncontrollable highs and lows that can come on unexpectedly throughout your life, so your physician may recommend that you continue to take lithium (or another mood stabilizing drug) for the rest of your life.

If you have experienced abnormal mood elevations along with your depression, your physician will almost definitely prescribe lithium or another comparable mood-stabilizing drug. Some studies suggest that if you are depressed and have a definite family history of mania, you might benefit from lithium even if you have never been manic yourself. However, most physicians would first prescribe a standard antidepressant and observe you carefully. Although antidepressants do not usually cause euphoria or mania in people with depression, they can occasionally have this effect in individuals with bipolar manic-depressive illness. The mania can begin as quickly as twenty-four to forty-eight hours after starting the antidepressant.

In my clinical practice, the development of a sudden and dangerous manic episode after starting an antidepressant has been quite rare, even in patients with bipolar illness. Nevertheless, if you have a personal or family history of mania, it is conceivable that you could experience this side effect. Be sure to tell your doctor about this so you can receive careful follow-up after starting an antidepressant. Your family, too, should be alerted to this possibility. Family members are often aware of the development of a manic episode before the patient realizes what is happening, and can alert the doctor that a problem has developed. This is because the distinction between normal happiness and the beginning of the mania may be unclear to the patient. Furthermore, mania feels so good at first that you may not recognize it as a dangerous side effect of the medication you are taking.

Doses of Lithium. As you will see in Table 20–1, lithium comes in 300-mg dosages, and normally three to six pills per day in divided doses are required. Your physician will guide you. Initially, you may take the lithium three or four times per day. Once you are stabilized on lithium, you may be able to take half your total daily dose in the morning and half before you go to bed. This twice-a-day schedule will be more convenient.

Sustained-release capsules containing 450 mg are also available. Because these drugs are released more slowly in the stomach and gastrointestinal tract, they may cause fewer side effects and they are more convenient because you don't have to take them so often. However, their increased cost, as compared with generic lithium, may not justify taking them. Furthermore, many patients have reported that the side effects of the inexpensive, generic brands of lithium are no different from the more expensive slow-release brands.

Like the other drugs used for treating mood disorders, lithium usually requires between two and three weeks to become effective. When taken for a prolonged period of time, its clinical effectiveness seems to increase. Thus, if you take it for a period of years, it may help you more and more.

Unfortunately, there appears to be a group of individuals

who do well on lithium, stop taking it, become symptomatic again, and then find that the lithium is less effective when they start taking it again. This is one reason why you should not stop taking lithium, or any other medication, without first consulting with your doctor.

Lithium Blood Testing. Too much lithium in your blood can cause dangerous side effects. In contrast, if your blood level is too low, the drug will not help you. Because there is a narrow "window" of effectiveness of lithium, blood-level testing is required to make sure that your dose is neither too high nor too low. Initially, your doctor will order more frequent blood tests so that she or he can determine what the proper dose should be. Later on, when your dose and symptoms have stabilized, you will not need the blood tests nearly as frequently.

If you are an outpatient and you are not experiencing severe mania, your doctor may order lithium blood tests once or twice a week for the first couple weeks, then once a month. Eventually, blood tests every three months may be sufficient.

If you are being treated for a more severe episode of mania, more frequent blood tests will be required. This is because higher blood levels of lithium are usually needed to control the severe symptoms. In addition, your body tends to get rid of lithium more rapidly during an episode of mania, so larger doses may be needed to maintain the proper blood level. As noted above, during a manic episode your doctor will almost definitely want to combine lithium with more potent drugs for the first few weeks until your symptoms have subsided.

Your blood must be drawn eight to twelve hours after your last lithium pill. The best time for a blood test is first thing in the morning. If you forget and take your lithium pill the morning of a blood test, *don't take the test!* Try again another day. Otherwise, the results will be misleading to your doctor.

Body size, kidney function, weather conditions, and

other factors can influence your lithium dose requirement, so blood tests should be performed on a regular basis when you are on lithium maintenance. Your doctor will probably try to maintain your blood level at somewhere between 0.6 and 1.2 mg per cc, but this will vary with your symptom level. During an episode of acute mania, your doctor will probably want to keep your blood level closer to the top of the therapeutic range. Some doctors feel that levels as low as 0.4 to 0.6 mg per cc can be effective to help prevent an episode of depression or mania when you are feeling good.

Patients with chronic irritability and anger may also respond to lithium at these lower blood levels, even if they don't suffer from clear symptoms of manic-depressive illness. The advantage of these lower levels is that there are fewer side effects.

Other Medical Tests. Prior to treatment, the doctor will evaluate your medical condition and order a series of blood tests and a urinalysis. These blood tests will usually include a complete blood count, tests of thyroid and kidney function, electrolytes, and blood sugar. Your thyroid functioning should be tested at six-month or yearly intervals while you are taking lithium because some patients on lithium develop goiters (a swelling or lump on the thyroid gland) or underactive thyroid glands. Your kidney function must also be evaluated from time to time because of kidney abnormalities reported in some patients taking lithium. Your doctor may order an electrocardiogram (ECG) before you start taking the lithium, especially if you are over forty or if you have a history of heart problems. Your doctor will also need to know about any other drugs you may be taking, because some of them may cause elevations in your blood lithium level. These include certain diuretics as well as some anti-inflammatory drugs such as ibuprofen, naproxen, and indomethacin. You will learn below that some drugs can have the opposite effect of causing your lithium level to fall.

Side Effects of Lithium. The side effects of lithium are listed in Table 20–12 on pages 624–625 and compared with the side effects of two other mood stabilizers I will discuss below. As you can see, lithium tends to have many side effects. Most of them are mildly uncomfortable but not serious.

Starting with the effects on the muscles and nervous system first, you will see that lithium can cause a fine tremor of the hands and fingers in 30 percent to 50 percent of patients. This tremor will be present when your hands are resting and often worsens when you do something purposeful with your hands. For example, the tremor can make it more difficult to hold a cup of coffee or to write clearly. The severity of the tremor is related to the dose and may be more severe when lithium is prescribed along with one of the tricyclic antidepressants, which can also cause tremor.

This tremor is one of the major reasons that some patients stop taking their lithium. An antitremor drug called propranolol (Inderal) can be given if the tremor is especially severe and troublesome, but it is my policy to avoid prescribing an additional drug if possible. A reduction in dose can also help.

If your doctor does prescribe propranolol, the usual dose to reduce a lithium tremor is 20 to 160 mg per day, given in divided doses. It is best to start with small doses and increase gradually. The smallest effective dose is best. This is because propranolol can have other effects, including a slowing of the heart, a drop in blood pressure, weakness and fatigue, mental confusion, and upset stomach. Propranolol can also cause breathing difficulties and must not be given to patients with asthma. It is also contraindicated for patients with Raynaud's disease. Metoprolol (25 to 50 mg) or nadolol (20 to 40 mg), drugs similar to propranolol, have also been used to treat lithium tremor.

Lithium may cause tiredness and fatigue initially, but these effects will generally disappear with time. Some patients complain of mental slowing or forgetfulness, partic-

Table 20–12. Side Effects of the Mood Stabilizers[a]

Category	*lithium*	*valproic acid*	*carbamazepine*
Muscles and Nervous System	tremor problems with coordination tiredness mental slowing or dulling memory loss	tremor problems with coordination tiredness weakness	dizziness problems with coordination tiredness weakness
Stomach and Gastrointestinal Tract	upset stomach weight gain diarrhea	upset stomach weight gain abnormalities in liver function pancreatitis	upset stomach abnormalities in liver function dry mouth
Kidneys	nephrogenic diabetes insipidus (excessive urination and thirst) interstitial nephritis, leading to (usually mild) renal insufficiency		syndrome of inappropriate secretion of antidiuretic hormone (SIADH)

Skin	rash hair loss acne	rash hair loss	rash
Heart	ECG changes		abnormal heart rhythms
Blood	increased white blood cell count	decreased platelets with bleeding problems	decreased platelets with bleeding problems bone marrow failure (rare)
Hormonal	hypothyroidism	menstrual changes	decreased levels of thyroid hormones (T3 and T4)

[a]Information in this table was obtained in part from the *Manual of Clinical Psychopharmacology*[1] and *Psychotropic Drugs Fast Facts*.[17] These excellent references are highly recommended.

ularly younger individuals. The forgetfulness has been confirmed by memory testing. Other antidepressants that have anticholinergic properties, such as Elavil, can also cause forgetfulness. Complaints about these mental changes are very common and cause many patients to stop taking their lithium. Memory difficulties seem to be more pronounced at higher lithium blood levels, as might be expected, and often improve when the dose is reduced.

Along the same lines, some patients complain of substantial weakness and fatigue. These symptoms often indicate an excessive lithium level, and a dose reduction may be indicated. Extreme sleepiness with mental confusion, a loss of coordination, or slurred speech suggests a dangerously elevated lithium level. Discontinue the drug and seek immediate medical attention if such symptoms appear.

Some patients express the fear that they may lose their creativity when they start taking lithium. This is especially of concern for artists and writers who have used their highs and lows as a source of painful inspiration for creative expression. Indeed, many well-known painters and poets through the centuries suffered from manic-depressive illness, and their moods were clearly reflected in their work. However, three quarters of patients on lithium report that it does not seem to reduce their creativity, and in some cases their creativity increases.[1]

Turning next to the digestive system, lithium can cause an upset stomach or diarrhea that is most troublesome during the first few days of treatment. These side effects will usually disappear with time. It may help to take the lithium with food or to take it in three or four divided doses throughout the day, so that your stomach isn't hit with a large dose all at once. It can also help to increase the dose of lithium more slowly. In rare cases lithium can cause vomiting as well as diarrhea, and your body may become dehydrated because of all the fluid loss. This can make your blood levels of lithium higher, and so the drug becomes more toxic. This, in turn, can cause more nausea and diarrhea, creating a vicious cycle. Medical attention may be needed to make sure you are adequately hydrated until the episode has passed.

Unfortunately, many patients on lithium experience weight gain; this is another common reason patients stop taking the drug. Dr. Alan Schatzberg[1] has suggested that this problem will be greater if you are already overweight. The weight gain results from the stimulation of your appetite. This is often very difficult to control. Obviously, if you exercise more and eat less, the weight gain can be prevented or reversed, but this is often much easier said than done! If the weight gain is excessive or troublesome, switching to an alternative mood stabilizer, such as carbamazepine, may be helpful.

Increased thirst and frequent urination can also occur when taking lithium. In some cases, patients develop intense thirst from urination that is so frequent and voluminous that the lithium must be stopped. This condition, known as nephrogenic diabetes insipidus (NDI), results from the effects of lithium on the kidneys. It is usually reversible when the lithium is stopped. In some cases, adding certain types of diuretics can also help. However, careful lithium monitoring must be performed, because these diuretics can cause increases in plasma lithium levels. Milder forms of increased urination probably occur in one half to three quarters of patients who take lithium.

Lithium can cause a form of kidney damage called "interstitial nephritis." This term simply means inflammation or irritation of the tissue. When first reported, psychiatrists were quite alarmed about this complication. Subsequent experience has indicated that although the problem may occur in 5 percent or more of patients who take lithium for many years, the degree of kidney impairment is usually mild. Your doctor will nevertheless want to monitor your kidney function periodically while you are on lithium. She or he will order two blood tests called the creatinine test and the blood urea nitrogen (BUN) test once or twice a year. These tests can be performed at the same time you are having your usual lithium blood test taken. If the tests indicate a change in kidney function, your doctor may request a consultation with a urologist and order a twenty-four hour creatinine clearance test. This is a more accurate test of kidney function and will involve saving all your urine for twenty-four hours in a special bottle that the clinical laboratory will give you. The results will help your doctor evaluate whether it will be safe for you to continue taking lithium.

An occasional patient will develop a rash, and patients with psoriasis who take lithium will often experience a flare-up of the condition. This may require consultation with a dermatologist, switching to another brand of lithium, going off lithium temporarily, or switching to one of the other mood-stabilizing medications. Acne may also worsen during lithium treatment. This can be treated with antibiotics or retinoic acid, but in some cases the lithium may have to be stopped. Some patients complain of hair loss, but the hair usually grows back, whether or not the patient continues taking lithium. It is interesting to note that lithium-related hair loss occurs primarily in women, and hair can disappear from anywhere on the body. Hair loss is sometimes a sign of hypothyroidism (see below) and so your doctor may order a thyroid blood test if the problem persists.

Lithium can cause a variety of changes in the electrocardiogram (ECG), but these are usually not serious. Older patients, as well as those with heart disease, should have an ECG taken before they start on lithium, as noted above. The ECG can be repeated once you are stabilized on lithium to see if there are any changes in heart rhythm that might be a cause for concern.

You can see in Table 20–12 that lithium can also cause an increase in your levels of white blood cells. These are the cells that normally fight infection. A normal white blood cell count is in the range of 6,000 to 10,000. The white blood cell count in patients on lithium typically increases to the range of 12,000 to 15,000 per cc, elevations that are not considered dangerous. However, if you go to a physician because you are ill, make sure you remind him or her that you are taking lithium and that the lithium may cause a false elevation of your white blood cell count. Otherwise, your doctor may falsely conclude that you have a serious infection, even if you actually do not.

Finally, lithium can affect thyroid functioning in as many as 20 percent of patients. As noted above, one common effect is an increase in the size of the thyroid gland (called a "goiter") without any changes in thyroid function. Other patients develop increases in the levels of thyroid stimulating hormone (TSH) in the blood. This indicates that the body is trying harder to stimulate the thyroid gland. As many as 5 percent of patients on lithium will develop hypothyroidism, and this may require treatment with thyroxine (0.05 to 0.2 mg per day), a thyroid hormone replacement. Hypothyroidism is more common in women than in men.

Lithium Drug Interactions. As you can see in Table 20–13 on pages 630–631, lithium interacts with many other drugs. Make sure you review this list with your physician if you are taking other medications at the same time you are taking lithium.

The drugs near the top of the table may cause lithium levels in the blood to increase. This can lead to more side effects, including lithium toxicity. The dose of lithium may need to be reduced to maintain blood levels in the proper range. These drugs that cause increased lithium levels include several drugs commonly used in the treatment of high blood pressure, such as the so-called ACE inhibitors, the calcium channel blocking agents, and methyldopa (Aldomet). The calcium channel blocking agents in particular may lead to greater lithium toxicity, with symptoms such as tremor, loss of coordination, nausea and vomiting, diarrhea, and ringing in the ears. Caution is required if you combine lithium with any of these drugs.

Many common non-steroidal anti-inflammatory drugs (NSAIDs) such as ibuprofen (Advil, Motrin, and other trade names) can also cause lithium levels to increase. Several

Table 20–13. Lithium Drug Interactions[a]

Note: This list is not exhaustive; new information about drug interactions comes out frequently. If you are taking lithium and any other medication, ask your doctor and pharmacist if there are any drug interactions.

Drugs Which Cause Blood Lithium Levels or Lithium Toxic Effects to Increase				
ACE (angiotensin-converting enzyme) inhibitors • benazepril (Lotensin) • catopril (Capoten) • enalapril (Vasotec) • fosinopril (Monopril) • lisinopril (Prinivil, Vestril) • quinapril (Accupril) • ramipril (Altace)	alcohol antibiotics • ampicillin (Omnipen) • spectinomycin (Trobicin) • tetracycline (Achromycin) anticonvulsants • carbamazepine (Tegretol) • phenytoin (Dilantin) • valproic acid (Depakene)	antifungal agents • metronidazole (Flagyl) calcium channel blockers • diltiazem (Cardizem) • nifedipine (Procardia) • verapamil (Isoptin) diuretics (loop type) • ethacrynic acid (Edecrin) • furosemide (Lasix)	diuretics (thiazides) • chlorothiazide (Diuril) • hydrochlorothiazide (Alsoril) diuretics (potassium-saving type) • amiloride (Midamor) • spironolactone (in Aldactazide) ketamine low-salt diet	mazindol (Sanorex) methyldopa (Aldomet) non-steroidal anti-inflammatory drugs • diclofenac (Voltaren) • ibuprofen (Advil) • indomethacin (Indocin) • ketoprofen (Orudis) • piroxicam (Feldene) • phenylbutazone (Butazolidin)

Drugs Which Cause Blood Lithium Levels or Lithium Toxic Effects to Decrease

acetazolamide (Diamox) bronchodilators • albuterol (Proventil) • aminophylline (Mudrane) • theophylline (Bronkaid)	caffeine (in coffee, tea, soda, chocolate) corticosteroids • hydrocortisone (Cortef) • methylprednisolone (Medrol)	osmotic diuretics sodium bicarbonate salty foods urea

Other Lithium Drug Interactions

Drug	*Effect*	*Drug*	*Effect*
antipsychotic agents • chlorpromazine (Thorazine) • haloperidol (Haldol) • thioridazine (Mellaril)	may cause increased lithium toxicity or increase the likelihood of neuroleptic malignant syndrome (NMS) (quite rare)	digitalis (Crystodigin; Lanoxin) hydroxyzine (Atarax, Vistaril) tricyclic antidepressants	abnormal heart rhythms and heart slowing abnormal heart rhythms increased likelihood of tremor

[a]Some information in this table was obtained from *Psychotropic Drugs Fast Facts*, pp. 213–215.[17] This book is an excellent source of information on psychiatric medications.

antibiotics raise lithium levels, as does the common anti-fungal agent metronidazole (Flagyl), which is often used to treat vaginal infections. Several anticonvulsants are also listed in the top portion of Table 20–13. If you are taking any of these medications, you might need lower doses of lithium.

If you have high blood pressure, you may also be treated with a diuretic (or water pill). Some diuretics cause lithium levels to increase. The loop diuretics and potassium-saving diuretics in Table 20–13 do not increase lithium levels as much as the thiazide diuretics that are listed there. Not all diuretics cause lithium levels to rise. For example, you can see in Table 20–13 that osmotic diuretics, which work a little differently from the others, can have the opposite effect of causing lithium levels to fall.

Your doctor may prescribe a low-salt diet if you have high blood pressure. However, a low-salt diet can cause lithium levels to rise. This is because your kidneys will excrete less salt in an attempt to preserve it. Since lithium is also a salt that is chemically very similar to table salt, your kidney will also excrete less lithium. By the same token, if you are sweating a great deal during the summer months, this can have the same effect of depleting your body of salt and causing your lithium levels to increase. Once again, your kidneys will try to preserve salt and lithium as well. Make sure you maintain an adequate intake of salt to compensate for the salt you will lose if you are sweating a great deal.

The opposite effect can also occur. You can also see in Table 20–13 that if you eat too much salt, it can cause lithium levels to fall. This is because your kidneys will sense that there is too much salt in your blood and will try to get rid of it. Your kidneys will excrete more lithium along with the extra salt.

In contrast, the drugs listed in the middle of Table 20–13 have the opposite effect of causing lithium levels in the blood to fall. As a result, lithium can lose its effectiveness. You can see that several drugs used in the treatment of asthma reduce serum lithium levels. Caffeine also has the same effect, so if you are a heavy coffee drinker, you may need to cut down on coffee or take higher doses of lithium. Corticosteroids, which are used in many conditions including poison ivy, can also cause lithium levels to fall. The dose of lithium may need to be increased to maintain blood levels in the proper range if you are taking any of these drugs.

A number of other drug interactions are listed in Table 20–13. Psychiatrists used to think that the combination of lithium with certain antipsychotic medications (especially haloperidol) greatly increased the risk of a toxic effect called NMS (neuroleptic malignant syndrome). NMS consists of severe muscle rigidity and confusion along with elevated temperature, profuse sweating, increases in blood pressure, rapid heartbeat and breathing, trouble swallowing, abnormal kidney and liver function, and other symptoms. However, although any patient on antipsychotic drugs runs a small risk of developing NMS, recent clinical experience has indicated that the likelihood of NMS may be increased only slightly when antipsychotics are combined with lithium. Lithium is now often used in combination with antipsychotic drugs and may enhance their effects in the treatment of schizophrenia, as described above.

As with most psychiatric drugs, pregnant women should avoid lithium, if possible, because its use has been associated with birth defects involving the heart. This is not an all-or-nothing issue, and the potential benefits must be weighed against the potential hazards. The risk of a heart defect known as Ebstein's anomaly is twenty times greater than normal in mothers who take lithium, but the likelihood is still less than 1 percent. Other birth defects can also occur, especially when lithium is used during the first trimester of pregnancy. In addition, lithium (as well as some other psychiatric drugs) is secreted in human milk and should be avoided by nursing mothers. If lithium is needed, breast-feeding should be avoided.

If you or your doctor have any questions about lithium

(as well as the other mood stabilizers described below), the lithium information center at the Madison Institute of Medicine, Madison, Wisconsin, can often help.[22]

Valproic Acid

Valproic acid is usually used in the treatment of epilepsy but was recently granted FDA approval for the treatment of bipolar disorder, especially acute mania. You can see in Table 20–1 on page 522 that this drug is prescribed in one of two forms: valproic acid (Depakene) or the slightly more expensive divalproex sodium form (Depakote). The two forms are equally effective. Studies comparing valproic acid with lithium indicate that the two drugs are comparably effective and both appear to be twice as effective as a placebo. Valproic acid, like lithium, also appears to be effective in preventing or reducing future manic episodes. The drug may be especially effective in the treatment of the rapid-cycling form of bipolar disorder. It can help patients who experience mania and depression at the same time (so-called "mixed states"), as well as patients who experience the more common forms of bipolar disorder. It is probably less effective in the prevention and treatment of depression than in the prevention and treatment of mania.

Doses for Valproic Acid. It is best to start valproic acid gradually, in order to minimize the side effects. The dose on the first day might be 250 mg administered with a meal. During the first week, the dosage can be gradually raised up to 250 mg given three times a day. As with any medication, the dose you receive may be slightly different depending on your size, gender, and clinical symptoms. For example, a man who weighs 160 pounds might be started on 500 mg twice a day.

During the second and third weeks, the dose may be slowly increased further. Most patients end up with a total daily dose in the range of 1,200 to 1,500 mg, given in divided doses (for example, 400 mg three times per day). Individual doses can vary widely. Some patients respond to as little as 750 mg per day and others need as much as 3,000 mg per day. As with any drug, doses outside the normal range are occasionally needed.

Some improvement should be observed within two weeks of attaining a therapeutic blood level. If you respond to valproic acid, your doctor may suggest that you remain on it for an extended period of time, just like lithium.

Blood Testing. Your doctor will order blood tests to adjust your dose of valproic acid. Initially your doctor may order a blood test once a week until your dose and blood level are stabilized. After that you will need a blood test only every month or two.

The blood should be drawn approximately twelve hours after your last dose, just like the lithium blood test. Most patients take valproic acid in divided doses twice a day. If so, the blood can be drawn in the morning, before you take your first daily dose. Most physicians think that a blood level of 50 to 100 micrograms per ml is therapeutic, but others are comfortable with blood levels up to 125 mcg per ml, especially if the patient is acutely manic. Of course, more side effects are observed at the higher blood levels.

Prior to treatment, your doctor will probably order a blood test to check your liver enzymes, a bleeding test, and a complete blood count (which includes a platelet count). These additional blood tests are performed because in rare cases valproic acid can cause hepatitis (an inflammation of the liver) as well as bleeding problems. From time to time after you have been on valproic acid, your doctor will repeat these tests to make sure that no changes have occurred. Many physicians feel that it is probably necessary to check the blood count and liver enzymes only every six to twelve months, especially if the patient has been educated to report immediately any signs or symptoms that indicate a liver inflammation, as described below. You should also tell your doctor if you notice any excessive bleeding or easy bruising.

Temporary increases in liver enzymes have been reported in as many as 15 percent to 20 percent of patients during the first three months of treatment. In most cases, these elevations are not considered serious. Nevertheless, if your liver enzymes do change, your doctor will probably reduce the dose of valproic acid and continue to monitor the liver enzymes. Your doctor will also want you to be educated about the symptoms of hepatitis so you can contact him or her immediately if they develop. Jaundice is the classic symptom. Jaundice is a condition in which your urine becomes dark and your skin and eyes become yellow in color. In addition, your bowel movements become pale. When the liver becomes inflamed, the pigment that normally causes your bowel movements to become brown gets backed up in your blood, staining your eyes, skin, and urine. Other symptoms of hepatitis include fatigue, nausea, a loss of appetite, tiredness, and weakness. Fortunately, hepatitis only rarely complicates treatment with valproic acid and can usually be treated successfully, especially if you notify your physician right away.

Although the liver inflammation is nearly always mild, it is important to watch carefully for these symptoms because they could, in theory, progress to fatal liver failure. This complication has been observed in infants and is rarely seen in adults. It usually occurs in individuals taking other anticonvulsants at the same time. In fact, some experts assert that it has not been seen in adults who take only one anticonvulsant.[17]

Side Effects of Valproic Acid. The side effects of valproic acid are listed in Table 20–12 on pages 624–625. On the average, valproic acid is usually better tolerated by patients than lithium because it has fewer side effects. Sleepiness is a common side effect. Taking more of your daily dose in the evening before you go to bed can often prevent the sleepiness from being problematic. Valproic acid can also cause stomach upset which can take the form of nausea, vomiting, cramping, or diarrhea. These effects on the gastrointestinal tract are less common and can often be helped by taking a drug like Pepcid twice a day. Drs. J. S. Maxmen and N. G. Ward indicate that the frequency of stomach upset is greater with valproic acid (15 percent to 20 percent) than with the enteric-coated divalproex sodium (10 percent) tablets, and so a switch to divalproex sodium may help if these symptoms are troublesome.[17]

You can see in Table 20–12 that valproic acid can also cause tremor. As with lithium, this effect can sometimes be helped by reducing the dose or by adding one of the beta-blocking drugs (see the discussion of lithium tremor above). Other uncommon side effects include a loss of coordination and weight gain.

Valproic acid can cause a rash in 5 percent of patients, much like the two other mood stabilizers listed in Table 20–12. Some patients have also reported hair loss, and if this develops you should discontinue the drug (after discussing this with your doctor, or course) because it can take several months for the hair to grow back. The hair loss is thought to be due to the fact that valproic acid can interfere with the metabolism of zinc and selenium. Vitamin supplements containing these two metals can be taken to try to prevent this. Dr. Alan Schatzberg and his colleagues recommend the vitamin supplement Centrum Silver for this purpose.[1]

As many as 20 percent of women have reported menstrual irregularities while on valproic acid. This may be due to the fact that valproic acid can cause blood levels of the relevant hormones to fall, resulting in impaired ovulation. Paradoxically, valproic acid can also cause certain oral contraceptives to fail, so in theory you could become pregnant. Make sure you discuss this possibility with your doctor if you are taking oral contraceptives.

Valproic acid, like a number of other anticonvulsants, may lead to birth defects and should usually not be taken during pregnancy. The deformities include a cleft lip, clotting abnormalities, spina bifida, and others. During the latter phases of pregnancy (the third trimester) valproic acid can cause liver toxicity for the developing baby, especially when blood levels are greater than 60 mcg per ml. Make sure you inform your doctor if you think there is any chance you could become pregnant while taking this drug.

Special precautions are indicated for women under twenty who receive long-term treatment with valproic acid. Some studies have suggested that they may be more likely to develop polycystic ovaries and increased levels of male sex hormones, but the actual incidence of this complication is not known.[17]

Drug Interactions for Valproic Acid. Valproic acid does not seem to have as many drug interactions as lithium or carbamazepine. Because valproic acid can cause sleepiness, it can enhance the effects of other sedative drugs such as alcohol, major and minor tranquilizers, barbiturates, or sleeping pills. These combinations could be hazardous, especially when driving or operating dangerous machinery. In addition, valproic acid can cause substantial increases in blood levels of barbiturates, causing extreme sedation or intoxication. Valproic acid may also cause levels of diazepam (Valium) to rise. The resulting depression of the central nervous system can be serious, and so great caution must be exercised if these drugs are combined with valproic acid.

As noted above, valproic acid can interfere with bleeding and clotting, and so caution needs to be exercised if it is combined with other drugs that interfere with bleeding or clotting, such as warfarin (Coumadin) or aspirin. In addition, valproic acid can lead to increased blood levels of warfarin. This can also enhance the tendency to bleed.

Some caution should be exercised when valproic acid is combined with a tricyclic antidepressant (especially nortriptyline and amitriptyline) because the blood levels of the antidepressant may increase. Your doctor may want to order a blood test to check the level of the antidepressant so the dose can be adjusted if necessary.

Several types of drugs can cause levels of valproic acid to increase. These include:

- antacids;
- non-steroidal anti-inflammatory drugs such as aspirin, ibuprofen (Advil, Motrin), and others;
- cimetidine (Tagamet);
- erythromycin (Erythrocin);
- felbamate (Felbatol), an anticonvulsant;
- lithium. Valproic acid also causes lithium levels to rise, and so the toxic effects of both drugs can increase;
- some antipsychotic drugs, especially phenothiazines such as chlorpromazine (Thorazine);
- SSRI antidepressants such as fluoxetine (Prozac) and fluvoxamine (Luvox).

If you are taking any of these drugs with valproic acid, your doctor may need to reduce your dose of valproic acid.

Some anticonvulsants, such as carbamazepine (Tegretol), ethosuximide (Zarontin), phenytoin (Dilantin) and possibly phenobarbital (Donnatal) can cause blood levels of valproic acid to fall, and so doses of valproic acid may need to be increased. At the same time, valproic acid can cause the levels of carbamazepine, phenytoin, phenobarbital, and primidone (Mysoline) to increase, and so the doses of these drugs may need to be reduced when they are combined with valproic acid. Patients with difficult cases of bipolar illness may be treated with more than one mood stabilizer, and some careful attention to these complex drug interactions will be needed.

Finally, the antibiotic rifampin (Rifadin) can cause blood levels of valproic acid to fall. This antibiotic is used in the treatment of tuberculosis, and it is also used as a two-to-four-day preventative treatment for individuals who have been exposed to patients with certain types of meningitis.

Carbamazepine

Carbamazepine (Tegretol) was introduced in the 1960s as a treatment for a certain type of epilepsy that originates in the temporal lobes of the brain. In the 1970s, Japanese investigators discovered that carbamazepine was helpful in treating manic-depressive patients who did not respond to lithium. Although the FDA has not yet officially approved carbamazepine for the treatment of mania and depression, it appears to be helpful for 50 percent of bipolar (manic-depressive) patients who have failed to respond to lithium. Carbamazepine can be combined with lithium or with one of the major tranquilizers (also known as neuroleptics) in order to enhance the effects of these drugs in the treatment of mania.

Carbamazepine can also be helpful for some rapidly cycling manic-depressives. These individuals have more than four manic episodes per year and can sometimes be challenging to treat. Some studies have also suggested that carbamazepine may be helpful for manic-depressive patients who experience anger and paranoia during their "high" phases. Finally, some psychiatrists report that carbamazepine may be helpful in the treatment of patients with borderline personality disorder when severe anxiety, depression and anger coexist with impulsive, self-destructive behavior such as wrist-slashing. However, in one study the therapists but not the patients reported that the carbamazepine was helpful. It is difficult to know how to interpret such findings.

Many of the studies of carbamazepine have been conducted on patients who were also taking other drugs at the same time, such as lithium or a neuroleptic. These drugs can also have effects on mania. Dr. Alan Schatzberg and his colleagues have pointed out that this makes it difficult to tease out the true effects of the carbamazepine.[1] The limited data and patent issues may explain why the drug is not yet approved as a primary treatment for mania—because the safety and effectiveness of the drug in the treatment of mania have not yet been convincingly demonstrated through large, well-controlled studies.

Doses for Carbamazepine. The beginning dose of carbamazepine is 200 mg twice daily for two days. It may then be raised to 200 mg three times a day for five days. After this, the dose is gradually increased by 200 mg per day every five days up to a total daily maximum of 1,200 mg to 1,600 mg.

Carbamazepine usually takes at least one to two weeks to be effective, as do many psychiatric medications. If it is helpful, your doctor will probably suggest you stay on the drug for a longer period of time to prevent a relapse of the mania.

Blood Testing. Carbamazepine blood testing is required, just as it is for the two mood stabilizers discussed above (lithium and valproic acid). You will need a blood test every week for the first two months. After that, you will need a blood test every one or two months. The results will guide your doctor in the amount she or he prescribes. The usual effective blood level for carbamazepine is in the range of 6 mg to 12 mcg per ml, but some experts recommend blood levels in the range of 6 mg to 8 mcg per ml for most patients with depression or mania. Like any drug, there are fewer side effects at lower doses, but if the blood level gets too low, the drug will lose its effectiveness.

Levels of other drugs in your blood may fall if you are taking carbamazepine. This is because carbamazepine stimulates certain liver enzymes, and so your liver clears these drugs out of your system faster than usual. One of the drugs that is affected by carbamazepine is carbamazepine! In other words, after you have been on the drug for several weeks, you may find that you need a larger dose to maintain the same blood level. This is because your liver begins to metabolize the carbamazepine more rapidly, so it leaves your body faster.

Your doctor will probably want to check the blood levels of certain liver enzymes before you start the carbamazepine, and from time to time when you are on it. This is because carbamazepine may cause an elevation of liver enzymes in your blood, indicating possible liver inflammation or damage. Earlier you learned that valproic acid can have similar effects on the liver. Some elevation of liver enzymes occurs in most patients taking carbamazepine, but this is not usually a cause of concern. However, you will still want to watch out for any signs of hepatitis described in the previous section on valproic acid.

Your doctor will also order frequent complete blood counts while you are taking carbamazepine. This is because carbamazepine may cause a drop in your red blood cells, white blood cells, or platelets. These cells are all produced by your bone marrow, and carbamazepine can sometimes make the bone marrow less active. Each type of blood cell serves a different function. The white cells help to fight infections. If you did not have enough white cells, you would be more vulnerable to infections. As noted above, a normal white blood cell count is in the range of 6,000 to 10,000. If your white cell count falls below 3,000, your physician will immediately consult with a hematologist (blood specialist). Roughly 10 percent of patients taking carbamazepine experience a drop in the white blood cell count, and levels below 3,500 are common. You should be reassured to know that a drop in the white blood cell count rarely develops into a serious problem. If carbamazepine is helping you, most doctors will continue prescribing it as long as your white cell counts are above 1,000. However, white cell counts below this level can be extremely dangerous, so your physician will monitor your blood count more frequently if your white cell count starts to drop.

Levels of red blood cells and blood platelets may also fall if you are taking carbamazepine. The red blood cells carry oxygen, and the platelets cause bleeding to stop. If your red blood cells fell to very low levels, you would experience anemia. You might appear pale and feel fatigued. If your platelets fell to low levels, you might experience an increased tendency to bleed. Dr. Alan Schatzberg and colleagues[1] state that these changes in the blood count are expected. They emphasize that good patient education and routine bloodcounts are the best ways to monitor them.[1] If you are taking carbamazepine, make sure you let your doctor know immediately if you develop any symptoms suggesting a change in your white cells, platelets, or red blood cells. These include fever, sore throat or sores in your mouth (indicating possible infection), bruising or bleeding (indicating a possible drop in the platelets in your blood), or fatigue along with pale lips and finger nails (suggesting anemia).

On extremely rare occasions, carbamazepine can cause a dangerous and potentially fatal failure of the bone marrow. In these cases, all your blood cells may drop to dangerously low levels. Recent estimates of severe and dangerous bone marrow failure range from approximately one patient in 10,000 to one in 125,000, so you can see that this complication is very rare.

When carbamazepine was first introduced, this possibility frightened many physicians, who were understandably reluctant to use the drug. Neurologists have been by far the largest group of doctors prescribing carbamazepine because it can be so valuable in the treatment of epilepsy as well as trigeminal neuralgia (facial nerve pain). Neurologists have now had vast experience with this drug and are quite comfortable with its use. More psychiatrists are also starting to recognize that this medication can be used safely.

Side Effects of Carbamazepine. A number of common or significant side effects of carbamazepine are listed in Table 20–12 on pages 624–625. Tiredness is the most common side effect, especially at the start of treatment. A third of patients experience tiredness, and some (5 percent) also complain of weakness. Raising the dose more slowly can minimize these effects. Usually the drowsiness will wear off over time. The drowsiness is usually not due to anemia, but just to the sedative properties of the drug.

Approximately 10 percent of patients report dizziness, especially when standing. This is due to a temporary drop in blood pressure because blood tends to pool in your legs when you rise. As a result, there is not enough blood for your heart to pump to your brain, and you get dizzy. This can usually be minimized by standing more slowly and exercising your legs (such as walking in place) immediately when you stand up. This "squeezes" blood from your legs to your heart so your heart can pump the blood to your brain.

You will see that carbamazepine can sometimes cause problems with coordination. This has been reported in as many as 25 percent of patients. Patients may appear a bit intoxicated and tend to stagger when walking. This sometimes indicates that the dose is too high. Other symptoms of an excess dose include double vision, slurred speech, mental confusion, muscle twitches, tremor, restlessness, and nausea, along with slowed or irregular breathing, a rapid heartbeat and changes in blood pressure. Immediate medical attention is required if these symptoms occur, because in extreme cases overdoses can lead to stupor, coma, and death.

You may also experience some nausea and vomiting at first. These effects are usually temporary and can usually be managed by raising the dose more slowly and by taking the medication with food. These effects are probably less common than with valproic acid or lithium. Most patients who have been on carbamazepine for several weeks do not report these effects.

Like the tricyclic antidepressants, carbamazepine can sometimes cause dry mouth or blurred vision. This is because carbamazepine blocks the cholinergic receptors (also called muscarinic receptors) in the brain. These anticholinergic effects are of special concern to patients with glaucoma, who have increased pressure in their eyes, because the carbamazepine may cause the glaucoma to worsen. If you have glaucoma you should have your intraocular pressures monitored closely while taking carbamazepine (or any drug with anticholinergic properties).

A side effect that involves the kidneys is called the syndrome of inappropriate secretion of antidiuretic hormone (SIADH), or water intoxication. Patients develop a great increase in thirst along with mental confusion and a fall in the levels of sodium in the blood. This side effect has been reported in as many as 5 percent of patients taking carbamazepine. If you develop excessive thirst, your doctor may order an electrolyte test to see if your sodium has dropped. She or he may want to reduce the dose, change to a different medication, or treat you with a drug called demeclocycline (Declomycin). This drug can often correct the problem of low sodium levels in your blood. Your doctor will probably monitor your kidney function from time to time by checking your levels of blood urea nitrogen (BUN) and creatinine.

Carbamazepine can have some adverse effects on the heart. If you are over fifty years of age you should have an ECG before starting the drug. The ECG should be repeated after you have been stabilized on the drug to make sure no changes of a serious nature have occurred. Carbamazepine often causes a slowing of the heart. These changes appear to be more common in older women. If you have a history of heart disease you may do better to take another mood-stabilizing drug with fewer effects on the heart, such as valproic acid.

As many as 5 percent to 10 percent of patients taking carbamazepine may develop a rash. You will see in Table 20–12 that any of the mood stabilizers (as well as many antidepressants) can cause a rash, but this is somewhat more common with carbamazepine. It can sometimes help to avoid direct sunlight (which may provoke the rash in some cases), to take an antihistamine, or to change to a different brand of carbamazepine. This is because you may be allergic to an ingredient in the pill other than the carbamazepine itself. On extremely rare occasions, two severe and potentially fatal skin rashes (called Lyell's syndrome and the Stevens-Johnson syndrome) have been reported in patients taking carbamazepine. Make sure you report any severe skin changes to your doctor immediately.

Like many other psychiatric drugs, carbamazepine can cause birth defects, especially spina bifida. A number of other fetal abnormalities have also been reported recently, especially when the drug is taken during the first trimester of pregnancy. Therefore, the potential benefit must clearly outweigh this risk if the drug is taken during pregnancy. The risk appears to be significantly higher when carbamazepine is combined with other anticonvulsants. If a pregnant woman definitely needs the drug, some experts recommend folic acid supplements that may reduce the likelihood of birth defects.

Carbamazepine is secreted in mother's milk. The concentration of carbamazepine in the milk is approximately 60 percent of the concentration in the mother's blood, and so the issue of nursing must be discussed with the pediatrician.

Drug Interactions for Carbamazepine. You can see in Table 20–14 on pages 648–650 that many drugs can influence the blood level of carbamazepine, and vice versa, so you and your physician will have to be very careful in this regard. At the top of the table, drugs are listed that cause carbamazepine level and toxicity to increase. If you are taking any of these drugs, your doctor may need to reduce the dose of carbamazepine. For example, many of the macrolide antibiotics (erythromycin is a common example) can double the blood level and toxicity of carbamazepine.

You can also see in Table 20–14 that some drugs, such as diuretics (water pills) and other anticonvulsant medications can cause the level of carbamazepine to fall. Your physician may have to give you a larger dose of carbamazepine to compensate for this.

Just as certain drugs can cause blood levels of carbamazepine to rise or fall, carbamazepine can change the levels of other drugs you are taking. Blood levels of the drugs that are listed next on the table may fall when combined with carbamazepine. This is because carbamazepine stimulates the liver enzymes that metabolize these drugs. As a result, the liver gets rid of these drugs more rapidly than usual. This would be equivalent to pulling out the plug while you are trying to fill the bath; the water may not rise to the proper level.

One important example would be birth control pills. The consequence of the decreased blood level is that the birth control pills may become ineffective, and you might become pregnant even though you are taking your birth control pills consistently. Levels of other drugs listed in the table that may fall when combined with carbamazepine include some antidepressants, antipsychotic drugs, anticonvulsants, antibiotics, thyroid hormones, and others.

Sometimes the drug interactions work in both directions. A drug may cause the blood level of carbamazepine to fall, and carbamazepine may in turn cause the blood level of the other drug to fall. For example, if you are taking an antipsychotic medication like haloperidol (Haldol), which is often also given for mania, the haloperidol may cause the level of carbamazepine to fall. At the same time the carbamazepine may cause the blood level of haloperidol to drop substantially. As a result, it may seem that neither drug is working properly, and the mania may not be controlled adequately. Your physician may need to do blood tests to determine the levels of both drugs so that the doses can be adjusted properly. Carbamazepine probably has similar effects on other antipsychotic drugs as well.

Finally, several other potentially dangerous drug interactions with carbamazepine are listed at the bottom of the table. In particular, carbamazepine must not be combined with any of the MAOIs discussed on page 564 because of

Table 20–14. Carbamazepine Drug Interactions[a]

Note: This list is not exhaustive; new information about drug interactions comes out frequently. If you are taking lithium and any other medication, ask your doctor and pharmacist if there are any drug interactions.

Drugs Which Can Cause Carbamazepine Levels or Toxic Effects to Increase				
acetazolamide (Diamox) antibiotics (macrolides) • azithromycin (Zithromax) • clarithromycin (Biaxin) • erythromycin (Pediazole) • troleandomycin (Tao) • other macrolides	antibiotics (other) • doxycycline (Vibramycin) • tetracycline (Achromycin) • ketoconazole (Nizoral) • isoniazid (INH) anticonvulsants • valproic acid (Depakene, Depakote)	antidepressants (SSRIs) • fluoxetine (Prozac) • fluvoxamine (Luvox) • sertraline (Zoloft) • others antidepressants (other) • nefazodone (Serzone) cimetidine (Tagamet)	calcium channel blockers • diltiazem (Cardizem) • verapamil (Calan) danazol (Danocrine) dextroprophoxyphene (Darvon) lipid lowering drugs • gemfibrozil (Lobid) • isonicotinic acid • niacinamide • nicotinamide	lithium mexiletene (Mexitil) prednisolone (Delta-Cortef) propoxyphene (Darvon) terfenadine (Seldane) viloxazine

Drugs Which Can Cause Carbamazepine Levels to Decrease

anticonvulsants	diuretics
• ethosuximide (Zarontin) • phenytoin (Dilantin) • primidone (Mysoline)	fentanyl (Duragesic)
barbiturates	major tranquilizers (neuroleptics)
• phenobarbital • others	• haloperidol (Haldol)
	methadone

Table 20-14. Cont.

Blood Levels of the Following Drugs May Fall When Combined with Carbamazepine			
acetaminophen (Tylenol) antibiotics • doxycycline (Vibramycin) • cyclosporine (Sandimmune; Neoral) anticonvulsants • phenobarbital • primidone (Mysoline) • phenytoin (Dilantin) • valproic acid (Depakene, Depakote)	antidepressants • bupropion (Wellbutrin) • imipramine (Tofranil) • others antipsychotics (neuroleptics) • haloperidol (Haldol) • others	benzodiazepines (minor tranquilizers) • alprazolam (Xanax) • clonazepam (Klonopin) • others corticosteroids • dexamethasone (Decadron) • methylprednisolone (Medrol) • prednisolone (Delta-Cortef)	emergency intubation drugs • pancuronium (Pavulon) • vecuronium (Norcuron) fenantyl (Duragesic) mebendazole (Vermox) methadone (Dolophine) oral contraceptives theophylline (Theo-Dur) thyroid hormones warfarin (Coumadin)

Other Carbamazepine Drug Interactions

Drug	*Effect*
clozapine (Clozaril)	increased possibility of bone marrow suppression
digitalis, digoxin (Lanoxin)	levels rise, may cause toxicity including slowing of the heart
MAOI antidepressants	serotonin syndrome (fever, seizures, coma)

[a]Some information in this table was obtained from *Psychotropic Drugs Fast Facts*, pp. 213–215.[17] This book is an excellent source of information on psychiatric medications.

the risk of the potentially fatal serotonin syndrome.

Although Table 20–14 is lengthy, it is not comprehensive because new drugs and new information about drug interactions are constantly emerging. As noted previously, only a small percentage of the potential drug interactions have been studied, and our knowledge about them is rapidly expanding. Other drugs may have important interactions with carbamazepine, so make sure your physician knows of all the medications you are taking. Ask specifically if any of them interact with carbamazepine.

Other Mood Stabilizing Agents

Until recently, lithium, valproic acid, and carbamazepine were the main drugs used for the treatment of bipolar illness. Recently, new drugs have been synthesized which may soon be available to treat patients with this disorder. Many of these new drugs are actually anticonvulsants that were designed for the treatment of epilepsy. At least two of them are already being used in the treatment of bipolar (manic-depressive) illness, and many others will undoubtedly become available in the next several years. It seems likely that at least some of them will provide powerful new tools for treating bipolar illness and possibly other psychiatric disorders as well.

These new drugs (as well as the three mood stabilizers discussed previously) are quite different from the antidepressants because they do not significantly increase levels of serotonin, dopamine, and norepinephrine in the brain. Instead, they seem to stimulate a transmitter substance called GABA (gamma-amino butyric acid) or inhibit a transmitter substance known as glutamate. GABA and glutamate are used by a large percentage of the nerves in the brain. The anticonvulsants that stimulate GABA tend to cause sleepiness. Medications in this category include valproic acid, discussed above, as well as gabapentin (Neurontin), tiagabine (Gabitril), vigabatrin (Sabril), and several others. The anticonvulsants that inhibit glutamate tend to cause stimulation and anxiety. Medications in this category include felbamate (Felbatol), lamotrigine (Lamictal), topiramate (Topamax), and several others.

Although it is not known for certain why or how these drugs prevent epilepsy or stabilize manic-depressive illness, it is known that the GABA system and the glutamate system in the brain tend to compete with one another. This may be why drugs that stimulate GABA or inhibit glutamate are helpful for epilepsy and for bipolar illness.

Most anticonvulsant drugs also inhibit sodium transport across nerve membranes in the brain. Sodium, as you know, is present in table salt. It is known as an ion, because it carries a tiny positive electrical charge when it is dissolved in a fluid. The electrical impulses of nerves result when ion channels in the nerve membranes open up and positively charged ions like sodium and potassium suddenly rush across the membrane. These ion fluxes create the electrical impulses in the nerves. Because these drugs inhibit the sodium channels, they may stabilize nerve conduction in the brain by making nerves less excitable. Because nearly all anticonvulsants have this property, they are sometimes classified as "sodium blockers." The sodium-blocking effects may also explain why these new drugs can prevent seizures and stabilize manic-depressive illness.

Of course, all new drugs have unforeseen benefits and hazards, and the new anticonvulsant drugs are no exception. Quite a bit of testing will be necessary before we can identify which ones have most promise for patients with epilepsy and bipolar illness. There is considerable excitement about one of the new drugs, called gabapentin (Neurontin), because it seems to have very few side effects, an excellent safety record, and few if any toxic interactions with other drugs. In addition, it does not require blood testing like the three mood stabilizers discussed above.

So far, the FDA has approved gabapentin only for the treatment of epilepsy. Although it has not yet been officially approved for psychiatric disorders, many psychiatrists are beginning to prescribe gabapentin for patients with difficult bipolar illness who have not responded to other medications. Its eventual role will have to be determined by clinical experience and by controlled outcome studies.

At least eight studies of the use of gabapentin in mood disorders were published in 1997, and many more will undoubtedly be published in subsequent years. In these studies, gabapentin was reported to be effective for many patients with bipolar illness. Gabapentin also appeared to have antidepressant and antianxiety properties, and it may be useful in the treatment of chronic pain (including migraine headaches), as well as PMS (premenstrual syndrome), panic disorder, and social phobia.

Doses for Gabapentin. The current dose of gabapentin for epilepsy is 300 mg to 600 mg three times daily, for a total dose range of approximately 900 to 2000 mg per day. In studies of bipolar patients, the average dose was about 1700 mg per day, with some investigators giving doses as high as 3600 mg per day.

The absorption of gabapentin from the stomach and intestinal tract is not affected by food. However, the antacid Maalox can reduce the absorption of gabapentin from the stomach by about 20 percent. Therefore, you should wait at least two hours after taking Maalox before you take gabapentin.

About half of a dose of gabapentin disappears from the body within five to seven hours, so it must be taken several times per day rather than all at once. If you take a high dose of gabapentin on a single occasion, a smaller proportion of the dose will be absorbed from your stomach and intestinal tract into your blood. For example, only 75 percent of a single 400-mg dose is absorbed, as compared with 100 percent of a 100-mg dose. From a practical point of view, this should not be a concern if you are taking gabapentin since you will be taking the medication several times per day in divided doses.

There is no evidence that men and women require different doses because of differences in metabolism, but individuals over seventy years of age may need only about half the doses used for younger people. This is because of changes in kidney function that occur with aging. Because the kidneys excrete gabapentin, individuals with impaired kidney function will require smaller doses.

Unlike lithium, carbamazepine, and valproic acid, blood testing does not appear necessary with gabapentin. This is another advantage of this medication.

Side Effects of Gabapentin. The main side effects are listed in Table 20–15 on pages 656–657. You can see that they include sleepiness, noted above, along with dizziness, tremor, problems with coordination, weight gain, and some visual side effects. All of these side effects will be more pronounced at higher doses and less noticeable at lower doses. Overall, the side effect profile of gabapentin is very favorable, especially when compared with the other currently available mood stabilizers.

In the studies cited in Table 20–15, gabapentin was given to patients with epilepsy who were already receiving one or more other anticonvulsants. Therefore, the side effects that were actually due to the gabapentin were lower. The best way to get a more realistic estimate of any side effect is to subtract the percentage seen in the placebo group from the percentage seen in the gabapentin group. For example, 11.0 percent of the gabapentin group experienced fatigue, whereas 5.0 percent of the placebo group experienced this side effect. The difference in these two numbers is 6.0 percent. This is a better estimate of the true incidence of fatigue that can be attributed to gabapentin.

Like nearly all psychiatric drugs, gabapentin should be used with great caution in pregnant women. Although there are no well-controlled studies of the effects of gabapentin on the developing fetus in pregnant women, fetal abnormalities have been observed when gabapentin was administered to pregnant mice and rabbits. Although animal studies do not always predict human responses, gabapentin should be used in pregnancy only if the need is great and if the potential benefit outweighs the potential risk to the developing fetus. Although it is not yet known whether gabapentin is secreted into human milk, many drugs are secreted into human milk; consequently, gabapentin should probably not be used by mothers who are nursing. Certainly, you should discuss this risk with your physician.

Drug Interactions for Gabapentin. Gabapentin has one unusual and desirable property; it is not metabolized by the liver, but is excreted unchanged by the kidneys directly into the urine. For this reason, it does not seem to interact in adverse ways with other drugs. You will recall from previous discussions that all the antidepressants and mood stabilizers have fairly complicated interactions with lots of other drugs. This is because these drugs compete with each other for certain metabolic enzymes in the liver. With gabapentin, this is not a problem, so it is much safer to combine gabapentin with other medications. In fact, many experts believe that gabapentin has no metabolic interactions at all with other drugs. One benefit is that gabapentin can be combined with other mood stabilizers for patients with difficult cases of bipolar illness or epilepsy who have not responded to other medications.

The properties of gabapentin are certainly very appealing. Is there a downside? Sometimes problems with new medications surface after the medication has been in widespread use for a period of time and the initial excitement has worn off. Gabapentin may be no exception. One concern already voiced by some neurologists and psychiatrists is that the drug may not be particularly effective for either epilepsy or bipolar illness. This would be disappointing, since the drug has so few side effects or interactions with other drugs. A colleague with considerable experience with gabapentin told me she is using it primarily to help anxious

Table 20–15. Side Effects of Gabapentin (Neurontin)

Note: The information in this table was adapted from the 1998 *Physician's Desk Reference* (*PDR*). In these studies, gabapentin or placebo was given to individuals with epilepsy who were already taking at least one other drug for epilepsy. The side effects in individuals not taking other drugs are likely to be less. Only the more common side effects are listed.

	Gabapentin (n = 543)	Placebo (n = 378)
	Digestive System	
weight gain	2.9%	1.6%
dry mouth	1.7%	0.5%
upset stomach	2.2%	0.5%
	Energy	
fatigue	11.0%	5.0%
sleepiness	19.3%	8.7%

Nervous System		
dizziness	17.1%	6.9%
trouble with coordination	12.5%	5.6%
tremor	6.8%	3.2%
slurred speech	2.4%	0.5%
memory problems	2.2%	0.0%
Eyes		
nystagmus (tremor of the eyes)	8.3%	4.0%
double vision	5.9%	1.9%
blurred vision	4.2%	1.1%

patients with insomnia, because it has excellent sedative and relaxing properties and is not habit-forming. Unfortunately, she feels it may not be powerful enough to be a primary mood stabilizer for bipolar patients, but it may have value when it is used in combination with other medications.

Another new anticonvulsant, lamotrigine (Lamictal) has also been approved by the FDA for the treatment of epilepsy. Like gabapentin, lamotrigine has been used in the treatment of treatment-resistant bipolar illness. Dr. Alan F. Schatzberg and colleagues[1] point out that very few formal studies of lamotrigine have been conducted in psychiatric patients, and so the reports of its effectiveness are still mainly anecdotal. In addition, lamotrigine has some significant and troubling side effects. In particular, rashes and skin reactions occur in as many as 5 percent or more of the adults taking lamotrigine. While most of these rashes are not dangerous, lamotrigine can cause a severe and life-threatening skin reaction known as the Stevens-Johnson syndrome in 1 percent to 2 percent of cases. These skin reactions are more common in pediatric patients than in adults, and so lamotrigine should not be given to individuals under sixteen years of age. Taking lamotrigine at higher doses or in combination with other drugs, such as valproic acid, may make these feared skin reactions more likely. In premarketing trials, five patients taking lamotrigine died from liver failure or multiorgan failure.

Lamotrigine causes many other side effects such as headache and neck pain, nausea and vomiting, dizziness, loss of coordination, sleepiness, trouble sleeping, tremor, depression, anxiety, irritability, seizures, speech problems, memory difficulties, runny nose, rashes, itching, double vision, blurred vision, vaginal infections, and others. Lamotrigine also has a number of interactions with other drugs because it is metabolized by the liver. Because it has many side effects, including some dangerous ones, lamotrigine must be used with great caution. Until we learn more about it, it should probably be reserved for patients who have failed to respond to the better-established mood stabilizers discussed above.

What If My Antidepressant Does Not Work?

As I have emphasized, I would recommend taking a mood test like the one in Chapter 2 to monitor your response to any treatment, including medications or psychotherapy. You can take the test once a week or even more frequently, and keep track of your scores. Your scores will show whether and to what extent the treatment is working. The goal of treatment is to get these scores reduced substantially. Ultimately, you want your scores to be in the range considered normal and ideally in the range considered happy.

If a drug doesn't help, or helps only somewhat, what should you do?

1. Make sure you have given the drug a fair trial. Ask yourself:
    - Is the dose adequate?
    - Have you taken the drug for an adequate period of time?
2. Make sure there are no drug interactions that are preventing the antidepressant from being effective. Remember that some other drugs can cause your blood level of an antidepressant to fall, even if you are taking the correct dose of the antidepressant. Inform your doctor about any other drugs you are taking.
3. You and your doctor may want to consider one of the augmentation strategies discussed below.
4. If these procedures are not successful, you and your doctor can discontinue the medication and try another type of antidepressant.

5. Psychotherapy along the lines described in this book, either alone or in combination with an antidepressant, can often be far more effective than treatment with drugs alone.

Let's examine each of these principles. First, you need to be certain the dose is sufficient. If for any reason your blood level of an antidepressant is too low, then the probability of a positive drug response will be diminished. However, a dose that is too high might also be less effective. This is because the side effects at excessively high doses may counteract the antidepressant effects. Concerns about the doses of antidepressant drugs are important because different people can metabolize these drugs quite differently. In other words, given a particular drug at a particular dose, different people can have dramatically different levels of the drug in their blood. In fact, the levels of a tricyclic antidepressant may differ by as much as thirty times in two different people who both receive comparable doses of the same drug. This can happen even if the two people are the same sex, height, and weight.

These differences in blood levels can result from differences in the ways people absorb a drug from their gastrointestinal tracts and from differences in how fast people get rid of a drug from their blood. Genetics can play a role. For example, approximately 5 percent to 10 percent of the Caucasian population in western Europe and the United States lack the liver enzyme called CYP2D6 (in the P450 family), and 20 percent of the Asian population lack the enzyme called CYP2C19.[23] These enzymes help to metabolize a wide variety of drugs including many antidepressants. Individuals who lack either of these enzymes may develop dramatically higher blood levels of certain antidepressants because their liver enzymes cannot get rid of these drugs nearly as rapidly as the average individual.

Medical conditions such as liver, kidney, or heart disease can have an impact on the blood level of antidepressants. Age can also be important. On the average, children and elderly individuals require lower doses of most medications including antidepressants. You may recall, for example, that individuals over sixty-five may develop blood levels of several SSRIs, citalopram (Celexa), fluoxetine (Prozac) and paroxetine (Paxil), that are approximately 100 percent greater than the blood levels of younger individuals taking identical doses. Sometimes gender can play a role as well. As noted previously, men may develop blood levels of fluoxetine (Prozac) or sertraline (Zoloft) that are 30 percent to 50 percent lower than women taking similar doses of these medications.

Weather, your personal habits, or other medications you are taking can sometimes influence blood levels of antidepressants or mood stabilizers. For example, if you are sweating a great deal during the summer, your blood level of lithium may rise, so your doctor may need to reduce the dose. If you are a smoker, your body will break down tricyclic antidepressants more rapidly because of the effects of the nicotine. Consequently, you may need a higher dose of these antidepressants. Many other drugs that can also cause a rapid breakdown of tricyclic antidepressants are listed in Table 20–5. In contrast, some drugs on this table can slow the metabolism of tricyclic antidepressant drugs by the liver, leading to excessively high blood levels of the antidepressants. Remember that these drug interactions can work both ways: an antidepressant or mood stabilizer may affect the level or activity of other drugs you are taking, and vice versa.

Before you and your doctor decide that a particular drug is not working, make sure that you review the dose with him or her. Ask about the possibility of drug interactions if you are taking more than one drug. Your doctor may want to order a blood test to ensure that the level in your blood is adequate. Blood-level testing is more commonly done for the mood stabilizers and for the tricyclic and tetracyclic drugs than for other types of antidepressants listed in Table 20–1.

If the blood level is adequate and you have been taking the medication for a sufficient period of time but your antidepressant is still not working, your doctor may try switching you to a different type of antidepressant or may try an augmentation strategy. This involves adding a small dose of a different drug to try to boost the effect of the antidepressant. Several kinds of augmentation strategies currently in vogue are listed in Table 20–16 on pages 664–669. A complete discussion is beyond the scope of this book; I will describe just a couple of them to give you a feel for this approach. Interested readers may want to consult the excellent reference by Schatzberg and his colleagues.[1]

Two drugs commonly used for antidepressant augmentation are lithium, a drug you've learned about in this chapter, and a thyroid hormone called liothyronine (also known as Cytomel, or T_3). Your doctor may add 600 mg to 1,200 mg per day of lithium carbonate or 25 to 50 micrograms per day of liothyronine to your antidepressant for several weeks if the antidepressant has not been working adequately. As noted above, lithium is usually used to treat bipolar (manic-depressive) illness, and liothyronine is used to treat people with underactive thyroid glands. However, in this case, the goal is different—the purpose of adding a small dose of lithium or liothyronine is to make the antidepressant more effective. It is not clear why lithium and liothyronine sometimes have this effect of boosting the effectiveness of antidepressants.

A liothyronine trial will usually last for one to four weeks. If you respond positively, your physician may continue the liothyronine for two more months. Then she or he will probably taper you off the augmentation medication over one to two weeks.

The dose of lithium used for augmentation will be adjusted with a blood test so that your blood level will remain in the range of around 0.5 to 0.8 mEq per L. These levels are a little lower than the levels used to treat patients who are experiencing mania. The lower levels have the advantage of having fewer side effects. The lithium augmentation trial will generally last for two weeks. Positive results have been reported when lithium was combined with tricyclics, SSRIs, and MAOIs. Research studies suggest that as many as 50 to 70 percent of patients who do not respond to an antidepressant will respond more favorably when lithium is added. If there is no improvement in your depression, your doctor will probably discontinue the lithium as well as the antidepressant and try another medication.

Some doctors use antidepressant combination therapy for patients with difficult depressions. For example, one new approach is to add an SSRI when a tricyclic does not work, or to add a tricyclic when an SSRI does not work. This combination can cause large increases in the blood level of the tricyclic medication, and so your doctor may decrease the tricyclic first and then check your tricyclic level with a blood test after you start the SSRI. Your doctor may also order an ECG to make sure there are no adverse effects on your heart.

An MAOI might also be combined with a tricyclic antidepressant as a combination antidepressant strategy. This is an advanced form of treatment for the specialist and requires careful teamwork between you and your doctor. You will recall that dangerous reactions can result from combining MAOIs with other antidepressant drugs or with lithium. Although the *Physician's Desk Reference* advises against such drug combinations, Schatzberg and colleagues report that the combination can be safe and helpful to some patients who fail to respond to single medications.[1] To maximize safety, these investigators recommend: (1) the MAOI and tricyclic should be started at the same time; (2) clomipramine should be avoided; (3) the safest tricyclics to use in combination with MAOIs appear to be amitriptyline (Elavil) and trimipramine (Surmontil); (4) among the two commonly prescribed MAOIs, phenelzine (Nardil) appears to be safer than tranylcypromine (Parnate) to use in combination with a tricyclic.

Table 20–16. Antidepressant Augmentation Chart.

Note: The first and second columns list several types of drugs that have been added in small doses to antidepressants in attempts to make the antidepressants more effective. The next three columns list three major classes of antidepressants. A check mark (✓) indicates that at least some favorable research regarding this augmentation strategy has been published in psychiatric journals. Some combinations are dangerous and best administered by experts in research settings. Information in this table was obtained in part from the *Manual of Clinical Psychopharmacology*.[1]

		Type of Antidepressant			
Augmentation Drug	*Augmentation Dose*	*TCAs*	*SSRIs*	*MAOIs*	*Comments*
			Amino acids		
inositol	6 gm twice daily	**see comment**	**see comment**	**see comment**	Inositol is the precursor of the phosphatidylinisotol (PI) system in the brain. There are no reports yet using inositol as an augmentation drug, but it appears to have antidepressant properties and will probably soon be used for augmentation.[25]
L-tryptophan	2–6 gm per day	✓		✓	Tryptophan is the precursor of serotonin in the brain. L-tryptophan plus MAOIs or SSRIs could cause the serotonin syndrome.
Phenylalanine	500 mg–5 gm per day	✓			Phenylalanine is the precursor of dopamine and norepinephrine in the brain. At least one authority has not been impressed with the effects of this drug in augmenting antidepressants.[1]

Antidepressants					
bupropion (Wellbutrin)	low doses usually, but doses up to 300 mg per day have been used.[1]		✓		Bupropion may be used in an attempt to combat the sexual side effects of SSRIs. There are anecdotal reports it may enhance the SSRIs, but no controlled studies.[1] There is at least one documented case of seizures from bupropion plus fluoxetine (Prozac).
buspirone (BuSpar)	15 mg–45 mg per day		✓		Buspirone (BuSpar) has been shown to enhance the effects of fluoxetine (Prozac) in an open trial.[26] However, double-blind studies have not confirmed this.[1] Buspirone may also be used in an attempt to combat the sexual side effects of SSRIs.
MAOIs	15 mg (or more) of phenelzine (Nardil) daily and 150 mg (or more) of amitriptyline (Elavil)	✓			The *PDR* states that the combination of an MAOI and a trycyclic is prohibited, but the combination can be relatively safe in the hands of experts. Double-blind studies have not documented the effectiveness of this combination. Both drugs should be started at the same time. Amitriptyline (Elavil) and trimipramine (Surmontil) appear to be the safest TCAs to combine with MAOIs, and phenelzine (Nardil) and isocarboxazid (Marplan) appear to be the safest MAOIs.[1]

Table 20-16. Cont.

Augmentation Drug	*Augmentation Dose*	*Type of Antidepressant* *TCAs*	*SSRIs*	*MAOIs*	*Comments*
		Antidepressants cont.			
SSRIs	First ↓ dose of TCA; 30 mg per day of nortriptyline or 50–75 mg of imipramine recommended[1]	✓			See comments on TCAs below
TCAs	when adding a TCA and patient is on an SSRI, start with 25 mg nortriptyline or 50 mg of imipramine; increase by 25 mg after 3 days[1]		✓	✓	See note on MAOI and TCA combinations above. Several reports suggest that desipramine (Norpramin) will enhance the effects of SSRIs.[27,28] TCA blood level testing must be done because the SSRI may cause a large ↑ in desipramine levels, side effects, and toxicity. ECGs must be monitored closely during TCA and SSRI combinations.
trazodone (Desyrel)	25 mg–300 mg per day		✓		Trazodone (100 mg HS) is often added to fluoxetine (Prozac) or bupropion (Wellbutrin) to help with sleep, since these two drugs may cause insomnia. However, trazodone (Desyrel) may also enhance the effects of SSRIs.[1,29]

Appetite Suppressants					
fenfluramine (Pondimin)	20 mg–40 mg per day[1]		✓		This is an amphetamine-like drug that may enhance serotonin release in the brain. Some patients become overly stimulated.
Hormones					
estrogen					Estrogen has been used for years, alone or in combination with other antidepressants, to treat depressed women. The evidence for effectiveness is shaky at best, and the combination is unwarranted.[1]
liothyronine (Cytomel; T3)	12.5 mg–25 mg per day; slowly ↑ to 50 mg per day	✓	✓	✓	Some studies have reported positive results with TCAs[30] but other studies have reported negative results.[1] May be more effective for women than men. Case reports also suggest augmentation of SSRIs and MAOIs.[1] Responses should be seen within one to four weeks. If the response is positive, you can continue for two more months. Should be used with caution in patients with heart disease or high blood pressure.

Table 20-16. Cont.

Augmentation Drug	*Augmentation Dose*	*Type of Antidepressant* *TCAs*	*SSRIs*	*MAOIs*	*Comments*
Mood Stabilizers					
lithium	600 mg–1,200 mg per day in divided doses	✓	✓	✓	Several open studies and double-blind studies suggest that lithium in small doses may enhance the effects of antidepressants about 50% of the time. The trial will last about two or three weeks, but the combination may be continued if effective. The combination may also help prevent relapses. Lithium may also be combined with either carbamazepine (Tegretol) or valproic acid (Depakene) in patients with difficult cases of bipolar manic-depressive illness, especially those with ''rapid cycling'' (many episodes per year).
Stimulants					
amphetamine (Dexedrine)	start at 5 mg per day			✓	In one study, amphetamine or pemoline (see below) was added to an MAOI in patients with severe, treatment-refractory depressions.[31] Some patients responded, but 1 in 5 developed symptoms of mania (abnormal euphoria). Such combinations are potentially dangerous and could trigger hypertensive crises. Also see comment on methamphetamine.

methamphetamine (Desoxyn)	start at 5 mg per day	✓			Any stimulant can be addictive. This drug is intensely addictive and potentially dangerous in combination with antidepressants. Its use in any other psychiatric disorders, alone or in combination with other drugs, is controversial. Large doses over prolonged periods can produce rage and a psychosis resembling paranoid schizophrenia.
methylphenidate (Ritalin)	start at 5 mg per day	✓			This combination causes an ↑ in TCA blood levels, so blood level testing should be done. Also see comment on methamphetamine.
pemoline (Cylert)	start at 37.5 mg per day or at 18.75 mg per day			✓	See comment on methamphetamine.
Beta-Blockers					
pindolol (Visken)	2.5 mg twice daily for 1 week; then ↑ to 5 mg twice daily; continue for 3 weeks[1]		✓		Pindolol blocks beta receptors and stimulates 5-HT_{1A} receptors. It is used to treat hypertension, so blood pressure should be monitored. Side effects include dizziness, fatigue, and activation, with anxiety, irritability, and insomnia.

You will see quite a number of additional augmentation strategies listed in Table 20–16. My experience with these antidepressant combination and augmentation strategies has been limited, but I have not been impressed with the results. I have tried lithium or thyroid augmentation with a number of patients but none of them seemed to improve. I was not encouraged to continue with this approach. However, if a depressed patient has failed to respond to an adequate trial of several antidepressants, one at a time, from different chemical classes, then a combination of antidepressants or an augmentation strategy might be worth a try.

If you have received an adequate dose of an antidepressant for an appropriate period of time and you are not responding, what antidepressant should you try next? Many physicians will switch you to an antidepressant of a completely different class to maximize the chance of a positive response. This idea makes good sense, since the different antidepressants have slightly different effects on the brain. If you have failed to respond to an SSRI such as fluoxetine (Prozac), your doctor may want to try a tricyclic such as imipramine (Tofranil), for example. Prozac selectively activates the serotonin systems in the brain, whereas imipramine has effects on many different systems.

If you switch to another drug, you will usually need to taper off your current drug slowly so as to prevent any withdrawal effects. Antidepressants are not addictive and they do not cause craving when you stop taking them. However, they need to be discontinued slowly to prevent uncomfortable withdrawal reactions. For example, the tricyclics can cause insomnia and upset stomach if you go off them abruptly, as noted previously.

Further, as noted above, there may be a mandatory waiting period when you are switching from one drug to another. This is because the two drugs might be dangerous if mixed together, and the effects of the first drug may persist for a while after you have stopped taking it. The classic example would be switching from an SSRI, such as fluoxetine (Prozac), to an MAOI, such as tranylcypromine (Parnate). The combination of these two drugs can cause the previously described serotonin syndrome, which is occasionally fatal. In addition, both types of drugs clear out of the body slowly, and so a drug-free period is necessary before switching from one to the other. When switching from Prozac, an SSRI, to Parnate, an MAOI, this waiting period may be five weeks or more. When switching from Parnate to Prozac, the waiting period will be at least two weeks. With some combinations of drugs, however, a waiting period is not necessary. Check with your doctor about this.

Suppose that all these strategies fail to bring about an optimal antidepressant response. What then? In my experience this is not unusual. I have seen lots of patients who were treated for years with all kinds of medications and yet they were still severely depressed. Early in my career, I realized that drugs did not provide the answer for many people. That is why I devoted so much of my career to the development of new psychotherapeutic techniques, such as those described in this book. I wanted to have more tools available than just drugs.

In my experience, the idea that a pill alone will solve your problems and bring you joy is not productive. In contrast, the willingness to use these cognitive therapy tools, often in combination with a compassionate, persistent, and creative therapist will often lead to substantial improvement.

Other Drugs Your Doctor May Prescribe

The various types of antidepressants I have described are the ones that in my opinion have a clear-cut indication in the treatment of depression. I will describe several types of drugs that you might want to avoid, although there are exceptions to this rule.

Minor Tranquilizers (Benzodiazepines). Some doctors use minor tranquilizers (called benzodiazepines) or sedatives to treat nervousness and anxiety. The benzodiazepines include many familiar drugs such as alprazolam (Xanax), chlordiazepoxide (Librium), clonazepam (Klonopin), clorazepate (Tranxene), diazepam (Valium), lorazepam (Ativan), oxazepam (Serax) and prazepam (Centrax). Minor tranquilizers may be added to the mix of drugs your doctor prescribes if you are depressed. Because most depressed patients also experience anxiety, this practice is unfortunately quite common.

I usually do not recommend minor tranquilizers because they can be addictive, and the sedation they produce might make your depression worse. In my experience, anxiety can nearly always be treated successfully without using these drugs. Two highly esteemed colleagues from Canada, Dr. Henny A. Westra from the Queen Elizabeth II Health Sciences Center, and Dr. Sherry H. Stewart from Dalhousie University, recently reviewed the world literature on the treatment of anxiety disorders with cognitive behavioral therapy versus medications. Based on their careful review of many clinical outcome studies, the authors recommended treatment of anxiety disorders with cognitive behavior therapy instead of medications.[1] The authors concluded that cognitive behavioral therapy without drugs is a highly effective and long lasting treatment for anxiety. In comparison, they emphasize that benzodiazepines may give some limited relief but only for a short period of time, tend to lose their effectiveness over time, and are very difficult to discontinue. If you have a serious interest in this topic, the scholarly article by Drs. Westra and Stewart would be worth reading.

Although the benzodiazepines such as Ativan, Librium, Ritrovil (available in Canada), Valium, Xanax, and others can have wonderfully calming effects almost immediately after you take them, the main problem is that these relaxing effects do not last. As soon as the drug leaves your body a few hours later, there is a high likelihood you will feel nervous again. In addition, if you take these drugs daily for more than a few weeks, you may experience withdrawal effects when you try to go off them. The most common withdrawal symptoms are anxiety, nervousness and trouble sleeping. Ironically, these are the exact reasons you started taking the drug in the first place. These withdrawal symptoms trick you into thinking you still need the drug, and so you start taking it again. This is how the pattern of drug dependency develops. Fortunately, antidepressants are also effective in treating anxiety, as are the cognitive and behavioral therapy techniques described in this book, and these treatments are not addictive. This is why I avoid the benzodiazepines in the treatment of depressed or anxious individuals.

There are other reasons to avoid minor tranquilizers in the treatment of anxiety. One of the cardinal treatment principles is that anxious individuals must face their fears and surrender to their fears in order to overcome them. For example, if you have a fear of heights, you may have to climb to the top of a ladder and stand there until the anxiety goes away. I could give you dozens of examples of patients who have experienced dramatic improvements or even complete recoveries when they faced their fears in this way. Anxious individuals who face their fears often feel tremendous relief because they discover their fears were not realistic in the first place. This realization may not occur if you are simply taking tranquilizers and not facing your fears. Even if you do manage to face your fears with the help of tranquilizers, the medication will tend to reduce the effectiveness of your efforts. In fact, when doctors prescribe tranquilizers for anxious patients, there is the danger that this will reinforce the idea that the fears really are dangerous and must be avoided and that the uncomfortable symptoms must be suppressed. These messages are the very antithesis of the newer exposure therapies that have shown so much promise in the treatment of anxiety.

If your doctor has been prescribing a benzodiazepine, or suggests this type of medication, a discussion of the pros and cons would be indicated. Remember that you are the consumer, and your doctor is working for you. You have every right to discuss your treatment in a frank and respectful way. This sense of teamwork and collaboration is quite important.

Sedatives. Many prescription sleeping pills can also be addictive and are easily abused: They can lose their effectiveness after only a few days of regular use. Then greater and greater doses may be required to put you to sleep. This can lead to a pattern of drug tolerance and dependency. If you take them daily, these pills can disrupt your normal sleep pattern. Severe insomnia is a withdrawal symptom from sleeping pills, and so every time you try to stop taking the pills you will falsely conclude that you need them even more. Thus they can greatly worsen your sleeping difficulties.

In contrast, there are several sedative medications that enhance sleep without requiring increased doses. In my opinion, these drugs represent a superior approach to treating insomnia in depressed individuals. Three that are often prescribed for this purpose are 25 to 100 mg of trazodone (Desyrel) or doxepin (Sinequan) or 25 to 50 mg of diphenhydramine (Benadryl). The first two are antidepressants that require a prescription. Benadryl is an antiallergy medication that is now sold without a prescription. Make sure that you consult with your doctor before taking any medication, even one that is sold over the counter, to make sure there are no dangerous drug interactions with other medications you are taking. Remember that many over-the-counter drugs, like Benadryl, were once available only on prescription, so they can be just as dangerous as prescription drugs. The new anticonvulsant, gabapentin, also has sedative and antianxiety effects without being habit-forming, and some doctors are prescribing it for this purpose.

If you are having trouble sleeping, you may have personal problems that make it hard to get to sleep. It could be anything—a problem at school or work, or a conflict with a family member or friend. Some people sweep these problems under the carpet so they won't have to deal with them. Then they develop a variety of symptoms instead. Some people become anxious, others have trouble sleeping, and some develop aches and pains that have no organic causes.

I have always felt it is better to try to identify and solve the problem rather than masking it with tranquilizers or sleeping pills. In our culture, the idea of a quick cure is tremendously appealing to patients and physicians alike. It is easy to prescribe a drug that will make the problem go away. This contributes greatly to the enormous popularity of sleeping pills and minor tranquilizers.

Stimulants. How about the "pep pills" (stimulants) such as methylphenidate (Ritalin) and the amphetamines that used to be so commonly prescribed for weight loss? It is true that these drugs can produce a temporary stimulation or elation (much like cocaine), but they can also be dangerously habit-forming. When you come down from the temporary high state, you may tend to crash and experience an even more profound sense of despair. When given chronically, these drugs can sometimes produce an aggressive, violent, paranoid reaction resembling schizophrenia.

I have not prescribed stimulants for depressed patients (or for any other problem) because of my concerns about these drugs, but this is clearly an area of controversy. Some psychiatrists do prescribe stimulants for elderly depressed patients under certain circumstances, and they are quite popular for treating hyperactive children and adolescents. If your doctor recommends taking such pills, you should certainly discuss the pros and cons. You might also want to obtain a second opinion if you feel uncomfortable about the treatment.

There are exceptions to this rule, like any. Because of its energizing properties, some doctors add methylphenidate (Ritalin) to a tricyclic antidepressant. This combination may be helpful for some patients who are very sluggish and unmotivated. However, methylphenidate also inhibits the breakdown of most tricyclic antidepressants by the liver, and so the blood level of these other antidepressants will increase. This may lead to greater side effects and may require a reduction in the dose of the antidepressant.

Antipsychotic Medications (Neuroleptics). What about the antipsychotic medications (also called neuroleptics or "major tranquilizers")? Some of the older drugs in this category include chlorpromazine (Thorazine), chlorprothixene (Taractan), haloperidol (Haldol), fluphenazine (Prolixin), loxapine (Loxitane), mesoridazine (Serentil), molindone (Moban), perphenazine (Trilafon), pimozide (Orap), thiothixene (Navane), thioridazine (Mellaril), and trifluoperazine (Stelazine). Some of the newer drugs include clozapine (Clozaril), olanzapine (Zyprexa), quetiapine (Seroquel), risperidone (Risperdal), sertindole (Serlect), and ziprasidone (trade name not yet available). These agents are usually reserved for patients with schizophrenia, mania, or other psychotic disorders. They do not play a major role in the treatment of most depressed or anxious patients. Pills that combined an antidepressant with an antipsychotic medication were marketed and promoted in the past, but most clinical studies have not documented any superior efficacy of such preparations in the treatment of depression.

Only a minority of depressed individuals benefits from antipsychotic agents. These include depressed patients who are delusional—that is, patients who draw false and highly unrealistic conclusions about external reality. For example, a depressed patient might have the delusion that there are worms in his or her body or that there is a conspiracy against him or her. Elderly depressed patients seem more likely to develop paranoid delusions. Depressed patients who are extremely agitated and cannot stop pacing sometimes benefit from the antipsychotic agents as well. However, the major tranquilizers may also cause a worsening of the depression because of their tendency to cause sleepiness and fatigue.

In addition, unlike most antidepressants, many of the antipsychotic medications carry the risk of an irreversible side effect called tardive dyskinesia. Tardive dyskinesia is an abnormality of the face, lips, and tongue; it involves repetitious, involuntary movements, such as smacking the lips over and over or grimacing. The abnormal movements can also sometimes include the arms, legs, and torso. The major tranquilizers can also cause a number of other alarming but reversible side effects. Therefore, these drugs should be used only when they are clearly needed so that their potential benefit outweighs the potential risk.

Polypharmacy

Polypharmacy refers to the practice of prescribing more than one psychiatric drug at a time to a particular patient. The idea is that if one drug is good, two, three, or more will be even better. Doctors may combine antidepressant drugs with other types of antidepressants as well as with other types of drugs, such as minor and major tranquilizers. The patient ends up taking a cocktail of many different types of drugs.

Polypharmacy used to be frowned upon. Now the practice has become more accepted, and many psychiatrists routinely prescribe two or more drugs for many of their psychiatric patients. In contrast, if a family physician is treating your depression, then it is much less likely that she or he will prescribe more than one psychiatric medication at a time. This is because a family doctor is usually more concerned with your medical problems and much less aggressive in the treatment of emotional problems.

In some instances, polypharmacy can be helpful in the treatment of mood disorders. For example, I have described several augmentation strategies that might boost the effectiveness of an antidepressant. I have also described how the occasional use of a second medication can combat a drug side effect. Rational polypharmacy might also be helpful when a patient has separate disorders that both require treatment. For example, a patient with schizophrenia may also be depressed and may benefit from a combination of an antipsychotic medication along with an antidepressant. A bipolar (manic-depressive) patient may receive an antidepressant in addition to the lithium during an episode of depression. During an episode of mania, the doctor may prescribe a neuroleptic or a benzodiazepine in addition to lithium to combat the acute symptoms, as described previously.

Although there are specific instances like these when combinations of drugs are indicated, I am usually not in favor of polypharmacy in the treatment of depression or anxiety because of the increase in side effects, drug interactions, and costs. In addition, polypharmacy has the tendency to convey the message that all the patient's problems can be dealt with by drugs. The patient may take one or two drugs for depression, one or two additional drugs to treat the side effects of the antidepressants, one more drug to treat anxiety, and so on. And if the patient is angry, she or he may get yet another drug, such as a mood stabilizer, to treat the anger.

The patient may end up in a rather passive role as a kind of human test tube. You may think I am exaggerating, but I have seen numerous patients who were in just this position. They were taking lots of drugs with lots of side effects but were receiving very little benefit from any of them. I have treated many of these patients successfully with cognitive therapy and no drugs or cognitive therapy and only one antidepressant.

I believe that some psychiatrists rely too much on drugs. Why is this? One problem is that most psychiatric training programs strongly emphasize biological theories about depression and stress the importance of drug treatments for depression and other disorders. In addition, a great many of the continuing education programs for psychiatrists in practice are sponsored by drug companies, and the focus of these conferences is nearly always on medications. The psychiatric journals, too, are filled with expensive drug company advertisements promoting the benefits of the latest medications for depression or anxiety, but I have never seen an ad promoting the latest psychotherapy technique. This is because there is simply no money to pay for such an ad! Drug companies also fund a great deal of the research on medications that appears in psychiatric journals, and concerns have been voiced about the potential conflict of interest inherent in such arrangements.

I do not mean to sound like a rabble-rouser! This is not a black-or-white issue. Clearly, the excellent research conducted by the pharmaceutical industry has been an enormous boon to the psychiatric profession and to individuals suffering from psychiatric disorders. My concern is that the emphasis on drugs sometimes seems excessive. Unfortunately, some psychiatrists do not have good training in the newer forms of psychotherapy, including cognitive behavioral therapy, which can be so helpful for individuals suffering from depression and anxiety. When a patient does not respond to medications, the main response of the psychiatrist may be to increase the dose or add another medication because this is what the psychiatrist has been trained to do. And when a patient complains of an adverse side effect, the psychiatrist may decide to add some other additional drug as an antidote—because that is what she or he has been trained to do. The result in some cases is that patients end up taking more and more drugs in larger and larger doses—without any real benefits. This is when polypharmacy can get out of hand.

When I was a psychiatric resident, I used to have the idea that if only I could find the right ''magic bullet'' (in other words, the right pill), I could help every patient. In those days, we treated our patients with pill after pill after pill but very little psychotherapy. My clinical experience taught me over and over again that this model was not sufficient—too many of my patients simply did not recover, no matter how many drugs I used, singly or in combinations.

To make things worse, most psychiatrists do not require patients to take mood tests, like the one in Chapter 2, between therapy sessions to track progress. As a result, the psychiatrist may conclude that the patient is being ''helped'' by a drug when the patient has not really improved substantially. To my way of thinking, treating patients without session-by-session assessments is antiscientific and represents a barrier to good treatment and progress in the field.

Some psychiatrists and many patients are almost exclusively committed to these biological theories and treatments for depression. They may discount the value of other approaches, sometimes with a religious fervor. A number of well-known psychiatrists are quite outspoken in this regard. The intensity of these debates about psychotherapy versus drug therapy is sometimes more reminiscent of a power struggle for turf than an intellectual search for the truth. Fortunately, there is a growing and healthy trend to recognize that all of our current psychiatric drugs are limited in their effectiveness. In addition, there is an increasing recognition that a combination of medication with the newer forms of psychotherapy (including cognitive behavioral therapy and others) usually provides a more satisfactory outcome than does treatment with drugs alone.

It is clear that antidepressant drugs can help some individuals, but it is also clear that many patients do not respond adequately. When patients do not respond, I would prefer to switch into a different gear and use cognitive therapy or a combination of cognitive therapy and one antidepressant medication at a time. Most depressed people have real problems in their lives, and nearly all of us need a compassionate, healing relationship with another human being to talk things out at times. The idea that drugs alone should work to cure depression and anxiety may be appealing, but this approach is often ineffective.

To be fair, an exclusive focus on psychotherapy alone can be just as biased. I have seen patients who did not respond to many psychotherapeutic interventions that I personally administered—week after week their depression scores on the test in Chapter 2 did not change. Sometimes I prescribed an antidepressant while we continued working with a variety of psychotherapeutic strategies. Within several weeks, the depression and anxiety often began to improve, and the psychotherapy suddenly began to work better. In these cases, I was glad to have the medications available.

A final problem contributing to polypharmacy is that

many patients are unassertive. Even though they feel uncomfortable about all the drugs they are taking, they may sometimes assume that "the doctor knows best." This is understandable. The doctor does have a great deal of training, and the patient's knowledge is usually limited. In addition, the patient often admires the doctor and respects his or her advice. But in psychiatry and psychology, treatment approaches are far more subjective and varied than in internal medicine, where the treatments are far more precise and uniform. Your feelings about the treatment are important, and you have every right to share these feelings with your doctor.

This review of drug-prescribing practices obviously represents my own approach. Your physician's ideas might differ. Psychiatry is still a blend of art and science. Perhaps some day the "art" will no longer be such a prominent ingredient. If you feel uncertain about your treatment, ask your physician questions. State your concerns and urge your doctor to explain the treatment in simple terms you understand. After all, it's your brain and body that are at risk, not the doctor's. The sense of teamwork and collaboration are important to successful treatment. As long as the two of you agree to a rational, understandable, and mutually acceptable strategy for your therapy, you will have an excellent chance of benefiting from your doctor's efforts to help you.

Notes and References (Chapters 17 to 20)

1. Schatzberg, A. F., Cole, J. O., & DeBattista, C. (1997) *Manual of Clinical Psychopharmacology*. Third Edition. Washington, DC: American Psychiatric Press.

2. Some psychologists are lobbying for the right to prescribe drugs, and some psychologists in the armed services have already been licensed to prescribe drugs. There is intense controversy about the merits of this proposal. Some psychologists argue that the right to prescribe drugs is desirable because it will put them on an even footing to compete with psychiatrists for patients. Other psychologists argue that drug prescribing requires extensive medical training and that the profession will lose an important part of its identity if psychologists win the right to prescribe drugs. They also point out that the role of the psychiatrist, particularly in managed care situations, has become quite unappealing. Many psychiatrists who work for HMOs are now forced to see huge numbers of patients for extremely brief visits consisting only of discussions about medications without any time to do psychotherapy or to learn about the problems in their patients' lives.

3. Baxter, L. R., Schwartz, J. M., & Bergman, K. S., et al. (1992). Caudate glucose metabolic rate changes with both drug and behavioral therapy for obsessive-compulsive disorders. *Archives of General Psychiatry* 49, 681–689.

4. Simons, A. D., Garfield, S. L., & Murphy, G. E. (1984). The process of change in cognitive therapy and pharmacotherapy for depression. *Archives of General Psychiatry* 41, 45–51.

5. Antonuccio, D. O., Danton, W. G., & DeNelsky, G. Y. (1995). Psychotherapy versus medication for depression: Challenging the conventional wisdom with data. *Professional Psychology: Research and Practice* 26 (6) 574–585.

6. Dobson, K. S. (1989). A meta-analysis of the efficacy of cognitive therapy for depression. *Journal of Consulting and Clinical Psychology,* 57(3), 414–419.

7. Hollon, S. D., & Beck, A. T. (1994). Cognitive and cognitive behavioral therapies. Chapter 10 in A. E. Bergin & S. L. Garfield (Eds.), *Handbook of Psychotherapy and Behavioral Change* (pp. 428–466). New York: John Wiley & Sons, Inc.

8. Robinson, L. A., Berman, J. S., & Neimeyer, R. A. (1990). Psychotherapy for the treatment of depression: Comprehensive review of controlled outcome research. *Psychological Bulletin,* 108, 30–49.

9. Scogin, F., Jamison, C., & Gochneaut, K. (1989). The comparative efficacy of cognitive and behavioral bibliotherapy for mildly and moderately depressed older adults. *Journal of Consulting and Clinical Psychology* 57, 403–407.

10. Scogin, F., Hamblin, D., & Beutler, L. (1987). Bibliotherapy for depressed older adults: A self-help alternative. *The Gerontologist* 27, 383–387.

11. Scogin, F., Jamison, C., & Davis, N. (1990). A two-year follow-up of the effects of bibliotherapy for depressed older adults. *Journal of Consulting and Clinical Psychology* 58, 665–667.

12. Jamison, C., & Scogin, F. (1995). Outcome of cognitive bibliotherapy with depressed adults. *Journal of Consulting and Clinical Psychology* 63, 644–650.

13. Smith, N. M., Floyd, M. R., Jamison, C., & Scogin, F. (1997). Three-year follow-up of bibliotherapy for depression. *Journal of Consulting and Clinical Psychology* 65 (2), 324–327.

14. Burns, D. D., and Nolen-Hoeksema, S. (1991). Coping styles, homework compliance and the effectiveness of cognitive-behavioral therapy. *Journal of Consulting and Clinical Psychology* 59 (2), 305–311.

15. Burns, D. D., & Auerbach, A. H. (1992). Do self-help assignments enhance recovery from depression? *Psychiatric Annals* 22 (9), 464–469.

16. Dessain, E. C., Schatzberg, A. F., & Woods, B. T. (1986). Maprotiline treatment in depression: a perspective on seizures. *Archives of General Psychiatry 43,* 86–90.

17. Maxmen, J. S., & Ward, N. G. (1995). *Psychotropic Drugs Fast Facts*, Second Edition. New York: W. W. Norton & Company.

18. You will notice that the percentages of patients reporting stomach upset in Table 20–5 are a little lower than

20 percent to 30 percent on the average. This is because the percentages in the table represent the differences between the rates for the actual drug minus the rates for patients taking placebo medications.

19. You will learn below that the MAOIs can cause dangerous blood pressure elevations, but this is only if you take one of the forbidden foods or medications. Usually, the MAOIs can cause a mild drop in blood pressure.

20. A patient with a "difficult" or "resistant" depression is simply one who does not readily respond to the usual treatments. If your doctor tries many antidepressant drugs and you do not improve, your doctor will naturally conclude that your depression is more difficult than usual to treat. However, you may respond nicely to another type of treatment. I have treated large numbers of patients who had years and years of unsuccessful treatment with a wide variety of drugs prior to seeing me. Many of these "difficult" patients recovered when I used cognitive therapy techniques like those described in this book.

    No single treatment is a panacea for everyone. That's why it is important to have lots of approaches available, including many different kinds of medicines and many different kinds of psychotherapeutic methods as well. The term, "different strokes for different folks" is right on target in the context of depression treatment!

21. Arky, R. (Medical Consultant). (1998). *Physician's Desk Reference*, 52 Edition. Montvale, NJ: Medical Economics Company, Inc.

22. The telephone number of the Madison Institute of Medicine is 608-827-2470; their fax is 608-827-2479; their address is 7617 Mineral Point Road, Suite 300, Mad-

ison, Wisconsin, 53717; and their email is INFOCTRS@Healthtechsys.com. They can do literature searches and supply pamphlets, reprints and other information for a modest fee.

23. Preskorn, S. H. (1997). Clinically relevant pharmacology of selective serotonin reuptake inhibitors. *Clinical Pharmacokinetics* Suppl. 1, 1–21.

24. Westra, H. A., & Stewart, S. H. (1998). Cognitive behavioral therapy and pharmacotherapy: Complementary or contradictory approaches to the treatment of anxiety? *Clinical Psychology Review* 18 (3), 307–340.

25. Levine, J., Brak, Y., Gonzales, M., et al. (1995). Double-blind controlled trial of inositol treatment of major depression. *American Journal of Psychiatry* 152, 792–794.

26. Joffee, R. T., & Shuller, D. R. (1993). An open study of buspirone augmentation of serotonin reuptake inhibitors. *Journal of Clinical Psychiatry* 54, 269–271.

27. Nelson, J. C., & Price, L. H. (1995). Lithium or desipramine augmentation of fluoxetine treatment (letter). *American Journal of Psychiatry* 152, 1538–1539.

28. Weilburg, J. B., Rosenbaum, J. F., Biederman, J., et al. (1989). Fluoxetine added to non-MAOI antidepressants converts nonresponders to responders: a preliminary report. *Journal of Clinical Psychiatry* 50, 447–449.

29. Nirenberg, A. A., Cole, J. O., & Glass, L. (1992). Possible trazodone potentiation of fluoxetine: a case series. *Journal of Clinical Psychiatry* 53, 83–85.

30. Joffee, R. T., Levitt, A. J., Bagby, R. M. et al. (1993). Predictors of response to lithium and triodothyronine:

augmentation of antidepressants in tricycylic non-responders. *British Journal of Psychiatry* 163, 574–578.

31. Fawcett, J., Kravitz, H. M., Zajeda, J. M., et al. (1991). CNS stimulant potentiation of monoamine oxidase inhibitors in treatment of refractory depression. *Journal of Clinical Psychopharmacology* 11, 127–132.

Suggested Resources

Other Books by Dr. Burns

The Feeling Good Handbook (New York: Plume, 1990). Dr. Burns shows how you can use cognitive therapy to overcome a wide variety of mood problems such as depression, frustration, panic, chronic worry and phobias, as well as personal relationship problems such as marital conflict or difficulties at work.

Intimate Connections (New York: Signet, 1985). Dr. Burns shows you how to flirt, how to handle people who give you the run-a-round, and how to get people of the opposite sex (or the same sex, if that is your preference) to pursue you.

Ten Days to Self-Esteem and *Ten Days to Self-Esteem: The Leader's Manual* (New York: Quill, 1993). In this ten-step program, Dr. Burns provides a practical, workable blueprint for breaking out of the bad moods that rob us of self-esteem. He provides you with clear, easy-to-understand instructions and specific tools gleaned from twenty years of systematic research and psychiatric practice. The *Leader's Manual* shows you how to develop this program in hospitals, clinics, schools, and other institutional settings.

Workshops and Lectures by Dr. Burns

Dr. Burns offers workshops and lectures for mental health professionals and for general public audiences as well. For a list of dates and locations, you are invited to visit Dr. Burns' Web site at www.FeelingGood.com

Audiotapes for the General Public

Burns, *The Perfectionist's Script for Self-Defeat.*

Dr. Burns helps you identify perfectionistic tendencies and explains how they work against you. He shows you how to stop setting unrealistically high standards and increase productivity, creativity, and self-satisfaction.

Burns, *Feeling Good.*

Dr. Burns describes ten common self-defeating thinking patterns that lead to depression, anxiety, frustration, and anger. He explains how to replace them with more positive and realistic attitudes so you can break out of bad moods and enjoy greater self-esteem now and in the future.

Audiotapes for Mental Health Professionals

Strategies for Therapeutic Success: My Twenty Most Effective Techniques—Volumes I and II. 8 Cassettes

In this two-day intensive workshop, Dr. Burns illustrates the most valuable therapy techniques he has developed during two decades of clinical practice, training, and research.

Feeling Good: Fast & Effective Treatments for Depression, Anxiety, and Therapeutic Resistance. 4 Cassettes

Dr. Burns describes the basic principles of CBT and illustrates state-of-the-art treatment methods for depression and anxiety disorders. He also illustrates how to deal with difficult, angry patients who seem to sabotage the treatment because they feel mistrustful and unmotivated.

Feeling Good Together: Cognitive Interpersonal Therapy
4 Cassettes

In this workshop, Dr. Burns shows how to modify the attitudes that sabotage intimacy and lead to anger and mistrust. He also explains how to deal with patients who blame others for their personal relationship problems.

Rapid, Cost-Effective Treatments for Anxiety Disorders
4 Cassettes

In this workshop, Dr. David Burns shows you how to integrate three powerful models in the treatment of the entire spectrum of anxiety disorders, including generalized anxiety, panic disorder (with or without agoraphobia), phobias, social anxiety, obsessive-compulsive disorder, and post-traumatic stress disorder (including victims of childhood sexual abuse).

You may order the audiotapes for professionals or for the general public by visiting Dr. Burns' Web page at www.FeelingGood.com

Treatment and Assessment Tools for Mental Health Professionals

Therapist's Toolkit 2000

Includes hundreds of pages of state-of-the-art assessment and treatment tools for the mental health professional. Purchase includes licensure for unlimited reproduction in your clinical practice. Site licenses are available.

Feeling Good Web Site

You are invited to visit Dr. Burns' Web site at www.FeelingGood.com. This Web site contains information about:

- dates and locations for upcoming lectures and workshops by Dr. Burns
- audiotapes for the general public
- training tapes for mental health professionals (including CE credits)
- links for referrals to cognitive therapists around the country
- description of Dr. Burns' new *Therapist's Toolkit*
- links to other interesting sites
- new information of potential interest to patients, therapists, and researchers
- Ask The Guru. You can submit questions about any mental health topic. Answers to selected questions are posted in a column format.

Index

Page numbers in *italics* refer to figures and tables. A small "n" following a page number refers to a note on that page.

DAVID D. BURNS, M.D.,

graduated *magna cum laude* from Amherst College, received his M.D. degree from Stanford University School of Medicine, and completed his psychiatric residency at the University of Pennsylvania. He has served as Acting Chief of Psychiatry at the Presbyterian/University of Pennsylvania Medical Center and as Visiting Scholar at the Harvard Medical School. In 1995 Dr. Burns and his family returned to California. He is currently Clinical Associate Professor of Psychiatry and Behavioral Sciences at his *alma mater*, the Stanford University School of Medicine, where he is actively involved in research and teaching. Dr. Burns is certified by the National Board of Psychiatry and Neurology.

Dr. Burns has received numerous awards, including the A.E. Bennett Award from the Society for Biological Psychiatry for his research on brain chemistry and the Distinguished Contribution to Psychology through the Media Award from the Association of Applied and Preventive Psychology. In 1998 he received the Teacher of the Year Award from the graduating psychiatric residents at Stanford.